Nanotechnology

Volume 1: Principles and Fundamentals
Edited by Günter Schmid

Related Titles

Nanotechnologies for the Life Sciences

Challa S. S. R. Kumar (ed.)

Volume 1: Biofunctionalization of Nanomaterials
2005
978-3-527-31381-5

Volume 2: Biological and Pharmaceutical Nanomaterials
2005
978-3-427-31382

Volume 3: Nanosystem Characterization Tools in the Life Sciences
2005
978-3-527-31383-9

Volume 4: Nanodevices for the Life Sciences
2006
978-3-527-31384-6

Volume 5: Nanomaterials - Toxicity, Health and Environmental Issues
2006
978-3-527-31385-3

Volume 6: Nanomaterials for Cancer Therapy
2006
978-3-527-31386-0

Volume 7: Nanomaterials for Cancer Diagnosis
2006
978-3-527-31387-7

Volume 8: Nanomaterials for Biosensors
2006
978-3-527-31388-4

Volume 9: Tissue, Cell and Organ Engineering
2006
978-3-527-31389-1

Volume 10: Nanomaterials for Medical Diagnosis and Therapy
2007
978-3-527-31390-7

Nanotechnology

Günter Schmid (ed.)

Volume 1: Principles and Fundamentals
2008
978-3-527-31732-5

Harald Krug (ed.)

Volume 2: Environmental Aspects
2008
978-3-527-31735-6

Rainer Waser (ed.)

Volume 3: Information Technology I
2008
978-3-527-31738-7

Rainer Waser (ed.)

Volume 3: Information Technology II
2008
978-3-527-31737-0

Viola Vogel (ed.)

Volume 5: Nanomedicine and Nanobiotechnology
2009
978-3-527-31736-3

Harald Fuchs (ed.)

Volume 6: Nanoprobes
2009
978-3-527-31733-2

Michael Grätzel, Kuppuswamy Kalyanasundaram (eds.)

Volume 7: Light and Energy
2009
978-3-527-31734-9

Lifeng Chi (ed.)

Volume 8: Nanostructured Surfaces
2009
978-3-527-31739-4

www.wiley.com/go/nanotechnology

G. Schmid, H. Krug, R. Waser, V. Vogel, H. Fuchs,
M. Grätzel, K. Kalyanasundaram, L. Chi (Eds.)

Nanotechnology

Volume 1: Principles and Fundamentals

Edited by Günter Schmid

WILEY-VCH

WILEY-VCH Verlag GmbH & Co. KGaA

The Editor

Prof. Dr. Günter Schmid
Institute for Inorganic Chemistry
University of Duisburg - Essen
Universitätsstrasse 5-7
45117 Essen
Germany

Cover: Nanocar reproduced with kind permission of Y. Shirai/Rice University

Library of Congress Card No.: applied for

British Library Cataloguing-in-Publication Data
A catalogue record for this book is available from the British Library.

Bibliographic information published by the Deutsche Nationalbibliothek
Die Deutsche Nationalbibliothek lists this publication in the Deutsche Nationalbibliografie; detailed bibliographic data are available in the Internet at http://dnb.d-nb.de.

Typesetting Thomson Digital, Noida, India
Printing betz-druck GmbH, Darmstadt
Binding Litges & Dopf Buchbinderei GmbH, Heppenheim

Printed in the Federal Republic of Germany
Printed on acid-free paper

ISBN: 978-3-527-31732-5

Contents

Nanotechnology. Volume 1: Principles and Fundamentals. Edited by Günter Schmid

ISBN: 978-3-527-31732-5

List of Contributors

Jörg Engstler
Technische Universität Darmstadt
Fachbereich Chemie
Eduard-Zintl-Institut für Anorganische und Physikalische Chemie
Petersenstrasse 18
64287 Darmstadt
Germany

Armin Grunwald
ITAS
Forschungszentrum Karlsruhe in der Helmholtz-Gesellschaft
76021 Karlsruhe
Germany

Alexander W. Holleitner
Ludwig-Maximilians-Universität
Fachbereich
Physik Geschwister-
Scholl-Platz 1
80539 München
Germany

Liberato Manna
National Nanotechnology Lab.
Istituto Nazionale per la Fisica della Materia
CNR-INFM
Lecce, Italy

Thomas Nann
University of East Anglia
Nanoscale Science
Earlham Road
Norwich NR4 7TJ
UK

Alfred Nordmann
Technische Universität Darmstadt
Institut für Philosophie
Schloss
64283 Darmstadt
Germany

Wolfgang J. Parak
Philipps-Universität
Marburg
Physik, AG Biophotonik
Renthof 7
35032 Marburg
Germany

Alexander Popp
Technische Universität Darmstadt
Fachbereich Chemie
Eduard-Zintl-Institut für Anorganische und Physikalische Chemie
Petersenstrasse 18
64287 Darmstadt
Germany

Nanotechnology. Volume 1: Principles and Fundamentals. Edited by Günter Schmid

ISBN: 978-3-527-31732-5

Günter Schmid
University of Duisburg – Essen
Institut für Anorganische Chemie
Universitätsstrasse 5–7
45117 Essen
Germany

Jörg J. Schneider
Technische Universität Darmstadt
Fachbereich Chemie
Eduard-Zintl-Institut für Anorganische und Physikalische Chemie
Petersenstrasse 18
64287 Darmstadt
Germany

Friedrich C. Simmel
Ludwig-Maximilians-Universität
Physik
Geschwister-Scholl-Platz 1
80539 München
Germany

Itamar Willner
The Hebrew University of Jerusalem
Institute of Chemistry
Jerusalem 91904
Israel

Maya Zayats
The Hebrew University of Jerusalem
Institute of Chemistry
Jerusalem 91904
Israel

1
Introduction

Günter Schmid

The term "Nanotechnology" is nowadays commonplace not only in all relevant scientific and technical areas, but also to a considerable extent in the public domain, based on reports in newspapers, on television and, justified or not, in a series of commercially available products with "nano" as part of their names. On the one hand, this development could be considered in a positive sense, indicating nanotechnology as an accepted new technology. On the other hand, it contains some risks that should not be neglected. This is due to the rather complex definition of nanotechnology and nanoscience as a sectional science, involving natural and materials sciences, engineering and medicine. Especially it is the lack of a generally accepted definition of nanotechnology and nanoscience that is responsible for many misunderstandings. The relevant scientific communities have agreed that the term "nano" must always be linked with the appearance of a novel property. If "nano" is restricted just to a length scale, one would preferably speak of "technology on the nanoscale", usually only based on scaling effects ranging from micrometer to nanometer dimensions, without being linked with the appearance of really novel physical or chemical properties. This imprecise view of nanotechnology is frequently misused for products that are linked with the term "nano", but do not really offer a "nano-effect".

The following chapters therefore deal with the principles and fundamentals of nanotechnology, explaining what nanoscience and nanotechnology really means and what it does not mean. Furthermore, this book contains philosophical and ethical aspects, since any new technology opens up questions concerning social consequences. Therefore, first of all, a scientifically unambiguous definition of nanotechnology and nanoscience is discussed in Chapter 2, followed by a series of examples elucidating this definition in various fields, reaching from size effects up to complex biosystems. Chapter 3 deals with the principles of how to generate effective nanosystems. Top-down techniques are completed by bottom-up procedures that are currently becoming increasingly important due to the use of ultimately small building blocks: atoms and molecules. Chapters 4 and 5 consider two kinds of fundamental objects of nanoscience: quantum dots, and wires, rods and tubes, respectively. Those species represent the world of size-determined properties of

Nanotechnology. Volume 1: Principles and Fundamentals. Edited by Günter Schmid

ISBN: 978-3-527-31732-5

manifold materials and so stand for one of the fundamental principles of nanotechnology, in agreement with the definition. Spherical or one-dimensional matter of appropriate size can no longer be described by classical physical laws, but by quantum mechanical rules, indicating the decisive change from the macroscopic or microscopic world to the nanoworld.

An extremely important field of nanoscience and also of nanotechnology is dealing with the intelligent combination of artificial nanoscopic building blocks with biomolecular systems, which can anyway be considered as the most powerful "nanotechnological" inventions that we know. Most building blocks of living cells represent perfect nanosystems, the interplay of which results in the microscopic and macroscopic world of cells. We have learned to learn from Nature and consequently try to develop technologically applicable devices reaching from novel sensor systems up to diagnostic and therapeutic innovations. Chapter 6 gives an insight into this fascinating part of the nanoworld.

Philosophical and ethical questions are discussed in Chapters 7 and 8. What kind of knowledge is produced and communicated by nanotechnology? What is its place in relation to other sciences? These and related problems are discussed in Chapter 7. Studying current and future developments in nanotechnology from the viewpoint of ethics is an essential requirement in order to elaborate rules and concerted actions on how to deal with them in society. Such reflections should accompany any novel technological development, especially nanotechnology, the power of which has already been compared with the beginning of a new genesis.

This is the first of a series of books dealing with the various fields of nanotechnology. In addition to the principles and fundamentals, treated in this volume, information technology, medicine, energy, tools and analytics as well as toxicity will be the subjects of subsequent other books. In all cases, developed fields of nanotechnology and future areas of nanotechnological applications will be described and discussed.

2
The Nature of Nanotechnology

Günter Schmid

2.1
Definition

Numerous definitions of "nanotechnology" exist in the literature. Most of them simply say that nanotechnology considers materials and architectures on the nanoscale. In some of the definitions it is stated that nanotechnology dealings mainly with structures in the region between 1 and 100 nm. In any case, the dimension plays the dominant role. For a more detailed description, however, this is much too simple. First, nanotechnology follows nanoscience, where fundamental effects have been discovered before. In a long course of development, a technology may result from scientific findings in some cases, but not in all cases by a long way. So, to understand nanotechnology, one has to define properly what nanoscience is.

Among these numerous attempts to define nanotechnology, the definition given by the Royal Society and the Royal Academy of Engineering is fairly close to the author's opinion on the matter [1]:

> *"Nanoscience is the study of phenomena and manipulation of materials at atomic, molecular and macromolecular scales, where properties differ significantly from those at a larger scale. Nanotechnologies are the design, characterization, production and application of structures, devices and systems by controlling shape and size at the nanometre scale."*

A more recent definition, formulated by a team of scientists at the Europäische Akademie zur Erforschung wissenschaftlich-technischer Entwicklungen Bad Neuenahr-Ahrweiler GmbH, focuses still more the real intent of nanoscience and nanotechnology and will therefore be used in the following [2]:

> *"Nanotechnology comprises the emerging application of Nanoscience. Nanoscience deals with functional systems either based on the use of subunits with specific size-dependent properties or of individual or combined functionalized subunits."*

Nanotechnology. Volume 1: Principles and Fundamentals. Edited by Günter Schmid

ISBN: 978-3-527-31732-5

This is the only existing definition without naming a particular lateral scale. It excludes any kind of simple scaling effect (see later). The decisive aspect is the appearance of novel properties. These can in principle be observed below 1 or above 1000 nm (1 μm). Therefore, the strict limitation to a distinct length scale does not appropriate seem. Scientifically, it would even be absurd to set a limit to size-dependent properties or novel properties of a construct of functionalized subunits. In spite of this deliberate leaving out of a particular lateral scale, the expressions nanoscience and nanotechnology are meaningful for practical use since most of the known "nano-effects" happen on the nanoscale. The limitation of the definition to the nanoscale, however, would degrade nanotechnology to the simple continuation of microtechnology. Microtechnology was and still is overwhelmingly successful by the continuous reduction of materials or tools, aiming not at the creation of novel abilities, but at other advantages. Some examples in the context of the above definition will help us to understand better what it is meant.

Typical *size-dependent* nano-effects that spontaneously occur when a critical dimension is reached are observed when metal particles are downsized. Depending on the kind of change of property, the critical size may vary for the same element.

A very typical and well known nano-effect is observed when gold is downsized. In the bulk state, the beautiful color of gold results from a very fundamental phenomenon, the relativistic effect, based on Einstein's Special Theory of Relativity. One of the basic messages of this theory is that the speed of light in vacuum is an absolute constant everywhere in the universe. It never can be surpassed. If an object were theoretically to be accelerated close to or even exactly to the speed of light, its mass would continuously increase with increasing speed, finally even *ad infinitum.*

Electrons in most atoms are moving around the atomic nucleus with speeds usually far below that of light. However, in very heavy atoms, the high mass of the nuclei accelerates electrons to such an extent that relativistic effects become obvious [3–5]. For instance, such effects are known for the elements lead, tungsten, mercury, platinum and gold. The interesting question is whether relativistic effects influence the physical and chemical properties of such a heavy element. Indeed they do! The acceleration primarily affects the s-electrons which are in the nuclear region, linked with an increase in their mass, reducing the average distance from the nucleus. Consequently, the s-orbitals and their energy are shrinking. As a secondary effect, d-electrons, being farer from the nucleus, become electronically better shielded. Their orbitals are therefore extended and increase energetically. For the p-electrons the effects approximately equilibrate. In bulk metals, the individual orbitals of atoms are extended to electronic bands. The valence band, resulting from the d-orbitals, is energetically increased, whereas the conduction band, formed from the s-orbitals, is lowered. The reduced energy difference between the valence band and conduction band finally allows the low-energy photons of blue light to lift electrons from the valence to the conduction bands. Consequently, gold absorbs blue light and shows the complementary color yellow. Relativistic effects are revealed in various ways: tungsten exhibits an unusually high melting point and mercury is the only metal that is liquid at normal temperatures. Figure 2.1 illustrates the relativistic effect of gold by shrinking the energy difference between the s- and d-orbitals.

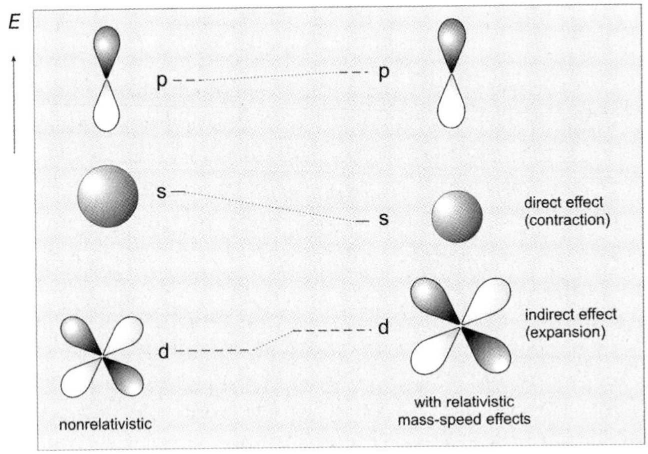

Figure 2.1 Influence of relativistic effects on the energy level of d-, s- and p-orbitals.

Since the relativistic effect is a property of each individual gold atom, it must also be present in nanosized particles, although their color is no longer golden. This metal changes its appearance at about 50 nm to become blue. Further reduction results in purple and finally, at about 15–20 nm, in bright red. This well-understood and long-known effect can be traced back to the existence of a so-called plasmon resonance. The phenomenon is quantitatively described by the Mie theory [6]. Qualitatively, the formation of a plasmon resonance can be explained by a collective electron oscillation with respect to the positive metal core of the particle, caused by the interaction of external electromagnetic radiation (visible light) with the confined electron gas of nanoparticles. The process is illustrated in Figure 2.2.

The energy taken up from light is responsible for the resulting color. This energy depends on, among others, the particle's size and shape and the surrounding medium. With decreasing size the color is shifted to shorter wavelengths. If the particles are not spherical but elongated, two plasmon bands may occur, one for the transverse and the other for the longitudinal resonance. Figure 2.3 shows the UV–visible spectra of spherical 18-nm gold nanoparticles (colloids) with an absorption maximum at 525 nm [7].

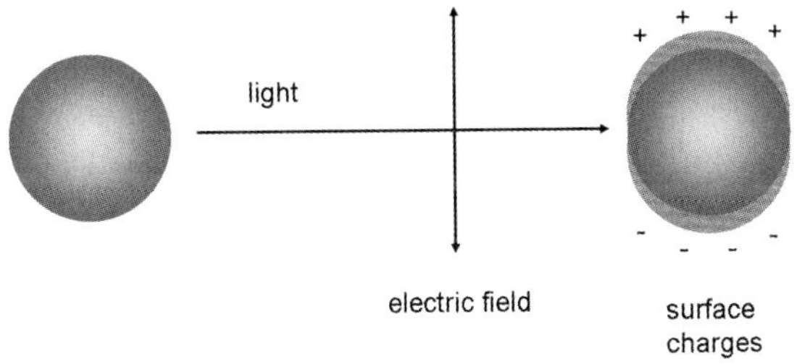

Figure 2.2 Formation of plasma resonance by the interaction of external electromagnetic radiation with the confined electron gas in a metallic nanoparticle.

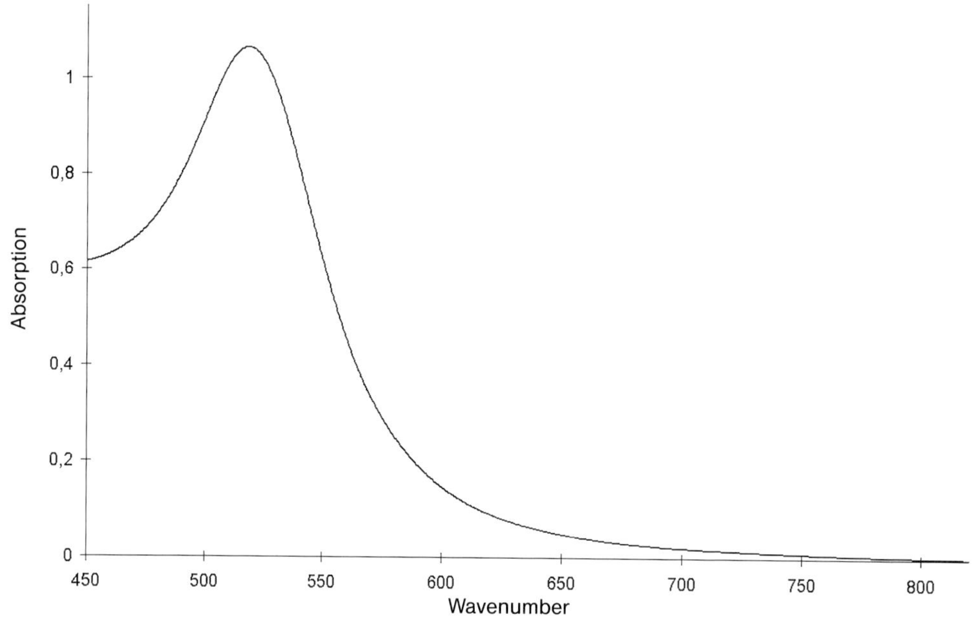

Figure 2.3 UV–visible spectrum of 18-nm gold particles.

Whereas the plasmon resonances of the three metals copper, silver and gold is in the visible region, for most other metals it is in the UV region and so cannot be observed with the naked eye.

The disappearance of the typical color in the case of gold when a critical size is reached and the appearance of blue or red colors simply means that the plasmon resonance superimposes the relativistic effect and so covers the typical color of bulk gold.

The second part of the above definition names the use of individual or combined functionalized subunits. As an *individual functionalized subunit* or building block, molecular switches may serve as an example. A molecular switch is a molecule that exists in two different states which can be adjusted by external stimuli. The two states must display different physical properties, each being stable with long lifetimes. If addressable by electrical or any other stimuli, such molecules could in principle serve as building blocks in future storage systems. For instance, catenanes consist of two interlocked rings equipped with electrochemically active parts (see Figure 2.4). Applied electric potentials cause Coulomb repulsion and make one of the rings move relative to the other, ending in another stable configuration. The sketch in Figure 2.4 elucidates the process [8].

Finally, an example of *combined functionalized subunits*, the last of the three conditions in the definition, can be given. Nature is a perfect nanoarchitect. Many parts of living cells can be considered as a combination of functionalized building blocks, although cells themselves have dimensions in the micro regime. The probably most exciting molecule in Nature is deoxyribonucleic acid (DNA) with its

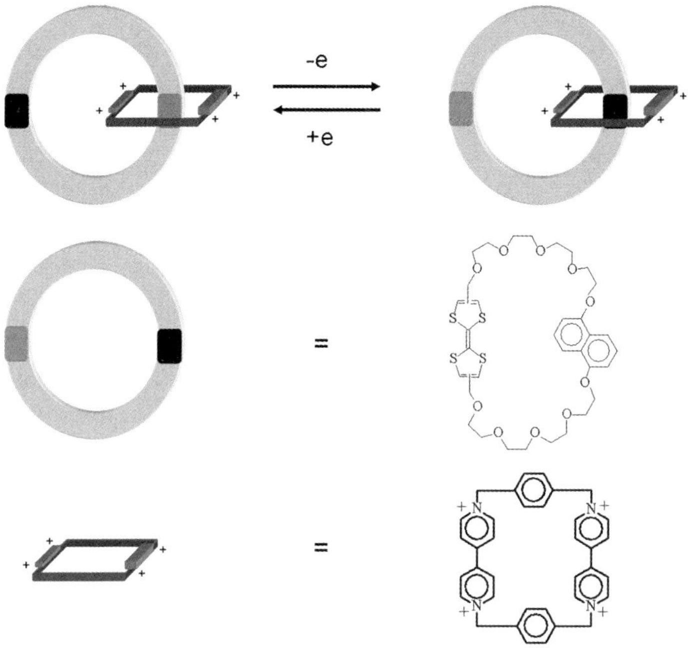

Figure 2.4 Sketch and molecular examples of a catenane-based switching device. The counteranions $[PF_6]^-$ have been omitted.

unique double helical structure. It consists of fairly simple subunits: four different heterocyclic bases, phosphate anions and pentose fragments. It is the shape complementarity of the bases that enables an almost infinite number of combinations to give base pairs, which finally encode the genome of any living system using hydrogen bonds to link complementary bases: thymine (T) combines only with adenine (A), cytosine (C) exclusively with guanine (G). The sugar fragments and the phosphates generate a backbone, holding the base pairs together. Figure 2.5 illustrates the decisive interaction between the four bases. The sequence of bases in one single strand determines the sequence in the other.

In practice, the number of different combinations in a human genome is infinite, considering the realistic length of a DNA double helix consisting of about 3.2 billion base pairs.

The above definition and the following elucidating examples clearly demonstrate that nanoscience and nanotechnology in a strictly scientific manner do not consider simple scaling effects. One speaks of scaling effects if a material is downsized from the macro/microscale even to the nanoscale and the properties, if at all, change continuously, but not spontaneously. In other words, characteristic properties are already present in the micro regime and only change gradually on reaching the nanoscale. A typical example will help to understand easily what a scaling effect is.

The well-known moth-eye effect is a well-developed natural system to avoid light reflection from the eyes of night-active insects [2]. The moth eye is built up from

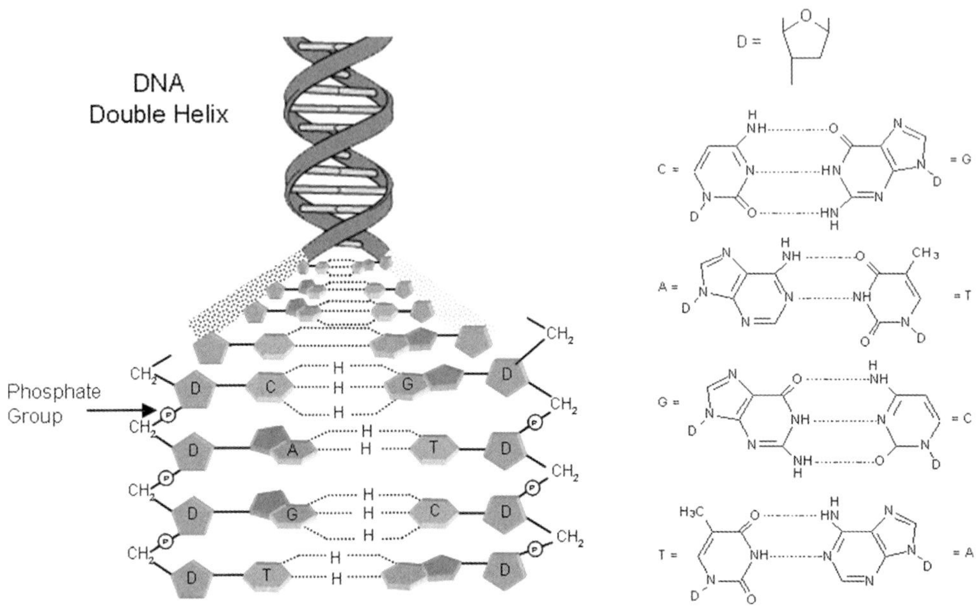

Figure 2.5 Illustration of the base pair interactions in DNA.

hemispheres 200–300 nm in diameter. Since they are smaller than the wavelength of visible light, a continuous increase in the refraction index follows, avoiding the strong reflection that occurs when light hits a flat and optically denser medium. The principle of the moth-eye effect can be used to structure surfaces artificially, for instance those of transparent materials, in order to avoid unintended reflection of light: windows, solar cells, spectacles, and so on. The techniques to nanostructure surfaces are manifold and will not be considered here. Antireflection does not start off from a distinct point. The only condition is that the structure units must be smaller than the wavelength of light. It works with 300-nm units and also with 50-nm building blocks, of course with varying results, but it works. Numerous such scaling effects have been developed into very important techniques. However, they are wrongly called "nanotechnology", since they are based only on scaling effects and not on real nano-effects as the definition demands.

2.2
From Nanoscience to Nanotechnology

Most of the currently known nano-effects are still deeply rooted in nanoscience, that is we cannot speak at all of a technology. In spite of the obvious contradiction, one usually speaks of nanotechnology even if a technology has not yet been realized. In the following, a careful differentiation will be put forward, not just between science and technology, but also with the usual intermediate step, called (nano)engineering. The development of a technique from a scientific finding never happens in a single

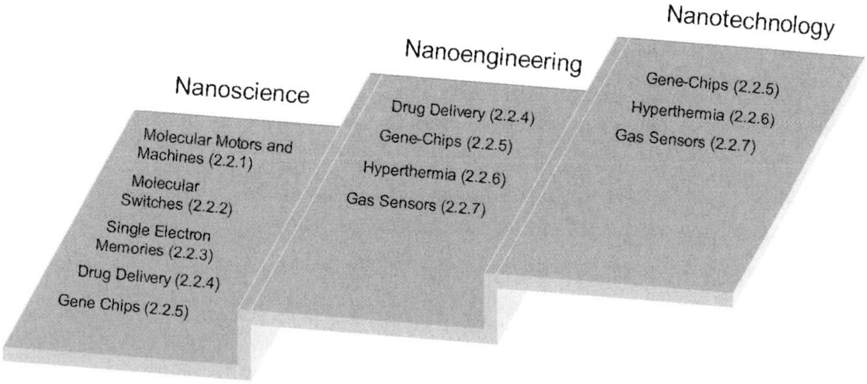

Figure 2.6 Examples of nanoscience, nanoengineering and nanotechnology.

step. Rather, a progressive procedure is necessary to develop a working device successfully. The "proof of principle" and provisional, but working, systems have to be generated before a final and commercially useful technique may result. Figure 2.6 illustrates in a rather simple manner what will be expressed.

Nanoscience, nanoengineering and nanotechnology are represented by three steps following each other. Each step contains several of very many possible examples, which will help to realize how nanoscience develops into nanotechnology. Some of the areas are still only to be found in "nanoscience", others have already developed into the "nanoengineering" step or even into "nanotechnology"; some are present in all three fields, indicating that there is still a further need for basic research to improve or to extend existing technologies. The examples presented for the three steps will be described below in order to elucidate the principles of development from basic research to nanotechnology.

2.2.1
Molecular Motors and Machines

Nature is a perfect nanotechnologist and we are well advised to learn from it. Distinct proteins and protein assemblies are known to perform special motions in response to biological stimuli [9–13]. Such systems are called molecular motors or molecular machines. Numerous attempts have been made during the last two decades to transfer the increasing knowledge of biological systems on a molecular level to devices consisting of completely artificial components or of hybrid systems where biomolecules and technical building blocks interact.

Myosins, kinesins and dyneins are frequently studied natural molecular motors [10–13]. Energetically fuelled by adenosine triphosphate (ATP), these proteins can move back and forth on actin filaments or microtubules transporting substrates. It is not the intension of this chapter to describe these natural molecular machines; rather, it is the discussion of man-made architectures in the sense of nanotechnology. Just one example illustrating Nature's principles will be briefly presented: the transport of actin filaments by myosins.

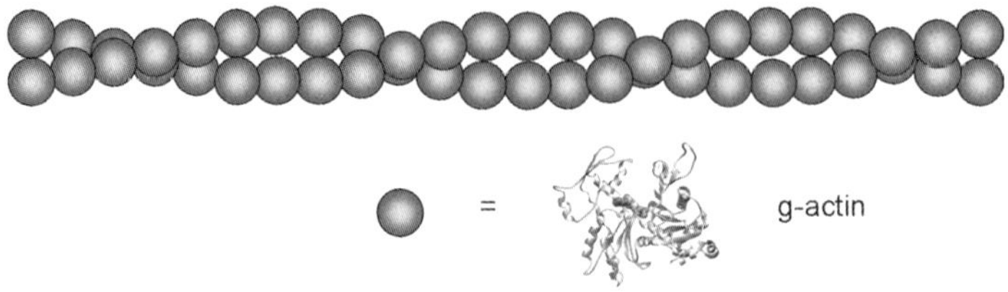

Figure 2.7 Sketch of an actin filament.

Actin filaments consist of double-stranded polymers, composed of globular actin monomers, as is indicated in Figure 2.7.

Eighteen different types of myosins exist, specialized for muscle contraction, signal transduction, vesicle transport, and so on. Myosins are composed of about 200-kDa "heavy chains" and 20-kDa "light chains", wound around the so-called neck domains. One myosin end is marked by the motor domain, as is indicated in Figure 2.8.

Figure 2.9 illustrates the formal transport of actin filaments on a myosin-modified surface. ATP is the energy supplier.

An impressive bio/artificial hybrid machine has been constructed using chaperonin systems. Chaperonins are proteins which can be isolated from *E. coli*, mitochondria and chloroplasts. A series of chaperonin crystal structures have been solved [14–18]. In Nature they make newly formed proteins, folding in their cylindrical cavities with the help of ATP as energy supplier [19]. A detailed description of the working mechanisms of different chaperonins has been published [20]. Instead of acting as host for natural proteins, chaperonins can also be used to capture and to release various nanoparticles with the help of ATP. In Figure 2.10 a primitive sketch of the uptake and the ATP-triggered release of 2–3-nm CdS nanoparticles is presented.

As mentioned above, kinesins are another class of natural molecular motors. They are capable of transporting cargo along intracellular microtubules [21]. In Nature they

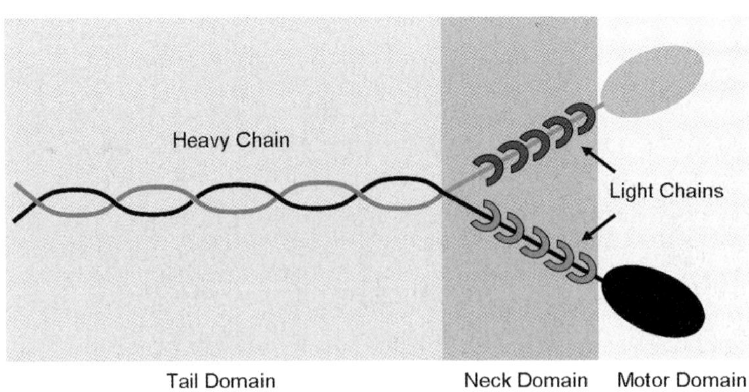

Figure 2.8 A double-headed myosin, consisting of heavy and light chains.

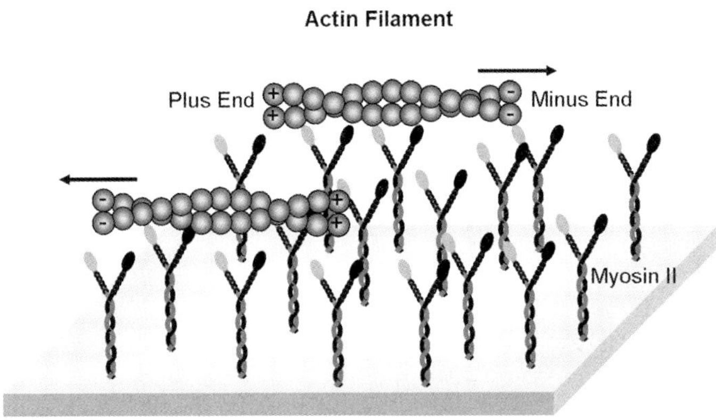

Figure 2.9 Myosin-coated surface transporting actin filaments from plus to minus. ATP serves as the energy supplier.

are part of the transport system for organelles, proteins and mRNAs. Conventional kinesin is composed of two 80-nm long 120-kDa chains, connected to two 64-kDa chains. The heavy chains are rod-like structures with two globular heads, a stalk and fan-like end [9, 22, 23]. One-headed kinesins are also known [24]. The mechanism of motion has been intensively studied using one-headed kinesins [25, 26], but will not be described here. Of course, it is also enabled by the energy of ATP hydrolysis. Microtubules are built up of 8-nm periodic building blocks of heterodimers of the subunits α- and β-tubulin, forming hollow tubes 24 nm in diameter.

Instead of a cargo of natural material such as vesicles or organelles, a recent example impressively demonstrates that artificial nanomaterial can also be transported by kinesin–microtubule systems; 7.6-nm core/shell CdSe/ZnS nanoparticles were functionalized with biotin–avidin. The as-modified quantum dot complexes were then bound to immobilized and fluorescently labeled microtubules. The movement of the loaded kinesin along the microtubules was observed by means of epifluorescence and total internal reflection fluorescence (TIRF) [27]. Figure 2.11 shows a sketch of the microtubule–kinesin–quantum dot hybrid system. The particle

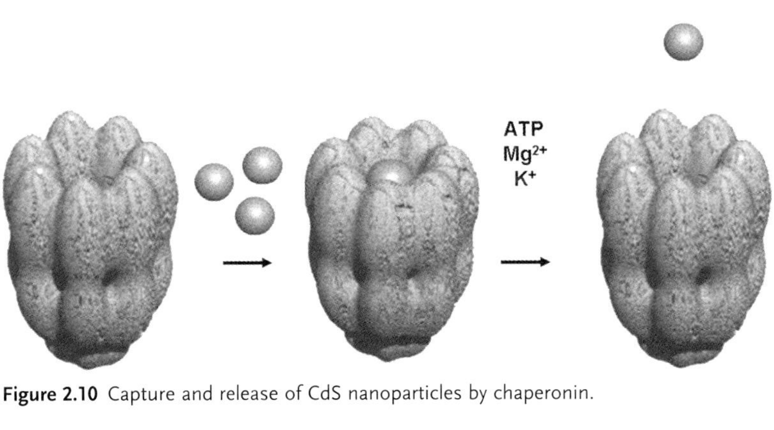

Figure 2.10 Capture and release of CdS nanoparticles by chaperonin.

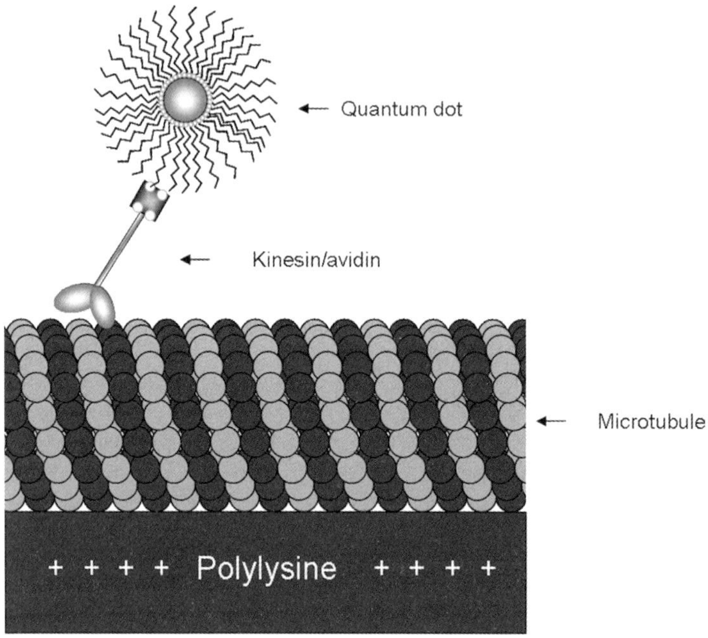

Figure 2.11 A kinesin–avidin–quantum dot hybrid system, moving on the surface of a microtubule.

transport could be visualized over 1200 s, considerably longer than in any comparable experiment before.

Numerous totally artificial molecular machines with non-biological components have also become known. For instance, the photoisomerization of substituted alkenes is another general method to create molecular motors, provided that the process is reversible and the two isomers are kinetically stable. In the example shown in Scheme 2.1, the four bulky substituents are responsible for an energy barrier at $-55\,^{\circ}$C between the *trans* and *cis* configurations [28].

Another light-driven system is the so-called "Irie" switch, a molecule with a light-sensitive C–C bond that can be opened or closed by two different frequencies (Scheme 2.2) [29].

Ring opening occurs under UV light and ring closure by light with a wavelength >600 nm. Switching processes induced by light are in practice much more easily

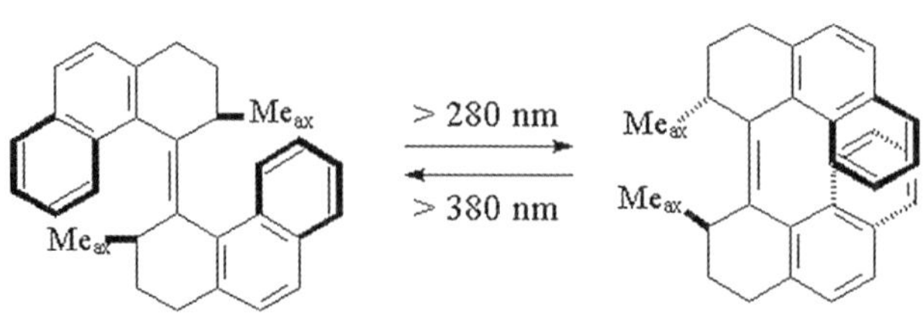

Scheme 2.1 A tetrasubstituted alkene acting as molecular motor by photoinduction.

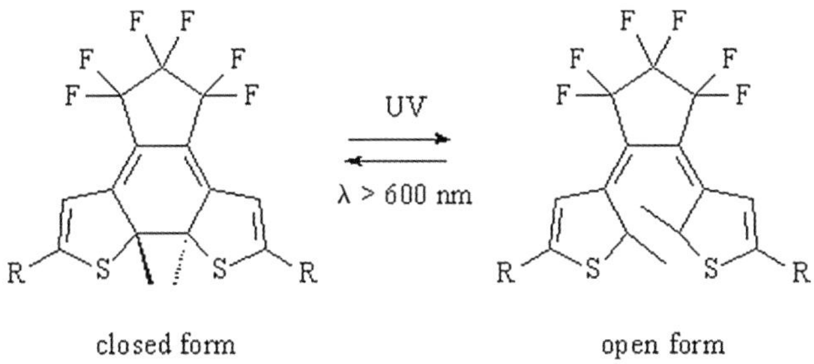

Scheme 2.2 Light-induced closure and opening of a C–C bond ("Irie" switch).

managed than those using extensive chemistry. Nevertheless, this and comparable objects are still part of basic research.

Rotaxanes consist of two parts: a stiff bar-like part and a ring-shaped part, arranged around the bar. Due to electrostatic interactions, the ring prefers a distinct position. By chemical oxidation of the interacting position in the bar, repulsion results and the ring is shifted to another position. Reduction brings the ring back to the former position. Figure 2.12 shows a rotaxane molecule with the ring in two different positions [30]. Both configurations are stable without any voltage applied.

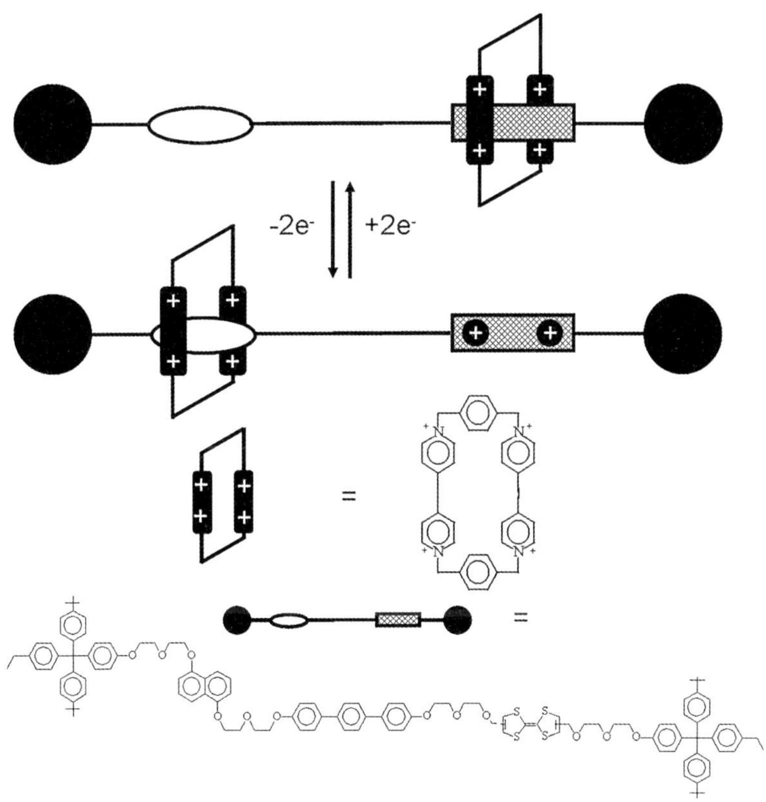

Figure 2.12 Illustration of a rotaxane system.

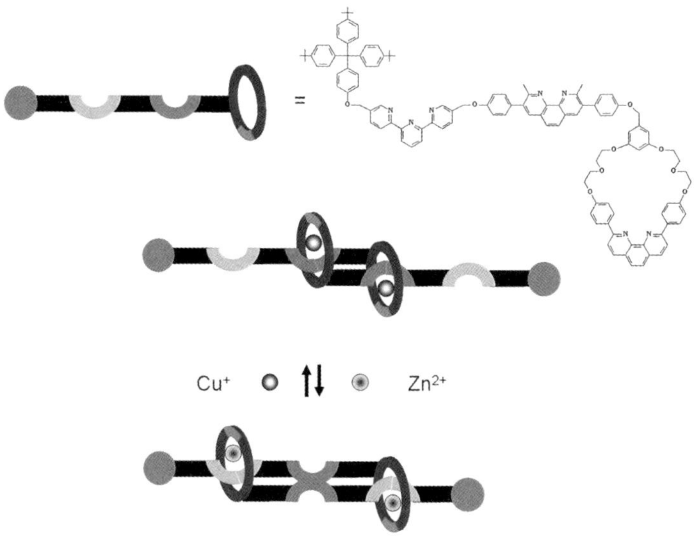

Figure 2.13 A redox-driven "molecular muscle" consisting of two combined rotaxane molecules.

Another elegant example of a rotaxane system is shown in Figure 2.13. A Cu^{+}-containing rotaxane dimer is contracted by the exchange of the Cu^{+} ions by Zn^{2+} ions and vice versa. Except as a molecular switch it can be considered as model of a molecular muscle [31].

A chemically driven rotaxane-like "molecular elevator" has been constructed and is illustrated in Figure 2.14. Protonation and deprotonation of the amine moieties of

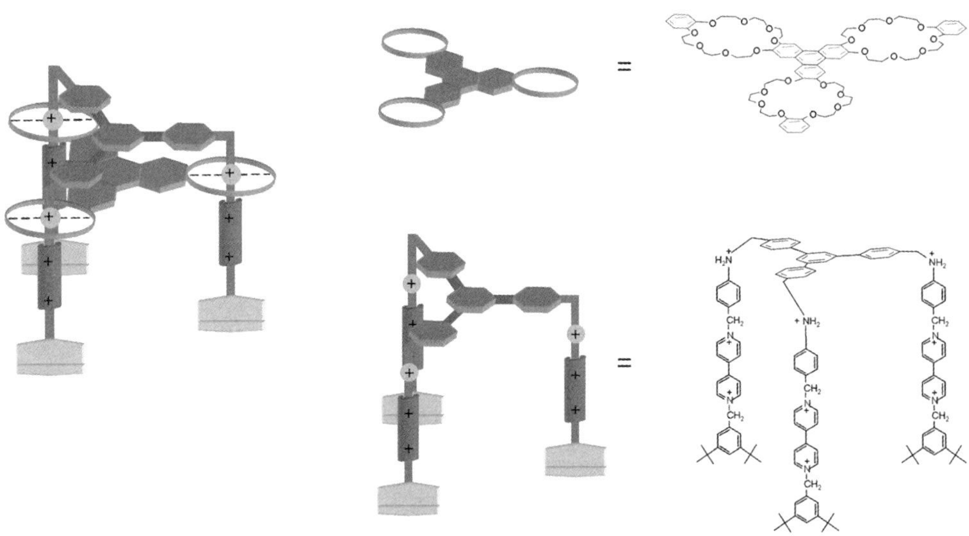

Figure 2.14 A rotaxane-based "molecular shuttle". The platform can be moved up and down by means of addition of acid or base.

part A makes the "platform" B move up and down depending on the NH/NH_2^+ situation in A [32, 33].

It is obvious that the above rotaxane examples are so far not really suited to work in devices, since complex chemistry is necessary to oxidize and to reduce the specific positions. However, the study of such or similar systems is of enormous importance in order to gather experience and to improve continuously the conditions to make such systems applicable.

2.2.2 Molecular Switches

The reasons for the worldwide and intensive search for novel generations of switches and transistors lie in the unavoidable fact that the limits of the present silicon technologies will eventually be reached. The famous Moore's law predicts that between 2010 and 2020 the two-year rhythm of doubling the capacity of computers will find a natural end due to a typical nano-effect [34]: below a not precisely known size barrier, silicon will lose its semiconductor properties and instead it will behave as an insulator. Other technologies which are based on very different nano-effects have to follow. For instance, magnetic data storage systems involving the so-called spintronics [35], magnetic recording systems using nanosized magnetic nanoparticles [36], magnetic domain walls in nanowires [37], and so on, are developing tremendously. All are still far from being of technological relevance in the near future and even the nanoengineering step has not really been reached so far. Hence they are still objects of intense basic research in nanoscience.

The term "molecular switch" is used for molecular systems which are stable in two different states. One state represents 1 and the other state 0. Different states may consist either of two different geometric conformations or of two different electronic states. Both states must be interconvertible by external stimuli. An example of molecular systems existing in two different geometric states has already been introduced in Section 2.1 with a catenane molecule that can in principle be switched by means of electrical pulses. Furthermore, all examples of artificial molecular motors described in Section 2.2.1 are at the same time molecular switches, since they exist in two different but convertible configurations. However, the use of systems, the switching of which is only based on more or less complex chemistry, looks not so much suited for application in future nanoelectronic devices; rather, it is the need to switch systems by light or electric pulses.

Molecular switches, based only on the change of the electronic spin situation in a molecule, are also promising candidates in this respect. Transition metal complexes with d^4–d^7 configurations can exist in either the high-spin (HS) or low-spin (LS) version. High-spin complexes are characterized by a maximum number of unpaired electrons, following Hund's rules. Low-spin complexes have zero or, in the case of an odd number of electrons, one unpaired electron. For instance, if an octahedral complex exists in the HS or the LS configuration depends on the energy gap between the t_{2g} and the e_g orbitals. The separation of the originally equivalent five d orbitals into three t_{2g} and two e_g orbitals (ligand field splitting) is due to the different influence of

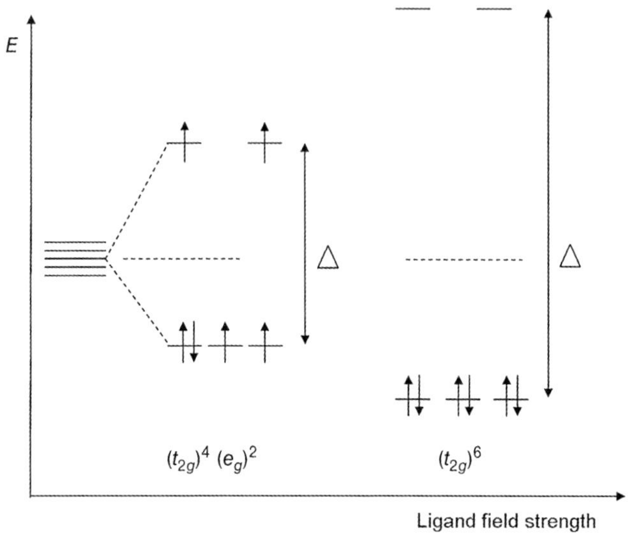

Figure 2.15 High- and low-spin configurations of a d^6 system.

the ligands coordinating the atom or ion along the *x*, *y* and *z* axes (e_g) or between the axes (t_{2g}). Small gaps Δ result in HS configurations and large Δ values in LS complexes. Figure 2.15 shows qualitatively the situation in both cases for a d^6 complex.

The Δ values are dominantly determined by the nature of the ligand molecules coordinating the corresponding transition metal atom or ion. So-called weak ligands (H_2O, halides X^-) cause small ligand field splitting and strong ligands (CN^-, CO, olefins) cause large Δ values. However, situations exist where the energy difference between the HS and LS configurations is small enough to be influenced by stimuli from outside and, consequently, switching between both electronic configurations becomes possible, provided that the transition between the two states is abrupt. Numerous such spin transition complexes have been identified and are of increasing interest with respect to molecular switching systems. It is of special relevance that HS–LS transitions can be induced by different stimuli such as temperature, pressure or light.

An example of a temperature-switchable complex is given in Figure 2.16. The two configurations of the d^7 Co(II) complex can be followed by the magnetic susceptibilities. The HS version has a total spin of $S = 3/2$ and the LS form of $S = 1/2$ [38].

The energetically higher lying states of the HS configuration are reached by increasing the temperature from about 130 to 150 K and vice versa.

The tetranuclear d^6 iron complex shown in Figure 2.17 can be switched by temperature, pressure or light [39]. The four iron centers allow switching over three magnetically different configurations: 3HS/1LS, 2HS/2LS, 1HS/3LS. This special situation allows manifold storage and switching varieties. The chance to switch this complex by light makes it a remarkable candidate for future applications. The

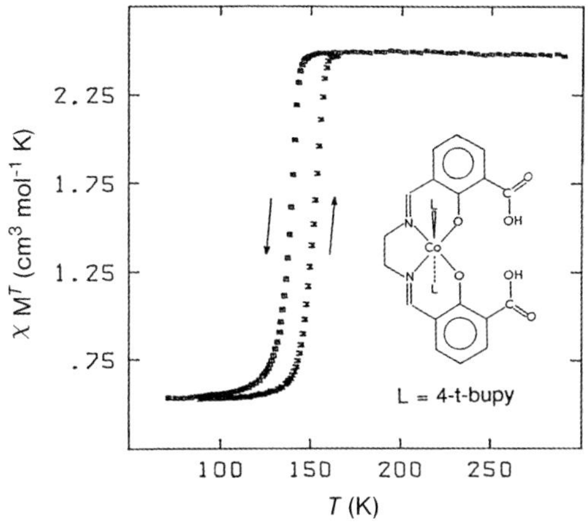

Figure 2.16 Temperature dependence of the magnetic susceptibility of an octahedral Co(II) complex. (© Elsevier, Amsterdam).

light-induced spin-state trapping (LIESST effect) however, could so far only be observed at 4.2 K with $\lambda = 514$ nm, but not the reverse effect.

Since HS and LS configurations are generally linked with a change of the bond length, pressure can be an alternative stimulus.

These few examples of existing molecular switches indicate a development in nanoscience opening up novel alternatives of storage systems on a level that can never be reached with traditional techniques. However, it is also obvious that giant efforts have to be made to reach the nanoengineering step or even the technology level. The route from the discovery of a novel fundamental effect to a product often fails due to unforeseen problems on the practical side. Nevertheless, the finding of fundamental effects in basic research is a presumption to install novel techniques, as history shows.

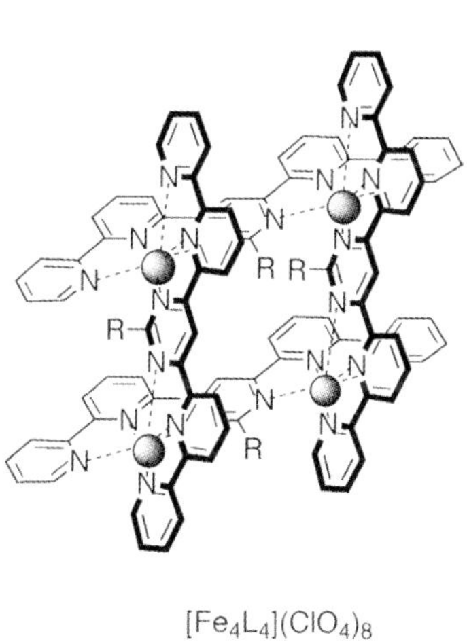

$[Fe_4L_4](ClO_4)_8$

Figure 2.17 Example of an LS ↔ HS switchable Fe(II) complex.

2.2.3 Single-Electron Memories

As already mentioned, the consequences of Moore's law demand alternatives, available on time. Single-electron memories might be a possible alternative to the present silicon technology if one succeeds in using the quantization of electric charge to handle digital information. Single-electron devices would be associated with enormous advantages compared with available techniques: extremely reduced power consumption, no or extremely reduced heat development, high density arrangements of building blocks and the principle possibility of generating three-dimensional memories. Single-electron memories would constitute the ultimate miniaturization of a memory device and it is worth intensively following up any chance to realize this goal. Single-electron memories, single-electron switches or single-electron transistors all require the transport of individual single electrons in a strictly controlled manner. This can only be realized by quantum mechanical tunneling. A device for single-electron tunneling (SET) must contain a unit that can be charged and discharged by single electrons. Charged and discharged units must be independently stable. To prevent uncontrolled transport of electrons, a charged unit must build up a so-called Coulomb blockade to prevent the transfer of a second, third, and so on, electron. Atoms would be ideal candidates to realize such conditions. The formation of an anion in chemistry is such an event, determined by the electron affinity of the neutral atom. However, atoms are still too small to be individually and routinely handled. Therefore, larger scaled, namely nanosized materials have to be found to overtake this "atomic capability". The ability of a unit to act as an atomic substitute depends decisively on the amount of energy to add an additional electron to an initially uncharged unit [40–44]. This charging energy scales roughly with $1/r$ (r = radius of the nanosized unit). What kind of material fulfils these conditions? – generally, all species that exhibit quantum size behavior. Nanotubes, nanowires and nanoparticles of conducting or semiconducting materials can have such extraordinary facilities. To observe quantum size behavior of a species, its dimensions in one, two or all three dimensions have to be reduced to such an extent that electrons are no longer freely mobile in all dimensions, but are confined to such an extent that they occupy more or less discrete energy levels. In Section 2.1, the phenomenon has already briefly been discussed in connection with the appearance of plasmon resonances. A detailed discussion of quantum size phenomena can be found in Chapter 4. A simplified demonstration of the consequences of electronic confinement of metals is depicted in Figure 2.18.

When reaching situation (c) in Figure 2.18, one speaks of quantum dots, since such particles no longer follow classical physical laws for bulk materials but obey quantum mechanical rules like atoms and molecules do, even at room temperature. Among many different ways to obtain information on whether the quantum dot situation has been reached or not, individual contacting of a particle by two electrodes and to study the current (I)–voltage (U) behavior gives the clearest answer. Figure 2.19 shows the principle of such an experimental arrangement.

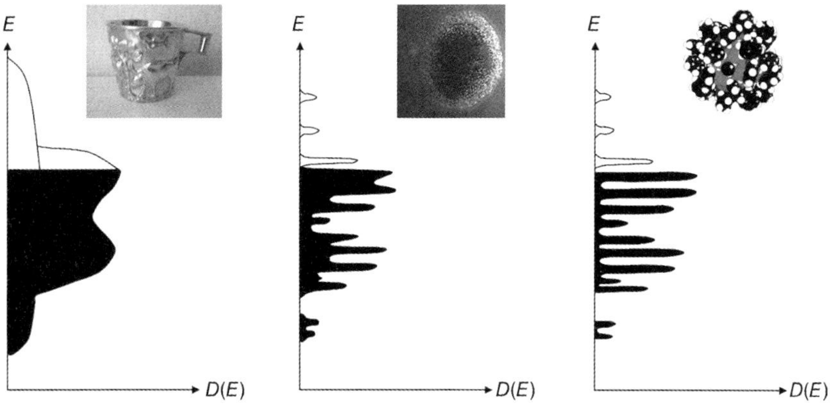

(a) Metallic State (b) 20 nm Gold particle at low temperature (c) Quantum dot at room temperature

Figure 2.18 The electronic transition from the bulk state to a quantum dot.

SET from one electrode into the nanoparticle leads to an increase in charge by e^- (1.6×10^{-19} C) linked with an increase in the electrostatic energy E_C by $E_C = e^2/2C$, where C is the capacity of the nanoparticle. As can be seen from Figure 2.19, the metal nanoparticle does not directly touch the electrodes, but an insulating envelope (or a respective distance) separates it from the contacts to gain an appropriate capacity of the system. In order to avoid uncontrolled thermal tunneling of electrons, E_C must be much larger than the thermal energy $E_T = k_BT$ (k_B = Boltzmann's constant $= 1.38\times10^{-23}$ J K^{-1}): $e^2/2C \gg k_BT$. The observation of an SET process will only be possible either at very low temperatures or with very small C values. Since $C = \varepsilon\varepsilon_0 A/d$ (ε = dielectric constant, d = electrode distance from metal core, A = surface area of the particle), small C values can be realized by very small particles having a sufficiently thick insulating shell. The charge generated on the particles causes a voltage $U = e/C$, linked with a current $I = U/R_T$ (R_T = tunneling resistance).

The temperature dependence of SET has been convincingly demonstrated by the study of a 17-nm Pd nanoparticle, covered by a shell of $H_2NC_6H_4SO_3Na$ molecules. As can be seen from Figure 2.20, the I–U characteristic at 295 K is a straight line, following Ohm's law. At 4.2 K, however, a well-expressed Coulomb blockade is observed, indicating that between about −55 and +55 mV the current is interrupted due to the existence of an additional electron in the particle, blocking the transport of a second one [45].

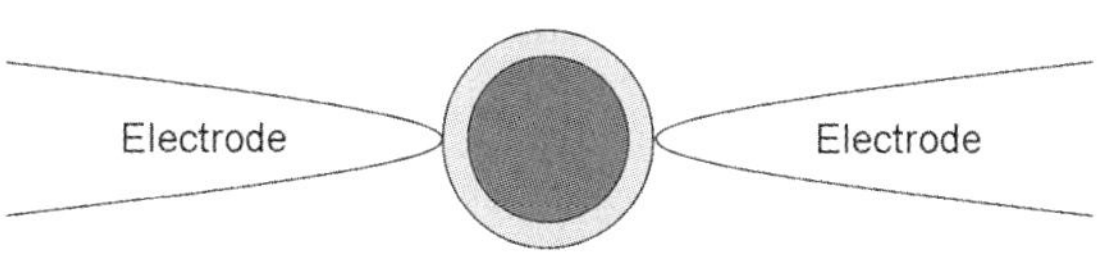

Figure 2.19 Experimental arrangement for the measurement of the I–U characteristic of a metal nanoparticle.

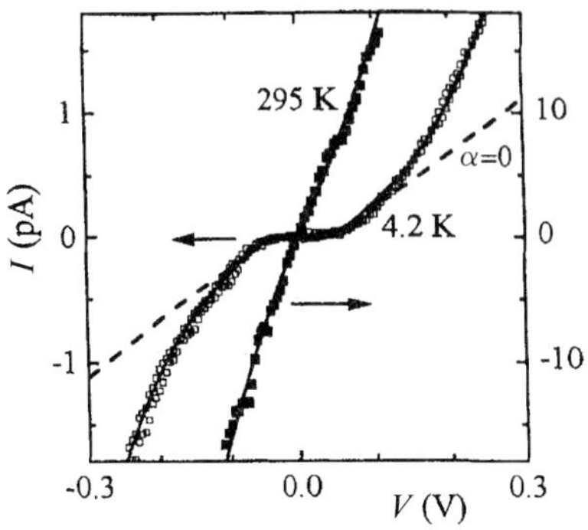

Figure 2.20 *I–U* characteristics of a 17-nm Pd particle at 295 K (Ohm's law behavior) and 4.2 K (Coulomb blockade). (© American Institute of Physics).

Instead of working at very low temperatures, in practice it is more attractive to decrease *C* sufficiently and to enable Coulomb blockade at room temperature. This aim has indeed been achieved by using the nanocluster $Au_{55}(PPh_3)_{12}Cl_6$ with its metal nucleus of only 1.4 nm and a 0.35-nm thick ligand shell [46, 47]. The *I–U* characteristic, shown in Figure 2.21, has been measured at room temperature. In spite of the high temperature, a Coulomb blockade between about −500 and +500 mV is registered [48].

The electric contacting of a single nanocluster has been performed using a tip of a scanning tunneling microscope (STM) and a conductive substrate on which the particles had been deposited from a very dilute solution.

A deeper insight into the electronic situation in these Au_{55} quantum dots was possible by studying the *I–U* behavior at 7 K [49]. Due to the low temperature, the

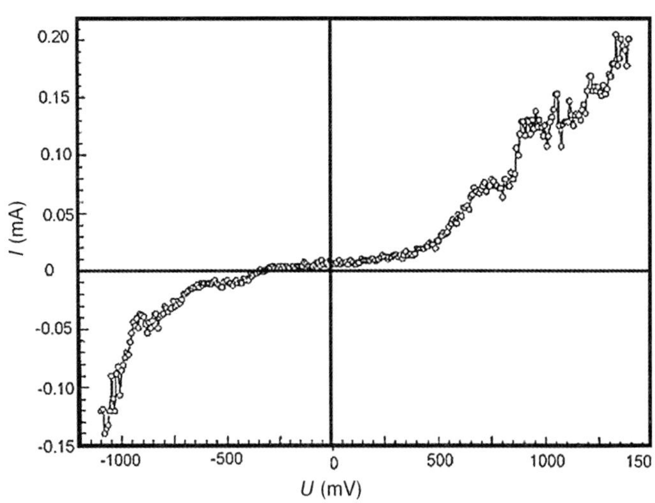

Figure 2.21 Room temperature Coulomb blockade of $Au_{55}(PPh_3)_{12}Cl_6$. (© Springer, Berlin).

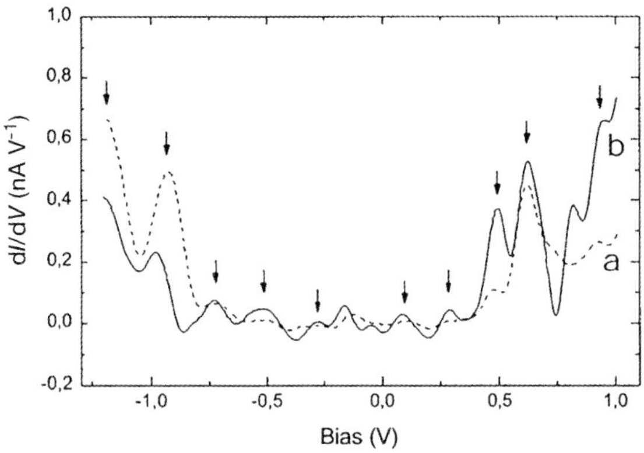

Figure 2.22 dI/dV diagram for $Au_{55}(PPh_3)_{12}Cl_6$ at 7 K indicating level splitting of 135 meV. (© American Chemical Society).

Coulomb blockade is enlarged; however, the most informative knowledge can be seen from a diagram using dI/dV values instead of I (Figure 2.22). Generally, the blockade then turns into a minimum on the U axis. Due to the low temperature, discrete energy levels in the minimum become visible in terms of conductivity oscillations with an average level spacing of 135 meV. The dashed line and the solid line result from two measurements on the same particle, but at different positions, namely above a phenyl ring of PPh_3 and at a position above bare gold atoms. They agree fairly well and so indicate that the result does not depend on the matter between the tip and Au_{55} nucleus.

This result demonstrates the existence of a perfect quantum dot, working at room temperature and representing exactly position (c) in the sketch in Figure 2.18. Figure 2.22 also makes it clear why such working units are sometimes called "artificial atoms".

The Coulomb blockade, based on the transfer of single electrons, represents a perfect single-electron switch and can in principle also be used as a single-electron transistor, as indicated in Figure 2.23.

These fundamental findings make Au_{55} and metal particles of similar size excellent candidates for use in future storage systems. Intensive research and development are still necessary to reach this ultimate goal. The very first steps from the science level to the engineering step are in progress, but nevertheless, quantum dot memories are still deeply involved in basic research.

2.2.4 Drug Delivery

The controlled release of a drug has a significant influence on the therapeutic efficiency. For numerous drugs there exists an area of concentration with the greatest success. Compared with traditional tablets or injections, drug delivery systems have a profile with continuous release, as indicated in Figure 2.24 [50].

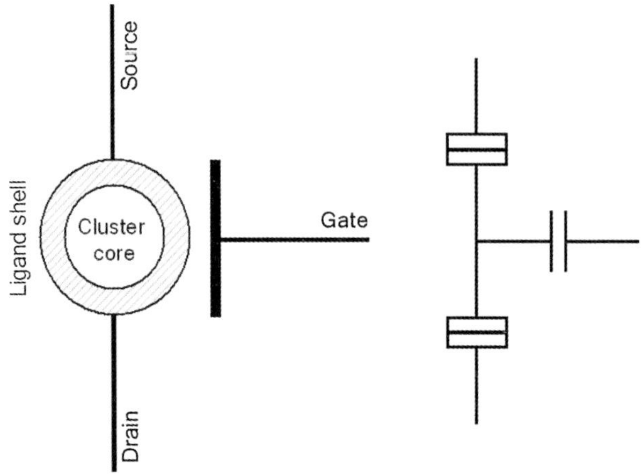

Figure 2.23 Circuit of a single-electron transistor.

Conventional methods are characterized by a rapid increase in release, followed by a fast decay (peak-and-trough cycle). Therefore, there is an urgent need for continuous drug release in the therapeutic area of concentration. The history of drug delivery based on implanted systems goes back to the 1960s [51]. Instruments at that time had considerable dimensions containing electrically driven pumps. In the course of the last two decades, numerous systems have been developed with novel principles [52–56]. Partially they also use novel principles such as release via the skin or the nasal mucous membrane. For longer times of therapy a drug delivery implant of reasonable size and high reliability would be the most effective system. Nanotechnology offers numerous chances to achieve that goal.

Entrapping and encapsulating drugs in nanostructured systems have been developed with remarkable success, releasing a drug uniformly over longer time periods. They are based on guest–host systems. Different kinds of chemical bonds between the guest (drug) and host system determine the speed of release. Hydrogen or van der Waals bonds and also electrostatic interactions can be used to combine the guest and

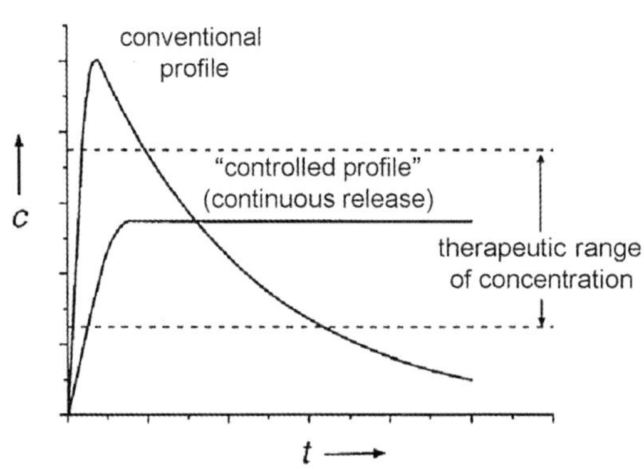

Figure 2.24 Concentration (c)–time (t) profile of a conventional and a controlled drug release.

host to such an extent that slow release is enabled. A host–guest system based on dendrimers as host molecules will be described briefly as an actual example of development. Dendrimers are highly branched molecules in the nanometer size regime. They consist of a core unit from where "branched branches" extend in different directions, forming a three-dimensional architecture bearing end groups of various functionality. Dendrimer structures unavoidably contain cavities inside the skeleton. These are able to take up guest molecules of appropriate size and to release them slowly depending on the surrounding conditions. With an increasing number of branches, the number and geometry of the cavities become variable and also increase [57]. Figure 2.25 shows a formal sketch of a dendrimer molecule taking up and releasing host particles.

Another interesting nano-based drug delivery system uses superparamagnetic iron oxide nanoparticles, usually embedded in a polymer matrix and attached to a drug system. External, high-gradient magnetic fields are applied to transport drug-loaded beads to the corresponding site in the body [58, 59]. Once the system has concentrated in the tumor, the drug is released using different techniques such as increase in temperature, change of pH value or enzymatic activity.

Another method uses superparamagnetic iron oxide nanoparticles, the surface of which is modified by DNA sequences. Those particles easily enter cells using receptor-mediated endocytosis mechanisms in combination with a magnetic field gradient [60]. Having entered the cell, the DNA is liberated and can enter the nucleus. This so-called non-viral transfection is of special interest for gene therapy.

A rapidly growing drug delivery development is based on the use of multifunctional nanoengineered capsules containing various kinds of active compounds. Attempts have been made to solve the general problem of treating only pathological cells and not healthy ones by using, for instance, functionalized polymer capsules having distinct release, permeability and adhesion properties. The inner volume can

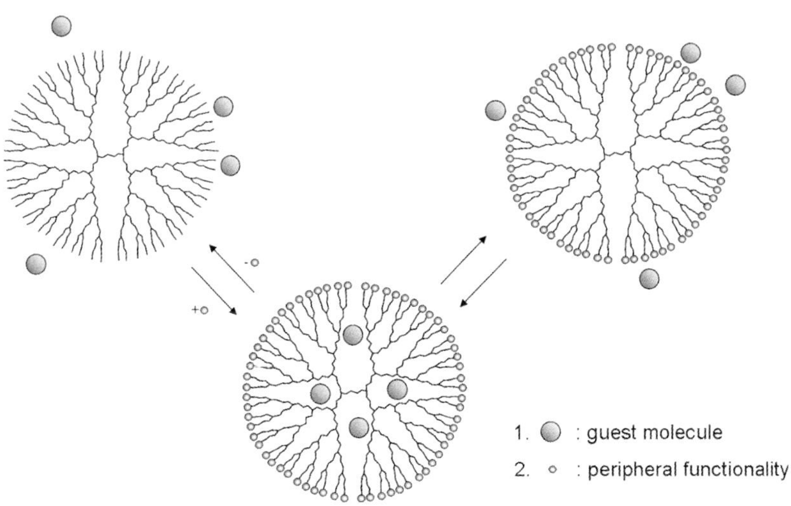

Figure 2.25 Inclusion of guest molecules in cavities of a dendrimer.

be filled with magnetic nanoparticles which allow aimed transport by outer magnetic fields, by modification of the capsule surface with specific receptors to target specifically diseased cells or by generating capsules acting as a nanoreactor producing products which are only toxic for diseased cells and cause selective apoptosis [61, 62].

Self-rupturing microcapsules consist of polyelectrolyte membranes, permeable to water, but not to the drug-containing degradable microgels, filling the holes of the capsules. The hydrolytic degradation of the microgels causes a swelling pressure, rupturing the outer membrane [63].

Polyelectrolyte capsules which degrade at physiological pH values open up novel prospects for drug delivery [64]. Intracellular targets such as nucleic acids or proteins can cause opening of the capsules. Using $CaCO_3$ particles as the carrying material for fluorescein isothiocyanate–dextran (FITC–dextran), assemblies of $CaCO_3$/FITC–dextran are formed by coprecipitation, followed by layer-by-layer polyelectrolyte membrane formation, for instance with poly-L-arginine as the polycation and dextran sulfate as the polyanion. Finally, the $CaCO_3$ particles are removed using buffered EDTA solution.

Finally, laser-induced opening of a polyelectrolyte membrane inside living cells can be mentioned [65].

A completely different drug delivery system has been developed using nanoporous alumina membranes for the controlled long-term release of drugs. Nanoporous alumina membranes with variable pore diameters between 10 and 200 nm are easy to prepare and are used to control the speed of release depending on the pore size [66]. Figure 2.26 shows a sketch of the implantable device and Figure 2.27 illustrates the influence of the speed of release of the same molecule depending on the pore diameter. Instead of a real drug, the system has been developed using crystal violet for the easy determination of concentrations by means of UV–visible spectroscopy. Of course, this system requires individual developments for each drug to optimize the pore size and solubility conditions, for instance with the help of surfactants.

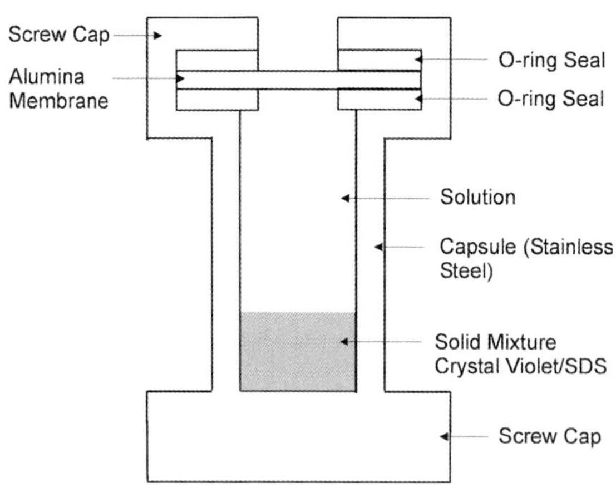

Figure 2.26 Sketch of a drug delivery system using nanoporous alumina membranes to control the speed of release.

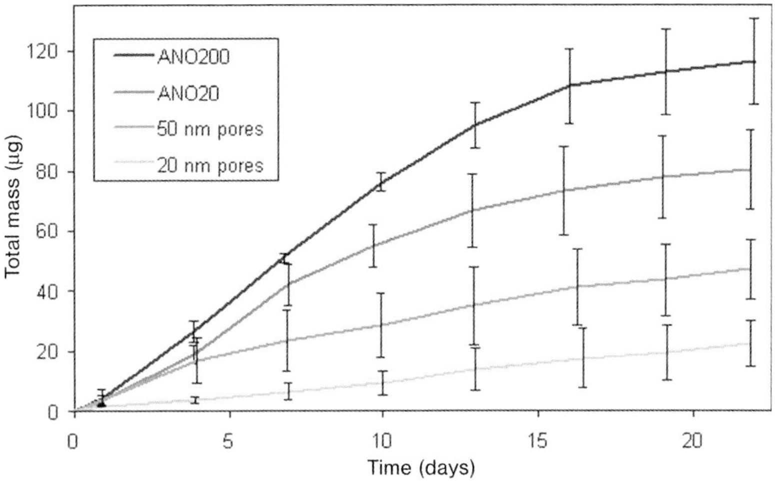

Figure 2.27 Dependence of the release of crystal violet on the pore size.

None of these nano-based techniques have yet reached the "technology standard". Rather, they are under development and have partially reached the "engineering state".

2.2.5
Gene Chips

Progress in the diagnostics of human diseases is of comparable importance to progress in therapy. Nanotechnology is, without doubt, the most promising field where improved developments can be expected. The scientific aim is to develop diagnostics on a molecular level, if possible with routine techniques. Gene chips, also called microarrays, represent the best way to fulfill such dreams. Gene chips are already commercially available, but are still also under investigation and development on the engineering level and also are still objects of intense scientific research. The early diagnosis of a cancer disease is crucial for successful therapy. So, this is a first example with a keyword existing in all three steps in Figure 2.6.

A gene chip consists of a collection of up to 400 000 single-stranded DNA segments (probes), fixed on a glass, silicon or plastic surface. Such DNA microarrays can be used to detect mRNAs which either interact with DNA fragments on the chip or not. Since there can be a huge number of reporters arriving, a microarray experiment can accomplish the equivalent number of genetic tests in parallel. Detection occurs by using organic fluorophore labels. So-called two-channel microarrays contain probes consisting of complementary DNA (cDNA), but also of oligonucleotides. This type of chip is then hybridized with cDNA from two different samples, labeled with two different fluorophores for comparison, for example from cancer cell lines and from a control. Mixing and hybridization on the microarray allow the visualization of the results. Gene chips working on that basis are commercially available and allow cheap and fast diagnosis on a level that was inconceivable one or two decades ago.

DNA microarrays are also used to analyze the sequence of particular genome sequences. Gene chips of this type have, in spite of their enormous contribution in diagnostics, inherent drawbacks due to the fluorophore labeling (in some cases even radioactive labeling is used). Recent advances in nanoscience open the door to increase the sensitivity of DNA detection to an unknown level. DNA–nanoparticle conjugates are powerful tools in this direction.

New developments of microarrays are based on the use of DNA, functionalized with quantum dots. Their excitation and emission properties make them powerful candidates for replacing fluorophore labeling techniques. Gold and silver nanoparticles, but also CdSe and ZnS quantum dots, have been successfully tested.

The surface plasmon resonance of gold nanoparticles and the resulting intense color allow the very sensitive and selective colorimetric detection of corresponding DNA sequences. Au nanoparticles in the size range from about 10 up to 100 nm, typically 10–20 nm, are linked with DNA probe strands by using 3′- or 5′-end mercapto-functionalized species. Due to the preferred formation of strong Au–S bonds, Au–DNA hybrid systems become easily available. Mirkin's group first reported the use of mercaptoalkyloligonucleotide-modified gold nanoparticles [67–71]. They used two samples, each complementary to one half of a target oligonucleotide, so that the formation of a polymeric network was induced by mixing the three species, as indicated in Figure 2.28. The purpose of this experiment was to perform a color change from red to blue, due to the aggregation of the nanoparticles.

Based on the colorimetric nanoparticle approach to DNA detection, microarrays using DNA–gold nanoparticle hybrid systems are being increasingly developed. A three-component system was first described by Mirkin and coworkers [69, 72, 73]. It consists of a glass chip, the surface of which is modified by capture DNA strands to recognize the DNA under investigation. The oligonucleotide-functionalized gold probe and the target DNA complete the system. Intense washing after the combination process results in a selectivity of 10 : 1 for single base-pair mutations. In a final special step, the gold nanoparticles are covered by a silver shell which is simply generated by the catalytic reduction of silver ions on the gold nanoparticles. The capture-strand–target–nanoparticle combination can then be visualized using a flatbed scanner. Due to the presence of silver shells on the gold particles, high

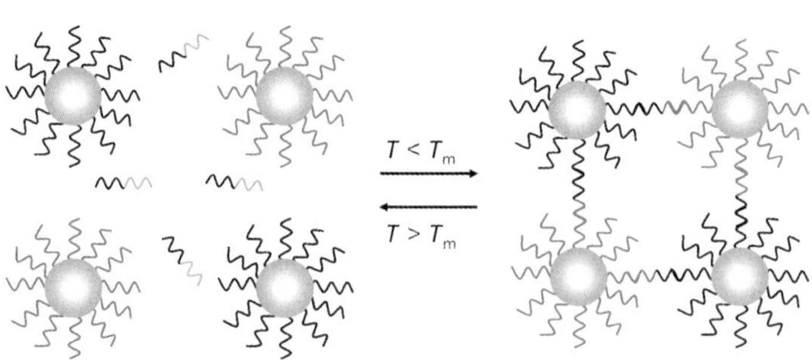

Figure 2.28 Aggregation of oligonucleotide-functionalized gold nanoparticles by means of complementary target DNA.

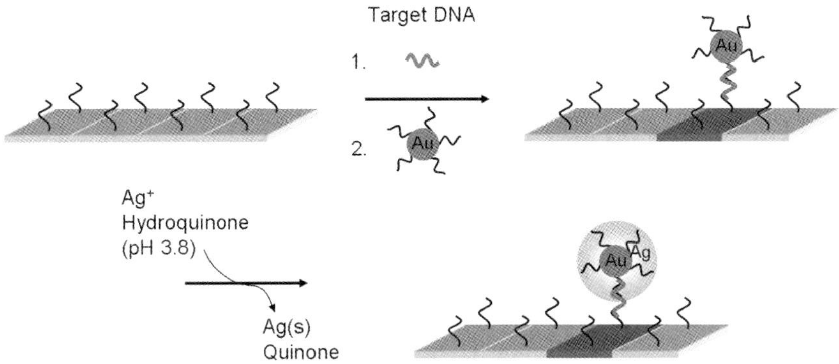

Figure 2.29 Principle of a scanometric DNA assay. A capture oligonucleotide on a surface binds one half of a target molecule and oligonucleotide gold nanoparticles bind the other half. A detectable signal results by the catalytic coverage of the gold nanoparticles by silver.

surface enhanced Raman scattering (SERS) is observed and can also be used for target DNA detection. Figure 2.29 elucidates the various steps.

The improvement of this technique, compared with conventional fluorophore-labeling techniques, is about 100-fold, namely as low as 50 fM. These remarkable nanotechnologically based developments will initiate great progress in diagnostics. As an example, first investigations of Alzheimer's disease (AD) can be mentioned [74].

Alivisatos and coworkers detected single base-pair mutations in DNA by the use of CdSe/ZnS quantum dots in chip-based assays [75]. The detection method was fluorescence microscopy and the detection limit was about 2 nM. This is not yet in the region of the detection limit described above, but it is likely that this technique can be improved considerably since it is known that even individual quantum dots can be detected under ideal conditions [76].

Gene chips based on the use of quantum dots belong to one of the most promising developments in nanotechnology. In an unusually short time, beginning with the very first experience with biomolecule–quantum dot interactions, a development started that has already led to commercially available devices. There are still simultaneous efforts to be made on all three levels. Further improvements of detection limits are still part of nanoscience. At the same time, improvements of routine detection processes are continuing in order to facilitate everyday clinical handling.

2.2.6
Hyperthermia

Hyperthermia, known for several decades, has more or less developed to a level of clinical applications based on nanotechnological attempts and can now be located in the field of "nanoengineering" and also "nanotechnology". It uses the fact that superparamagnetic nanoparticles can be warmed up by external alternate magnetic fields. As has long been known, tumor cells respond sensitively with apoptosis on temperature increases of only a few degrees (40–44 °C) [77–81].

The superparamagnetic state of a material at room temperature is reached when the thermal energy kT (k = Boltzmann's constant) overcomes the magnetostatic energy of a domain or particle. If the particle or domain is small enough, a hysteresis no longer exists or, in other words, the magnetic unit no longer exhibits the ability to store the larger particle's magnetization orientation; rather, the magnetic moments rotate and so induce superparamagnetic behavior. Typical particle sizes for the transition from ferro- to superparamagnetism are in the range 10–20 nm for oxides. Metal particles have to be downsized to 1–3 nm. A great advantage of superparamagnetic particles is the fact that they can be dispersed in various liquids without any tendency to agglomerate, an important condition for applications in medicine.

Several types of superparamagnetic oxide nanoparticles have been investigated for application in hyperthermia. The most promising candidates are magnetite and maghemite since their biocompatibility has already been shown. The amount of magnetic material to reach the necessary temperature depends on the concentration of the particles in the cells. Direct injection allows larger quantities than intravascular administration or antibody targeting. On the other hand, direct injections into the tumor involve a certain danger of the formation of metastases. In any case, the amount of magnetic nanoparticles necessary for a sufficient temperature increase depends on the magnetic properties of the particles and on the external radio-frequency field. Under then optimum conditions only 0.1 mg per mL of tissue is necessary to induce cell death.

2.2.7
Gas Sensors

Gas sensors have been known and applied since the early 1960s [82, 83]. The fields of application range from industrial and automotive needs (NO_x, NH_3, SO_2, hydrocarbons, etc.) via domestic gas determinations (CO_2, humidity) up to the security sector, where traces of explosives have to be detected. In a working sensor system, the information resulting from the chemical or physical interaction between a gas molecule and the sensor has to be transformed into a measurable signal. Numerous possibilities such as electrochemical, calorimetric, acoustic, chemoresistant and other effects are well established. Chemoresistors typically use metal oxides. These change their electrical resistance when they oxidize or reduce gases to be detected [82–85]. Continuous improvements have been elaborated concerning sensitivity, selectivity, stability, response and recovery time [86]. In connection with nanotechnological developments sensors based on metal oxides and metal nanoparticles have been intensively studied and have reached the state of engineering and even technology (see Figure 2.6) [86].

Sensors based on nanosized *metal oxides* provide both receptor and transducer functions [87]. The receptor must ensure a specific interaction of the sensor's surface with the target analyte. The transducer's task is to transform the molecular information into a measurable change of the electrical resistance. For instance, the conductivity of n-type semiconducting metal oxides increases on contact with reducing gases, whereas that of p-type oxides decreases. Oxidizing gases cause opposite effects. A frequently used wide-bandgap n-type semiconductor is SnO_2. A qualitative explanation (for details see [86]) of the working principle says that, in the presence of dry air, oxygen is ionosorbed on the oxide surface, depending on temperature as O_2^- (<420 K), as O^- (420–670 K) or as O^{2-} (>870 K). The electrons required for the reduction of O_2 come from the conduction band, so generating an electron-deficient region, the so-called space–charge layer Λ_{air} [88–91]. Λ_{air} depends on the Debye length L_D, a material- and temperature-dependent value. For SnO_2 at 523 K it is about 3 nm [92]. In real systems water is present to some extent, forming hydroxyl groups on the surface and affecting the sensor's properties. The influence of water has been discussed in detail [93].

Reducing gases interact with the ionosorbed oxygen species and are oxidized, for instance $CO \rightarrow CO_2$, which is desorbed. Even traces of reducing gases decrease the number of oxygen species to such an extent that, due to the release of surface-trapped electrons, the increase in conductance becomes measurable. In the case of oxidizing target gases, the process is inverse: additional electrons are removed from the semiconductor, resulting in an increase in Λ_{air}. Hence adsorption of oxidizing gases, for example NO_2 or O_3, causes a decrease in conductance.

The efficiency of a gas sensor depends not only on the material of which it is made, but decisively also on the size of the particles and their arrangement, since the relevant reactions occur on the particles' surface. In an ideal case all existing percolation paths are used, contributing to a maximum change in conductance. The response time depends on the equilibrium between the diffusion rate of the participating gases. Film thickness and porosity are therefore of special relevance for the quality of a sensor [94, 95].

A vital role in this connection is played by the size of the particles forming the macroscopic film. Since the analyte molecule–sensor interaction occurs on the particles surface, their surface:size ratio plays a dominant role. Since the relative proportion of the surface increases with decreasing particle size, smaller particles should be more efficient than larger particles. This can clearly be seen from Figure 2.30 [87].

SnO_2 particles with diameters below about 10 nm exponentially increase the sensor's response. In addition to the increase in surface area, particle radii in the range of the space–charge layer Λ_{air} decrease the Schottky barriers between depleted zones or even lead to an overlap, with the consequence that surface states dominate the electrical properties and so have a decisive influence on the sensor performance. Very small differences in the particle size can have crucial consequences for the sensor's ability. As has been shown for WO_3 nanoparticles, a reduction from 33 to 25 nm increases the sensitivity towards 10 ppm NO_2 at 573 K by a factor of 3–4 [96]. Several similar examples for other metal oxides are known [86].

Metal nanoparticles as building blocks for sensor systems have been under investigation since the late 1990s. Thin films of ligand-protected metal nanoparticles change their conductance when gas molecules are absorbed in the regions between the nanoparticles [97, 98]. Films of 2-nm gold nanoparticles, covered with octanethiol molecules, turned out to change the conductance reversibly if gas molecules such as toluene, 1-propanol or water became part of the interparticle sphere. This principle

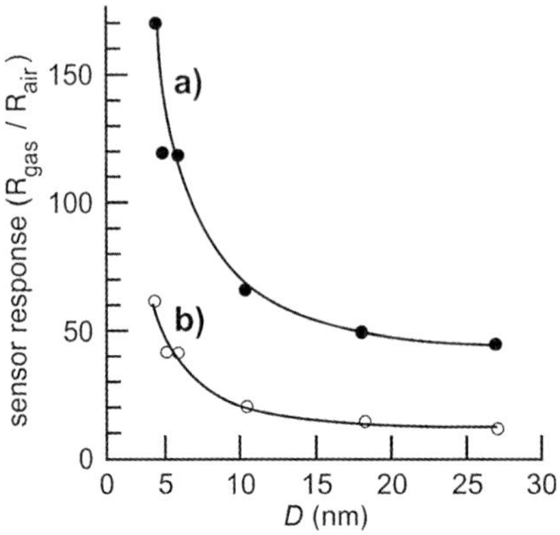

Figure 2.30 Dependence of the sensor response R on the particle diameter D (SnO_2), expressed as R_{gas}/R_{air}. (a) 800 ppm H_2 and (b) 800 ppm CO in air at 573 K. (© Springer, Berlin).

has since been improved by the introduction of specifically functionalized ligand molecules, increasing the sensitivity [99–102]. Self-assembled layers of gold nanoparticles and dendrimer molecules use the ability of dendrimers to host guest molecules in their cavities [101, 103].

Apart from gold, other noble metal nanoparticles have also been tested. Platinum nanoparticles, for instance, partially crosslinked by dithiol molecules, are active sensors towards toluene, H_2O, CO and NH_3 [104–107]. In the case of NH_3, traces down to 100 ppb could be detected. Figure 2.31 informs on the different sensitivities of such a chemoresistor towards different gases.

Whereas charge transport mechanisms between ligand-protected metal nanoparticles have been investigated in detail [86, 108–119], knowledge about the change in conductance due to the influence of various gases is still rather limited.

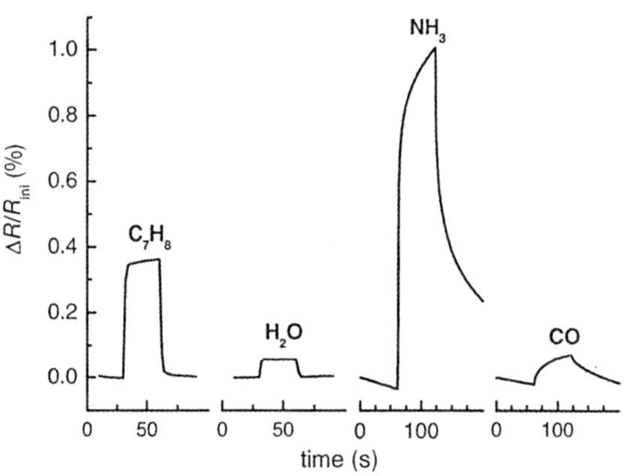

Figure 2.31 Response of a sensor system consisting of partially nonanedithiol-crosslinked Pt nanoparticles towards 400 ppm amounts of different gas molecules. ($\Delta R = R_{gas} - R_{ini}$; R_{ini} = resistance in dry air). (© Elsevier, Amsterdam).

2.3
Technologies on the Nanoscale

2.3.1
Introduction

This section deals with images of "nanotechnology" that do not fulfill the strict definition of nanoscience and nanotechnology, explained in Section 2.1. Why is this necessary? The term "nanotechnology" has now reached such a broad and thereby diffuse meaning that it seems helpful to discuss some examples of those already installed techniques that are falsely integrated into nanotechnology in a broader sense. Indeed, in some cases it is not trivial to decide if an effect and a thereby resulting technique follow the precise definition or not. From experience it can be seen that "wrong" nanotechnology means techniques that are settled on the nanoscale, but without the decisive size-dependent or functionality-determined nano-effect. What is usually meant by this common understanding is "technology(ies) on the nanoscale". The exclusion of those techniques from the scientifically exactly defined techniques is not discrimination. Rather, some of them became very important and others will follow. To conclude this introduction, one should try to differentiate clearly between nanotechnology and technologies on the nanoscale. The latter can also be considered as scaling effects without indicating nano-specific effects.

2.3.2
Structured Surfaces

It has long been known that structured surfaces change the physical and chemical properties of the corresponding material. Two property changes dominate the interest in structured surfaces: (i) change in wettability and (ii) change in optical properties. Both are of enormous importance both in nature and in technique. What kind of structure are we talking about? Let us consider first a natural "technique", that has been copied in many respects: the *wettability behavior.* Barthlott *et al.* have investigated since about 1990 the surface of lotus leaves for its special property of having a permanent clean surface [120–122]. Like all primary parts of plants, lotus leaves are covered by a layer of hydrophobic material. In case of lotus leaves, this layer consists of epicuticular wax crystals. Scanning Electron Microscopy (SEM) investigations additionally showed a structural design that is responsible for the super-hydrophobicity of lotus leaves. This special behavior has subsequently become known as the lotus effect. The SEM images showed that the surface of the leaves consists of a double structure of microsized cells decorated with nanosized waxy crystals.

The physical background for this phenomenon can be seen in the behavior of water droplets on a micro/nanostructured surface. It is important to state that the effect is not dependent on a distinct size of the structure units. As it turned out, the lotus combination of micro- and nanosized units is advantageous, but it is not a condition. Also, the absolute size of the nanostructure units is not decisive to observe the effect, it can only improve or worsen the hydrophobic nature.

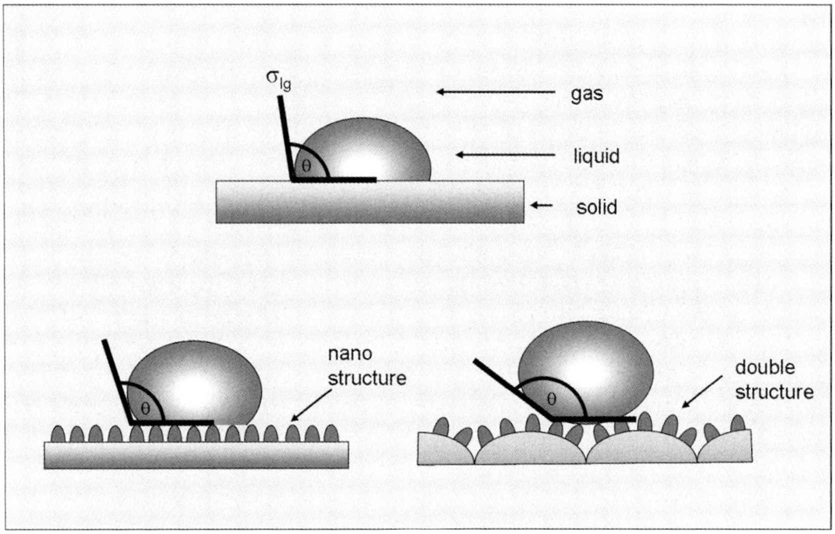

Figure 2.32 Sketch of the gas–liquid–solid three-phase system. (a) Water droplet on a smooth surface resulting in small contact angles θ, (b) on a nanostructured surface with increased θ value and (c) on a bimodal micro/nano structured surface with the largest contact angle.

The wettability generally describes the interaction of a liquid with a solid surface. It is described by the Young equation, $\sigma_{sg} - \sigma_{sl} = \sigma_{lg}\cos\theta$, where σ_{sg} = solid–gas interfacial tension, σ_{sl} = solid–liquid interfacial tension, σ_{lg} = liquid–gas interfacial tension and θ = solid–liquid contact angle [123]. The contact angle θ is the angle between the solid surface and the tangent applied at the surface of the droplet. Figure 2.32 illustrates the situation of a water droplet on a smooth surface, a nanostructured surface and a micro/nanostructured surface.

It is obvious that structured surfaces in any case cause larger contact angles than flat surfaces and so show an increased hydrophobicity. The reason is that the energy to distribute a water droplet on a structured surface extends the gain in energy by additional interactions of water molecules with the surface.

Soon after its recognition, the lotus effect led to the development of artificially micro/nanostructured surfaces. Lithographic techniques, self-assembly processes, controlled deposition, size reduction and replication by physical contact are applicable routes [124]. An elegant replication procedure will be briefly considered. It uses masks consisting of nanoporous alumina films. The advantage is their rather simple fabrication by anodization of aluminum surfaces, the easy adjustability of the pore diameters and the hardness and the temperature stability of alumina [125–129]. Using appropriate imprinting devices, various polymers and metals could be nanostructured [130]. The successful 1 : 1 polymer transfer from the mask to the surface is shown in Figure 2.33 by means of a poly(methyl methacrylate) (PMMA) surface.

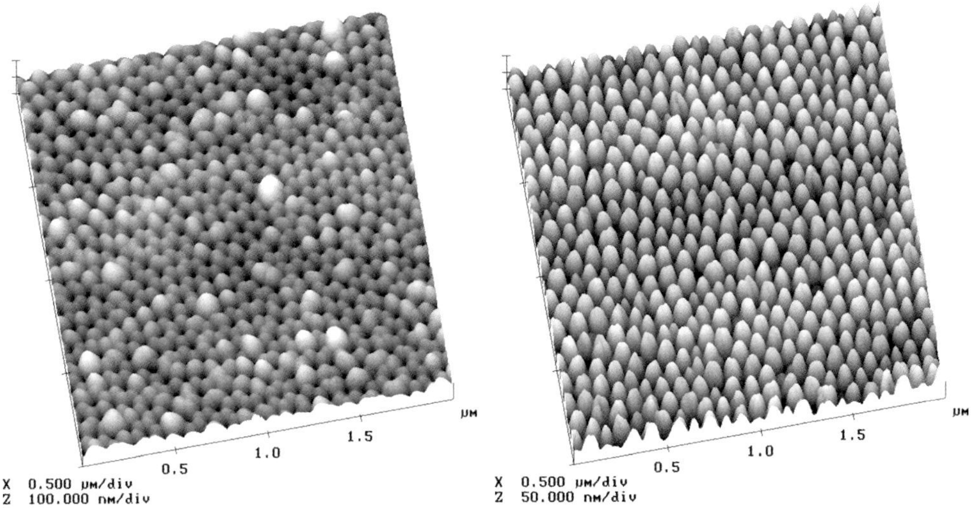

a) b)

Figure 2.33 Proof of 1 : 1 pattern transfer. (a) AFM image of an alumina surface with 50-nm pores; (b) imprinted PMMA surface indicating some defects from the mask at the same positions.

Polycarbonate (PC) and polytetrafluoroethylene (PTFE) could also be nanostructured with different pore widths. Aluminum, iron, nickel, palladium, platinum, copper, silver and brass are examples of successfully nanostructured metals. Figure 2.34 shows an AFM image of a nanostructured silver surface using a mask with 50-nm pores.

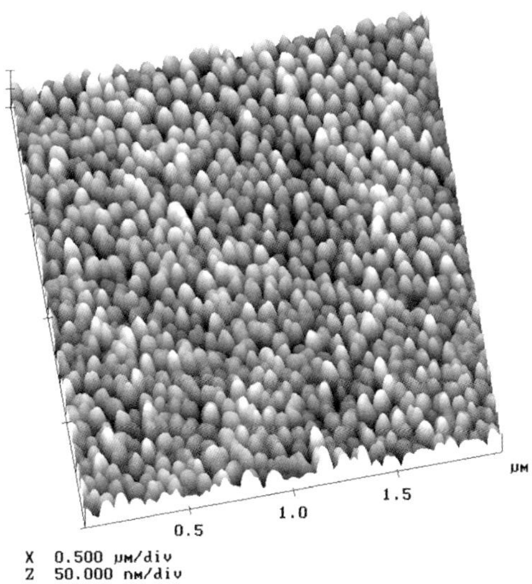

Figure 2.34 AFM image of an imprinted silver surface.

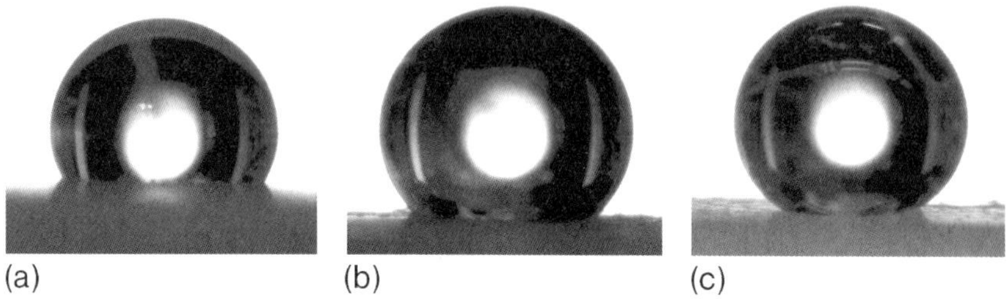

Figure 2.35 Light microscopic images of water droplets on (a) 50-nm, (b) 120-nm and (c) 170-nm structured PTFE surfaces.

By means of PTFE surfaces, imprinted with masks of 50, 120, 170 and 200 nm, it has been demonstrated that each of the surfaces makes the hydrophobicity increase, compared with an untreated surface. However, there is no spontaneous effect to be observed, rather it is a scaling phenomenon. In Figure 2.35, light microscopic images of water droplets on variously nanostructured PTFE surfaces are shown. There is an increasing contact angle to be registered if the pore width and thereby the pillars on the surfaces increase. This continuous development of the contact angles can also be followed from Figure 2.36.

In addition to the wettability properties, a second physical behavior changes with structure: the *light transmission* of transparent materials. In Figure 2.37, the increasing transmission of visible light through PMMA windows with decreasing structure size is demonstrated.

Improvements in the transparency of glasses, linked with a reduction in reflection, has important practical consequences in optical devices.

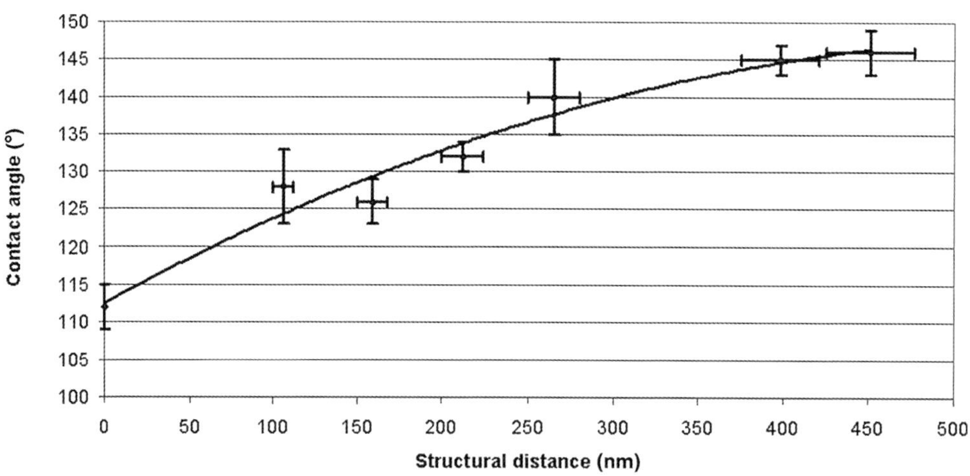

Figure 2.36 Dependence of contact angles of water droplets on PTFE surfaces on the structure size.

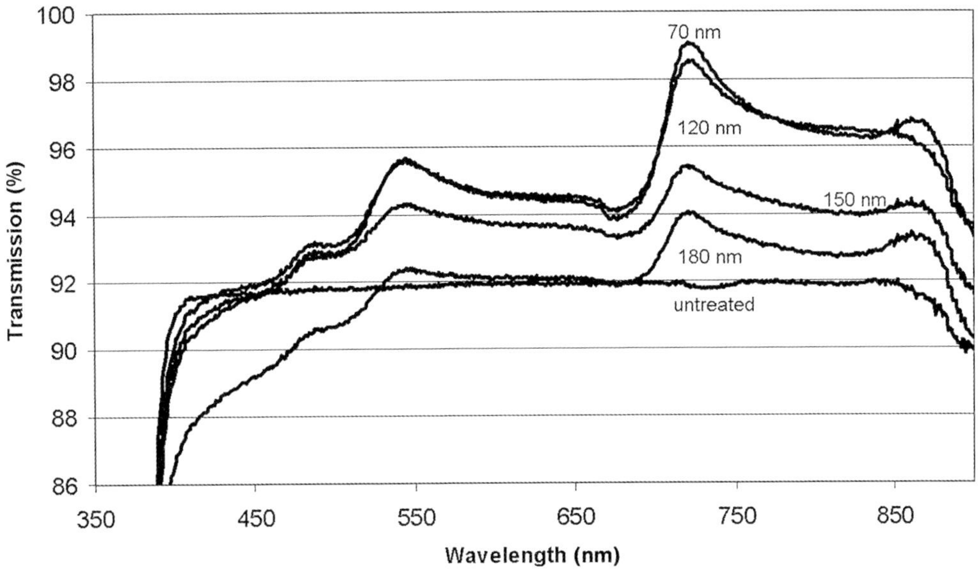

Figure 2.37 UV–visible transmission spectra of PMMA samples structured with 70-, 120-, 150- and 180-nm pillars and of a non-structured sample.

2.4 Final Remarks

Referring to Figure 2.6, only seven of a huge number of possible examples have been selected here to demonstrate the enormous diversity of nanoscience and nanotechnology. Nano-effects can occur everywhere, both in "simple" materials and in complex biological structures. This makes nanoscience and nanotechnology a unique field of research and development. The examples discussed illustrate the universality of this future-determining technology, which in many of the most attractive fields is still at the very beginning. However, it can be predicted that distinct fields that are still part of basic research will develop into techniques which will influence daily life dramatically. Others, usually those of easier and faster research and development, have already become routine techniques.

The selected examples also indicate that basic research is fundamental to developing nanotechnology further. Only by basic research will we discover nano-effects in the different fields named in the definition. However, also in basic science the research assignment should not just be to look for novel nano-effects: physics, chemistry, biology and materials science will also discover relevant effects when working in other fields.

References

1 Nanoscience and nanotechnologies: opportunities and uncertainties, Royal Society and The Royal Academy of Engineering, (2004) 5, *Science Policy Section, The Royal Society, London.*

2 Brune, H., Ernst, H., Grunwald, A., Grünwald, W., Hofmann, H., Krug, H., Janich, P., Mayor, M., Rathgeber, W., Schmid, G., Simon, U., Vogel, V. and Wyrwa, D. (2006) *Wissenschaftsethik und Technikfolgenabschätzung Vol. 27. Nanotechnology – Assessment and Perspectives*, Springer, Berlin.

3 Schwerdtfeger, P. (ed.) (2002) *Relativistic Electronic Structure Theory. Part 1: Fundamentals*, Elsevier, Amsterdam.

4 Schwertdfeger, P. (ed.) (2005) *Relativistic Electronic Structure Theory. Part 2: Applications*, Elsevier, Amsterdam.

5 Hess, B.A. (ed.) (2002) *Relativistic Effects in Heavy-element Chemistry and Physics*, Wiley, New York.

6 Mie, G. (1908) *Annals of Physics*, **25**, 377.

7 Schmid, G. and Giebel, U.unpublished work.

8 Collier, C.P., Mattersteig, G., Wong, E.W., Luo, Y., Beverly, K., Sampaio, J., Raymo, F.M., Stoddart, J.F. and Heath, J.R. (2000) *Science*, **289**, 1172.

9 Schliwa, M. (ed.) (2003) *Molecular Motors*, Wiley-VCH, Weinheim.

10 Tyreman, M.J.A. and Molloy, J.E. (2003) *IEE Proceedings Nanobiotechnology*, **150**, 95.

11 Vallee, R.B. and Hook, P. (2003) *Nature*, **421**, 701.

12 Schliwa, M. and Woehlke, G. (2003) *Nature*, **422**, 759.

13 Ball, P. (2002) *Nanotechnology*, **13**, R15.

14 Braig, K., Otwinowski, Z., Hegde, R., Boisvert, D.C., Joachimiak, A., Horwich, A.L. and Sigler, P.B. (1994) *Nature*, **371**, 578.

15 Ditzel, L., Lowe, J., Stock, D., Stetter, K.-O., Huber, H., Huber, R. and Steinbacher, S. (1998) *Cell*, **93**, 125.

16 Wang, J. and Boisvert, D.C. (2003) *Journal of Molecular Biology*, **327**, 843.

17 Shomura, Y., Yoshida, T., Izuka, R., Maruyama, T., Yohda, M. and Miki, K. (2004) *Journal of Molecular Biology*, **335**, 1265.

18 Shimamura, T., Koike-Takeshita, A., Yokoyama, K., Yoshida, M., Taguchi, H. and Iwatawa, S. (2003) *Acta Crystallographica Section D-Biological Crystallography*, **59**, 1632.

19 Saibil, H.R., Ranson, N.A. (2002) *Trends in Biochemical Sciences*, **27**, 627.

20 Kinbara, K. and Aida, T. (2005) *Chemical Reviews*, **105**, 1377.

21 Vale, R.D., Reese, T.S. and Sheetz, M.P. (1985) *Cell*, **42**, 39.

22 Kozielski, F., Sack, S., Marx, A., Thormahlen, M., Schonbrunn, E., Biou, V., Thompson, A., Mandelkow, E.M. and Mandelkow, E. (1997) *Cell*, **91**, 985.

23 Kikkawa, M., Okada, Y. and Hirokawa, N. (2000) *Cell*, **100**, 241.

24 Kikkawa, M., Sablin, E.P., Okada, Y., Yajima, H., Fletterick, R.J. and Hirokawa, N. (2001) *Nature*, **411**, 439.

25 Hancock, W.O. and Howard, J. (1999) *Proceedings of the National Academy of Sciences of the United States of America*, **96**, 13147.

26 Stewart, R.J., Thaler, J.P. and Goldstein, L.S. (1993) *Proceedings of the National Academy of Sciences of the United States of America*, **90**, 5209.

27 Muthukrishnan, G., Hutchins, B.M., Williams, M.E. and Hancock, W.O. (2006) *Small*, **2**, 626.

28 Koumura, N., Zijistra, R.W.J., van Delden, R.A., Harada, N. and Feringa, B.L. (1999) *Nature*, **401**, 152.

29 Irie, M. (1993) *Molecular Crystals and Liquid Crystals*, **227**, 263.

30 Tseng, H.-R., Vignon, S.A. and Stoddart, J.F. (2003) *Angewandte Chemie-International Edition*, **42**, 1491.

31 Jimenez, M.C., Dietrich-Buchecker, C.O. and Sauvage, J.-P. (2000) *Angewandte Chemie-International Edition*, **39**, 3248.

32 Badjic, J.D., Balzani, V., Credi, A., Silvi, S. and Stoddart, J.F. (2004) *Science* **303**. 1845.

33 Balzani, V. (2005) *Small*, **1**, 278.

34 Moore, G. (1965) *Electronics*, **38**, 114.

35 Wolf, S.A., Treger, D. and Chtchelkanova, A. (2006) *MRS Bulletin*, **31**, 400.

36 Richter, H.J. and Harkness, S.D. IV, (2006) *MRS Bulletin*, **31**, 384.

37 Allenspach, R. and Jubert, P.-O. (2006) *MRS Bulletin*, **31**, 395.

38 Zarembowitch, J. and Kahn, O. (1991) *New Journal of Chemistry*, **15**, 181.

39 Breuning, E., Ruben, M., Lehn, J.-M., Renz, F., Garcia, Y., Knesofontov, V., Gütlich, P., Wegelius, E. and Rissanen, K. (2000) *Angewandte Chemie-International Edition*, **39**, 2504.

40 Feldheim, D.L. and Keating, C.D. (1998) *Chemical Society Reviews*, **27**, 1.

41 Simon, U. (1998) *Advanced Materials*, **10**, 1487.

42 Simon, U. and Schön, G. (2000) in *Handbook of Nanostructured Materials and Nanotechnology* (ed. H.S. Nalwa), Academic Press, New York, Vol. 3, p. 131.

43 Simon, U., Schön, G. and Schmid, G. (1993) *Angewandte Chemie-International Edition in English*, **2**, 250.

44 Simon, U. (2004) in *Nanoparticles. From Theory to Application* (ed. G. Schmid), Wiley-VCH, Weinheim, p. 328.

45 Bezryadin, A., Dekker, C. and Schmid, G. (1977) *Applied Physics Letters*, **71**, 1273.

46 Schmid, G., Boese, R., Pfeil, R., Bandermann, F., Meyer, S., Calis, G.H.M. and van der Velden, J.W.A. (1981) *Chemische Berichte*, **114**, 3634.

47 Schmid, G. (1990) *Inorganic Syntheses, Vol 32*, **7**, 214.

48 Chi, L.F., Hartig, M., Drechsler, T., Schaak, Th., Seidel, C., Fuchs, H. and Schmid, G. (1998) *Applied Physics Letters A*, **66**, 187.

49 Zhang, H., Schmid, G. and Hartmann, U. (2003) *Nano Letters*, **3**, 305.

50 Santini, J.T., Jr. Richards, A.C., Scheidt, R., Cima, M.J. and Langer, R. (2000) *Angewandte Chemie-International Edition*, **39**, 2396.

51 Folkman, J. and Long, D.M. (1964) *Journal of Surgical Research*, **4**, 139.

52 Bar-Shalom, D., Bukh, N. and Larsen, T.K. (1991) *Annals, New York Academy of Sciences*, **618**, 578.

53 Florence, A.T. and Jani, P.U. (1994) *Drug Safety*, **10**, 233.

54 Pereswtoff-Morath, L. (1998) *Advanced Drug Delivery Reviews*, **29**, 185.

55 Davis, S.S. (1999) *Pharmaceutical Science & Technology Today*, **2**, 265.

56 Schmidt, U. (2003) *Spektrum der Wissenschaften*, **10**, 42.

57 Gestermann, S., Hesse, R., Windisch, B. and Vögtle, F. (2000) in *Stimulating Concepts in Chemistry* (eds F. Vögtle, J.F. Stoddart and M. Shibasaki), Wiley-VCH, Weineim. p. 187.

58 Mornet, S., Vasseur, S., Grasset, F. and Duguet, E. (2004) *Journal of Materials Chemistry*, **14**, 2161.

59 Pankhurst, Q.A., Connolly, J., Jones, S.K. and Dobson, J. (2003) *Journal of Physics D*, **36**, 167.

60 Huth, S., Lausier, J., Gersting, S.W., Rudolph, C., Plank, C., Welsch, U. and Rosenecker, J. (2004) *Journal of Gene Medicine*, **6**, 923.

61 Sukhorukov, G.B., Rogach, A.L., Zebli, B., Liedl, T., Skirtach, A.G., Köhler, K., Antipov, A.A., Gaponik, N., Susha, A.S., Winterhalter, M. and Parak, W. (2005) *Small*, **1**, 194.

62 Muños Javier, A., Kreft, O., Piera Alberola, A., Kirchner, C., Zebli, B., Susha, A.S., Horn, E., Kempner, S., Skirtach, A.G., Rogach, A.L., Rädler, J., Sukhorukov, G.B., Benoit, M. and Parak, W.J. (2006) *Small*, **2**, 394.

63 De Geest, B.G., Dégunat, C., Sukhorukov, G.B., Braeckmans, K., De Smedt, S.C. and Demeester, J. (2005) *Advanced Materials*, **17**, 2357.

64 De Geest, B.G., Vandenbroucke, R.E., Guenther, A.M., Sukhorukov, G.B., Hennink, W.E., Sanders, N.N., Demeester, J. and De Smedt, S.C. (2006) *Advanced Materials*, **18**, 1005.

65 Skirtach, A.G., Muños Javier, A., Kreft, O., Köhler, K., Piera Alberola, A., Möhwald, H., Parak, W.J. and Sukhorukov, G.B. (2006) *Angewandte Chemie-International Edition*, **45**, 4612.

66 Kipke, S. and Schmid, G. (2004) *Advanced Functional Materials*, **14**, 1184.

67 Mirkin, C.A., Letsinger, R.L., Mucic, R.C. and Storhoff, J.J. (1996) *Nature*, **382**, 607.

68 Elghanian, R., Storhoff, J.J., Mucic, R.C., Letsinger, R.L. and Mirkin, C.A. (1997) *Science*, **277**, 1078.

69 Rosi, N.L. and Mirkin, C.A. (2005) *Chemical Reviews*, **105**, 1547.

70 Jin, R., Wu, G., Li, Z., Mirkin, C.A. and Schatz, G.C. (2003) *Journal of the American Chemical Society*, **125**, 1643.

71 Storhoff, J.J., Lazarides, A.A., Mucic, R.C., Mirkin, C.A., Letsinger, R.L. and Schatz, G.C. (2000) *Journal of the American Chemical Society*, **122**, 4640.

72 Taton, T.A., Mirkin, C.A. and Letsinger, R.L. (2000) *Science*, **289**, 1757.

73 Cao, Y.W.C., Jin, R. and Mirkin, C.A. (2002) *Science*, **297**, 1536.

74 Georganopoulou, D.G., Chang, L., Nam, J.-M., Thaxton, C.S., Mufson, E.J., Klein, W.L. and Mirkin, C.A. (2005) *Proceedings of the National Academy of Science of USA*, **102**, 2273.

75 Gerion, D., Chen, F.Q., Kannan, B., Fu, A.H., Parak, W.J., Chen, D.J., Majumdar, A. and Alivisatos, A.P. (2003) *Analytical Chemistry*, **75**, 4766.

76 Empedocles, S. and Bawendi, M. (1999) *Accounts of Chemical Research*, **32**, 389.

77 Rosensweig, R.E. (2002) *Journal of Magnetism and Magnetic Materials*, **252**, 370.

78 Jordan, A., Scholz, R., Wust, P., Fahling, H. and Felix, R. (1999) *Journal of Magnetism and Magnetic Materials*, **201**, 413.

79 Jordan, A., Schmidt, W. and Scholz, R. (2000) *Radiation Research*, **154**, 600.

80 Jordan, A., Scholz, R., Maier-Hauff, K., Johannsen, M., Wust, P., Nadobny, P., Schirra, H., Schmidt, H., Deger, S., Loening, S., Lanksch, W. and Felix, R. (2001) *Journal of Magnetism and Magnetic Materials*, **225**, 118.

81 Jordan, A., Rheinlander, T., Waldofner, N. and Scholz, R. (2003) *Journal of Nanoparticle Research*, **5**, 597.

82 Seiyama, T., Kato, A., Fujiishi, K. and Nagatami, M. (1962) *Analytical Chemistry*, **34**, 1502.

83 Seiyama, T. and Kagawa, S. (1966) *Analytical Chemistry*, **38**, 1069.

84 Brattein, W.H. and Bardeen, J. (1953) *Bell System Technical Journal*, **32**, 1.

85 Heiland, G. (1954) *Zeitschrift Fur Physik*, **138**, 549.

86 Franke, M.E., Koplin, T.J. and Simon, U. (2006) *Small*, **2**, 36.

87 Yamazoe, N., Sakai, G. and Shimanoe, K. (2003) *Catalysis Surveys from Asia*, **7**, 63.

88 Samson, S. and Fonstad, C.G. (1973) *Journal of Applied Physics*, **44**, 4618.

89 Jarzebski, Z.M. and Marton, J.P. (1976) *Journal of the Electrochemical Society*, **123**, 299.

90 Maier, J. and Göpel, W. (1988) *Journal of Solid State Chemistry*, **72**, 293.

91 Göpel, W. and Schierbaum, K.D. (1995) *Sensors and Actuators B*, **26–27**, 1.

92 Ogawa, H., Nishikawa, M. and Abe, A. (1982) *Journal of Applied Physics*, **53**, 4448.

93 Bârsan, N. and Weimar, U. (2003) *Journal of Physics-Condensed Matter*, **15**, R813.

94 Sakai, G., Matsunaga, N., Shimanoe, K. and Yamazoe, N. (2001) *Sensors and Actuators B*, **80**, 125.

95 Matsunaga, N., Sakai, G., Shiman, K. and Yamazoe, N. (2003) *Sensors and Actuators B*, **96**, 226.

96 Lu, F., Li, Y., Dong, M. and Wang, X. (2000) *Sensors and Actuators B*, **66**, 225.

97 Wohltjen, H. and Snow, A.W. (1998) *Analytical Chemistry*, **70**, 2856.

98 Snow, A.W. and Wohltjen, H. (2001) US Patent 6 221 673.

99 Evans, S.D., Johnson, S.R., Cheng, Y.L. and Shen, T. (2000) *Journal of Materials Chemistry*, **10**, 183.

100 Han, L., Daniel, D.R., Mayer, M.M. and Zong, C.-J. (2001) *Analytical Chemistry*, **73**, 4441.

101 Krasteva, N., Besnard, I., Guse, B., Bauer, R.E., Muellen, K., Yasuda, A. and Vossmeyer, T. (2002) *Nano Letters*, **2**, 551.

102 Zang, H.-L., Evans, S.D., Henderson, J.R., Miles, R.E. and Shen, T.-H. (2002) *Nanotechnology*, **13**, 439.

103 Vossmeyer, T., Guse, B., Besnard, I., Bauer, R.E., Muellen, K. and Yasuda, A. (2002) *Advanced Materials*, **14**, 238.

104 Joseph, Y., Guse, B., Yasuda, A. and Vossmeyer, T. (2004) *Sensors and Actuators B*, **98**, 188.

105 Simon, U., Flesch, U., Maunz, W., Müller, R. and Plog, C. (1999) *Microporous and Mesoporous Materials*, **21**, 111.

106 Moos, R., Müller, R., Plog, C., Knezevic, A., Leye, H., Irion, E., Braun, T., Marquardt, K.-J. and Binder, K. (2002) *Sensors and Actuators B*, **83**, 181.

107 Franke, M.E., Simon, U., Moos, R., Knezevic, A., Müller, R. and Plog, C. (2003) *Physical Chemistry Chemical Physics*, **5**, 5195.

108 Kreibig, U., Fauth, K., Granquist, C.-G. and Schmid, G. (1990) *Zeitschrift Fur Physikalische Chemie-International Journal of Research in Physical Chemistry & Chemical Physics*, **169**, 11.

109 van Staveren, M.P.J., Brom, H.B. and De Jongh, L.J. (1991) *Physics Reports-Review Section of Physics Letters*, **208**, 1.

110 Simon, U., Schön, G. and Schmid, G. (1993) *Angewandte Chemie-International Edition in English*, **32**, 250.

111 Schön, G. and Simon, U. (1995) *Colloid and Polymer Science*, **273**, 101.

112 Schön, G. and Simon, U. (1995) *Colloid and Polymer Science*, **273**, 202.

113 Brust, M., Betel, D., Shiffrin, D.J. and Kieley, C.J. (1995) *Advanced Materials*, **7**, 795.

114 Andres, R.P., Bielefeld, J.D., Henderson, J.I., Janes, D.B., Kolagunta, V.R., Kubiak, C.P., Mahoney, W.J. and Osifchin, R.G. (1996) *Science*, **273**, 1690.

115 Simon, U. (1998) *Advanced Materials*, **10**, 1487.

116 Torma, V., Schmid, G. and Simon, U. (2001) *ChemPhysChem*, **2**, 321.

117 Torma, V., Vidoni, O., Simon, U. and Schmid, G. (2003) *European Journal of Inorganic Chemistry*, **6**, 1121.

118 Simon, U. (2004), in *Nanoparticles. From Theory to Application* (ed. G. Schmid), Wiley-VCH, Weinheim, p. 328.

119 Schmid, G. and Simon, U. (2005) *Chemical Communications*, 697.

120 Barthlott, W. (1990) in *Scanning Electron Microscopy in Taxonomy and Functional Morphology* (ed. D. Claugher), Clarendon Press, Oxford, p. 69.

121 Barthlott, W. (1993) in *Evolution and Systematics of the Caryophyllales*, (eds H.D. Behnke and T.J. Mabry), Springer, Berlin, p. 75.

122 Barthlott, W. and Neinhuis, C. (1997) *Planta*, **202**, 1.

123 Israelachvili, J. (1995) *Intermolecular and Surface Forces*, 2nd edn, Academic Press, London.

124 Xia, Y., Rogers, J.A., Paul, K.E. and Whitesides, G.M. (1999) *Chemical Reviews*, **99**, 1823.

125 O'Sullivan, J.P. and Wood, G.C. (1970) *Proceedings of the Royal Society, London*, **317**, 511.

126 Thompson, G.E. and Wood, G.C. (1983) *Treatise on Materials, Science and Technology*, **23**, 205.

127 Diggle, J.W., Downie, T.C. and Goulding, C.W. (1969) *Chemical Reviews*, **69**, 365.

128 Masuda, H., Yamada, H., Satoh, M., Asoh, H., Nakao, M. and Tamamura, T. (1997) *Applied Physics Letters*, **71**, 2770.

129 Pelzer, K., Philippot, K., Chaudret, B., Chaudret, W., Meyer-Zaika, W. and Schmid, G. (2003) *Zeitschrift fur Anorganische und Allgemeine Chemie*, **629**, 1217.

130 Schmid, G., Levering, M. and Sawitowski, T. (2007) *Zeitschrift fur Anorganische und Allgemeine Chemie*, **633**, 2147.

3
Top-Down Versus Bottom-Up

Wolfgang J. Parak, Friedrich C. Simmel, and Alexander W. Holleitner

3.1
Introduction

Today, the term "nanotechnology" is a widely used keyword occurring in a variety of different contexts [1, 2]. Semantically, "nano" is an official SI prefix for physical units, which is equivalent to the factor 10^{-9}. This prefactor can be combined with units of physical quantities such as time (nanoseconds), mass (nanograms) or length (nanometers). Historically, nanotechnology has evolved from different scientific fields such as physics, chemistry, molecular biology, microelectronics and material sciences. Generally, nanotechnology aims to study and to manipulate real-world structures with sizes ranging between 1 nm, that is one millionth part of a millimeter, and up to 100 nm. The set of typical "nano"-objects includes colloidal crystals, molecules, DNA-based structures and integrated semiconductor circuits.

We can approach the scale between 1 and 100 nm from different sides, for instance by emphasizing biological and chemical aspects. On the one hand, we can conceptually subdivide complex biological organisms, such as a human body, into smaller and smaller subunits. In a simplified way, the entire body can be classified into organs, each organ is composed out of cells and each cell is a dynamically organized assembly of biological molecules. Arguably, complex molecules, such as lipids, proteins and DNA, are the smallest biologically relevant subunits. These molecules typically have dimensions of a few nanometers up to several tens of nanometers. On the other hand, the smallest integral part of a molecule is the physical bond between individual atoms. The latter have a size of a few tenths of nanometers. Molecules can be very small (such as hydrogen, H_2) and very big (such as polymeric macromolecules). If we leave aside the combination of molecular building blocks to new macromolecules (as in polymerization reactions or in supramolecular chemistry [3]), the typical size range of molecules that can be synthesized from atomic building blocks and their derivatives is again the region from a few nanometers up to several tens of nanometers. It can be concluded that if macroscopic objects such as biological organisms are investigated on a smaller and smaller length scale, one ultimately ends

Nanotechnology. Volume 1: Principles and Fundamentals. Edited by Günter Schmid

ISBN: 978-3-527-31732-5

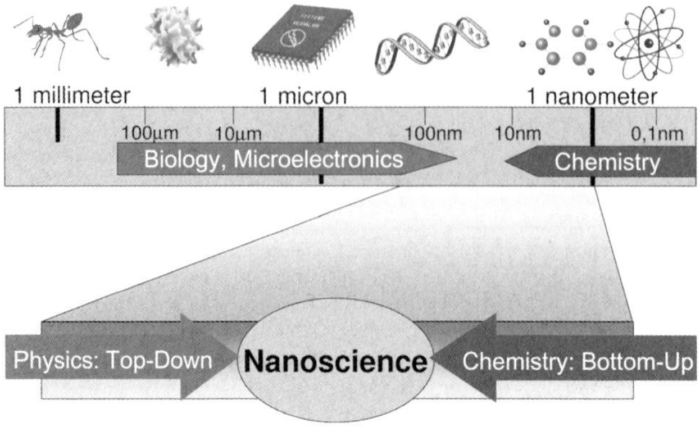

Figure 3.1 Some objects (ant, virus, circuits in computer chip, DNA, benzene, atom) are presented that are characteristic of different size scales ranging from milli- to nanometers. Adapted from a presentation by Professor Dr. J. P. Kotthaus.

up in the nanometer range as the smallest relevant scale for functional subunits. Then again, if materials are built up synthetically from their basic chemical building blocks, one will also first arrive at this length scale. In this sense, "nano" is the size scale where physics, chemistry and biology meet in a natural way (Figure 3.1).

The first scientist who pointed out that "there is plenty of room at the bottom" was Richard Feynman in 1959 [oral presentation given on 29 December 1959 at the Annual Meeting of the American Physical Society at the California Institute of Technology (Caltech)]. He envisioned scientific discoveries and new applications of miniature objects as soon as material systems could be assembled at the atomic scale. To this end, machines and imaging techniques would be necessary which can be controlled at the nanometer or subnanometer scale. Fifty years later, the scanning tunneling microscope and the atomic force microscope are ubiquitous in scientific laboratories, allowing to image structures with atomic resolution [4–12]. As a result, various disciplines within nanotechnology aim towards manufacturing materials for diverse products with new functionalities at the nanoscale.

As we have seen, we can approach the nanoscale from two sides: by making things smaller, that is, by downscaling, and by constructing things from small building blocks, that is, by upscaling. The first method is referred to as the "top-down" and the second as the "bottom-up" approach. The top-down approach follows the general trend of the microelectronic industry towards miniaturization of integrated semiconductor circuits. Modern lithographic techniques allow the patterning of nanoscale structures such as transistor circuits with a precision of only a few nanometers (see http://www.icknowledge.com and publications therein for more information). As we will see in this chapter, the industrial demand for ever smaller electronic circuits has provided several physical tools by which materials can be probed and manipulated at the nanometer scale. In contrast, the bottom-up approach is based on molecular recognition and chemical self-assembly of molecules [13]. In combination with chemical synthesis techniques, the bottom-up approach allows for the assembly of macromolecular complexes with a size of several nanometers. To visualize the different strategies of these two approaches is the topic of this chapter. We first will give a few simple explanations and arguments. Then we will illustrate the differences of the two strategies in the context of a variety of examples. Finally, we will discuss the limits of both approaches.

3.1.1 Top-Down Strategies

How can we get to smaller and smaller sculptures and objects? The idea of top-down-strategies is to take processes known from the macroscopic world and to adopt them in such a way that they can be used for doing the same thing on a smaller scale. Since ancient times, humans have created artwork and tools by structuring materials. Let us take the artist as an example, who carves and sculpts figures out of blocks of wood or stone, respectively. For this purpose, the artist needs tools, usually a carving knife or chisel, with which he can locally ablate parts from the original piece of material and thus give it its desired shape (Figure 3.2). If we continue with the idea of a sculptor, it is evident that in order to sculpt smaller objects, smaller tools are needed, such as miniature rasps and knives. Chinese artisans have used such tools to carve little pieces of wood and other materials into sculptures in the submillimeter regime, which can only be seen with magnifying glasses [some of these highly impressive sculptures can be seen in the National Taiwan Museum (http://www.ntm.gov.tw)].

If we want to fabricate even smaller structures or objects and study their physical properties, classical mechanical tools such as rasps will no longer work. For that purpose, scanning probe microscopes are powerful instruments for probing and manipulating materials at the nanometer scale and can thus be seen as one of the key inventions for nanotechnology. On the one hand, they allow imaging and probing of the characteristics of nanoscale objects with the highest resolution. Examples include topology, material configuration and electrical, chemical and magnetic properties of the studied objects. On the other hand, scanning probe microscopes allow local manipulation and even shaping of the nanostructures. In a seminal work in 1982, Binnig and Rohrer invented the scanning tunneling microscope (STM) [4], for which they were awarded the Nobel Prize in 1986. In such a microscope, piezoelectric

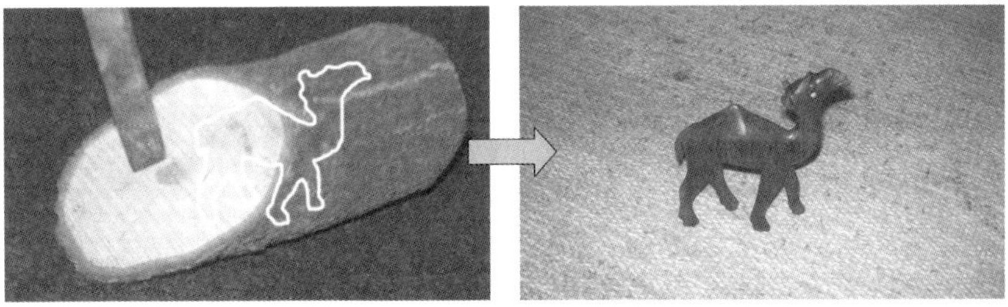

Figure 3.2 An artifact is made out of a piece of wood by carving. The wooden figure is created by locally ablating material from the initial block according to a building plan.

crystals move a scanning tip across the surface of a sample, while the electric current is recorded between the tip and the sample. If the tip is located very close to the surface of the sample, the electric current is composed of tunneling electrons. In the nanoscale quantum world, the wave character of electrons plays an important role. As soon as the distance between the tip and the surface is on the order of the electron wavelength, electrons can tunnel from the tip to the surface. The size of the tunneling current has an exponential dependence on the distance between the tip and surface. Therefore, the point of the tip which is closest to the surface predominantly contributes to the current. In principle, this point can be made up of only one atom, which allows for atomic resolution. The scanning tunneling microscope is sensitive to the electronic density at the surface of the sample, which allows for, for example, imaging of the electronic orbitals of atoms and molecules. The scanning tunneling microscope can operate not only under ultrahigh vacuum conditions, but also at atmospheric pressure and at various temperatures, which makes it a unique imaging and patterning tool for the nanosciences. For instance, scanning tunneling microscopes are utilized to pattern nanostructures by moving single atoms across surfaces, while the corresponding change of the quantum mechanical configuration can be recorded *in situ* [14].

The atomic force microscope (AFM) operates similarly to the scanning tunneling microscope [5]. Here, the force between the scanning tip and the sample surface is extracted by measuring the deflection of the tip towards the sample. Again, the atomic force microscope can be utilized as an instrument to image and to shape materials on the nanometer scale [15].

In experiments in which only a few nanostructures need to be patterned and probed, scanning probe microscopes can be exploited to structure and shape materials at the nanometer scale. To define an array of millions of nanoscale systems in parallel – such as in integrated electronic circuits – the top-down approach of microlithography is the technique of choice (Figure 3.3). Microlithography has been the technological backbone of the semiconductor industry for the last 45 years (see http://www.icknowledge.com and publications therein for more information). The minimum feature size of optically defined patterns depends on the wavelength of the utilized light and also factors that are due to, for example, the shape of the lenses and the quality of the photoresist. In 2003, the typical linewidth of semiconductor circuits fell below 100 nm, that is, the semiconductor industry can be literally seen as being part of the nanotechnologies. For this achievement, argon fluoride excimer lasers are applied with a laser wavelength of 193 nm in the deep ultraviolet region. For lithography in this optical range, tricks such as "optical proximity correction" and "phase shifting" were invented and successfully implemented. On the one hand, the miniaturization of semiconductor circuits is limited by economic costs for the semiconductor industry, since the implementation of new techniques for the realization of ever smaller feature sizes results in ever increasing costs. On the other hand, physical material properties, such as the high absorption level of refractive mirrors at short optical wavelengths, set a natural limit of a few tens of nanometers for the miniaturization process. Since the 1980s, the decline of optical lithography has been predicted as being only a few years away. However, each time optical

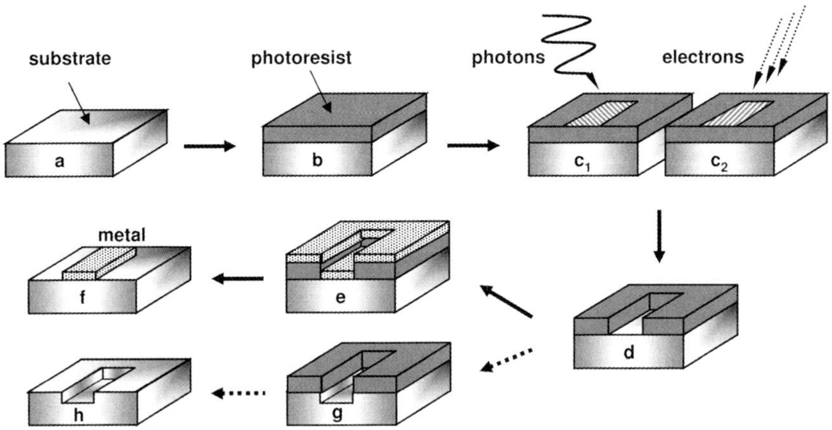

Figure 3.3 Top-down microlithography: in (a) and (b) a so-called photoresist – a liquid, photosensitive polymer – is spin-coated on to the polished surface of a substrate. (c) After hardening of the photoresist, for example, by heating, the substrate with the photoresist layer on top is locally exposed to photons or to electrons. Thereby, patterns can be defined in the photoresist. (d) After a photoresist specific developing process, the exposed (or the unexposed) parts of the original substrate are bared. By follow-up processes, such as metallization (e) or etching (g), the patterns in the photoresist can be translated to the substrate. To this end, the remaining parts of the photoresist are finally removed by a solvent, such as acetone [(f) and (h)]. Thereby, metallic conductor paths, logical circuits or memory cells can be defined.

lithography had reached traditional limits, new techniques extended the economically sustainable lifetime of top-down microlithography.

At present, there are several possible successor top-down nanotechnologies for industry, for example, extreme ultraviolet light lithography (EUV), electron beam lithography with multicolumn processing facilities [see also Figure 3.3(c_2)], the focused ion beam (FIB) technique and the ultraviolet nano-imprinting technique [16]. The implementation of each of these techniques requires enormous technical challenges to be overcome. One of the most promising techniques is that using EUV light with a wavelength of only 13 nm. For this technique, fabrication errors of the "optical" components need to be in the nanometer or subnanometer range. For comparison, state-of-the-art X-ray telescopes, such as the Zeiss XMM Newton telescope, exhibit a granularity of the mirror surfaces of 0.4 nm (see W. Egle, *Mission Impossible: XMM-Newton Proves the Opposite,* Innovation, Carl Zeiss, 2000, *8,* 12–14, ISSN 1431-8059; this file can be downloaded from http://www.zeiss.com). In addition to "state of the art" optical requirements, all metrological components of the EUV technique need to exhibit subnanometer resolution. As a result, the cost of a stepper machine for EUV exposure of photoresists is US$50 million per system, providing a linewidth of 35 nm (see http://www.icknowledge.com for more information).

For medium-sized businesses, the "ultraviolet nano-imprinting technique" seems to be the most promising method to fabricate nanoscale circuits [17]. Here, nanostructures are mechanically imprinted into a photoresist. The stamp with the nanoscale patterns is made out of fused quartz, a material which is transparent to ultraviolet light. As soon as the stamp has been plunged into the photoresist, a short ultraviolet light pulse causes the photoresist to polymerize along the patterns on the stamp. In line with optical lithography, follow-up processes allow for the definition of nanoscale circuits in various geometries [Figure 3.3(e)–(h)]. The minimum feature size of the nano-imprinting technique is about 10 nm (see the webpage of the Chou group at Princeton University: http://www.princeton.edu/~chouweb/newproject/page3.html). At the same time, the technique is applicable to metals and plastic materials. The costs of an industrial nano-imprinting machine are less than US$1 million. However, the throughput of nano-imprinting machines is much lower than that for stepper machines.

What we have illustrated above is the top-down approach. We take tools and methods known in the macroscopic world and scale them down to structure materials on increasingly smaller length scales. As explained above, this downscaling process is not simply a reduction in size of the tools – we cannot make a nanoscale carving knife that directly resembles a macroscopic knife to take away material on the nanometer scale. In most cases, the relative importance of the forces and interactions involved changes with decreasing system size – and this also has to be considered when constructing the tools themselves.

3.1.2
Bottom-Up Strategies

The antipode of the top-down approach is the so-called bottom-up technique. Here a complex structure is assembled from small building blocks. These building blocks possess specific binding capabilities – often termed "molecular recognition properties" – which allow them to arrange automatically in the correct way. Self-assembly is an essential component of bottom-up approaches [18]. The ultimate examples of molecular recognition are biological receptor–ligand pairs: molecules that recognize and bind to each other with very high specificity. Prominent examples of such pairs are antibodies and their corresponding antigens and complementary strands of deoxyribonucleic acid (DNA) [19].

We can visualize the bottom-up assembly of materials with the example of LEGO building blocks, a common toy for children. Again we take the example of a small sculpture as also used to describe top-down approaches. LEGO building blocks can have different functions, as symbolized by their color (Figure 3.4). Furthermore, there are building blocks of different size and each building block has a defined number of binding sites, realized here as knobs. In this way the blocks can only be attached to each other in a defined way. This can be seen as "molecular" recognition. However, we should point out that the example of the LEGO blocks fails to describe self-assembly. There is still a helping hand needed to assemble the individual bricks to form the complete structure – the assembly process is not spontaneous (cf. the self-assembly models of Whitesides [20]). The example of LEGO blocks hold the basic concept of a bottom-up strategy: the construction of a new material by assembling basic small building blocks. If instead of LEGO blocks nanoscopic building blocks are

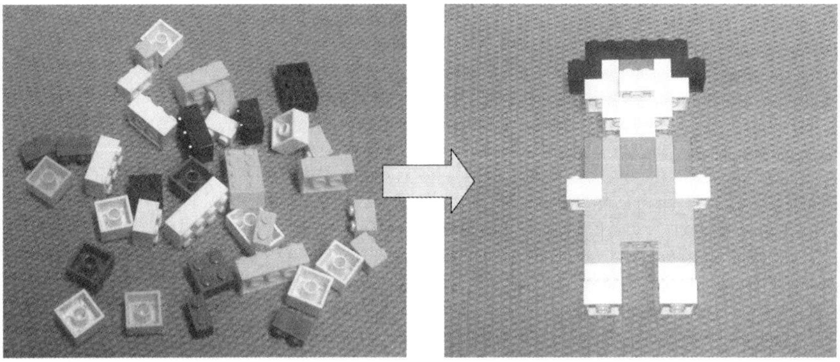

Figure 3.4 LEGO blocks can symbolize different functionalities (colors) and they can be assembled with their knobs in a defined way. Assembly of all building blocks (bricks) leads to the formation of the desired structure.

used, structures on the nanometer scale can be assembled in a very analogous way [21–24].

Generally, bottom-up assembly techniques seek to fabricate composite materials comprising nanoscale objects which are spatially ordered via molecular recognition. The prime examples of the technique are self-assembled monolayers (SAMs) of molecules [25]. A substrate is immersed in a dilute solution of a surface-active organic material that adsorbs on the surface and organizes via a self-assembly process. The result is a highly ordered and well-packed molecular monolayer. The method can be extended towards layer-by-layer (LBL) assembly, by which polymer light-emitting devices (LEDs) have already been fabricated [17]. The self-assembly technique also allows positioning of single molecules between two metal electrodes and subsequently into an experimental circuit. By this set-up, quantum mechanical transport characteristics of single molecules, such as photochromic switching behavior, can be studied in order to build electronic devices with new functionalities [26].

There are several combinations and variations of the SAM technique. It can be combined with nano-imprinting methods, the atomic force microscope or the focused ion beam technique, allowing the fabrication of geometric patterns of molecules with variable wetting properties, chemical functionality and/or topological characteristics [17]. Since most of the processes are performed in solution, electrical fields may be utilized to assemble charged compounds in a directed way, for example, for the creation of functionalized sensor electrodes. Electrodes modified with negatively charged gold nanoparticles and positively charged host molecules have proven highly sensitive for the detection of, for example, adrenaline (as the guest molecule) [27]. Another very promising field is the DNA-directed assembly of network materials [28]. Here, molecular recognition reactions between single DNA strands are translated into aggregate formation of nano-objects such as nanoparticles [21, 22]. Self-assembly can also be used to construct simple nanomechanical devices. For instance, the molecular recognition between DNA molecules has been exploited to build "nanotweezers" [29]. Further, networks of materials can be synthesized on top of so-called block copolymer templates [30]. A typical application of this technique is the formation of networks of metallic nanowires, that is, metals are vapor deposited on to a preformed template matrix made out of copolymers. The polymer networks can be used as two- or even three-dimensional templates. Generally, there is a wide range of materials which can be engineered by bottom-up techniques. The research involved has led to numerous sensing, electronic and optoelectronic interfaces in addition to devices.

3.2 First Example: Nanotweezers

Among the fundamental tools for mechanical work are tweezers or fingers. The basic function of tweezers is to hold and release things. In principle, tweezers comprise two fingers, which can close to hold something and open to release it again. So far only a few functional nanofingers are available. Although a variety of tools exist which can hold nanometer-sized objects, the problem is to release them again. This problem is of fundamental rather than of technological nature. The interested reader is referred to the excellent articles by Nobel Prize winner Richard Smalley, who describes the problem of nonspecific adhesion as the problem of "sticky fingers" [31, 32]. The basic module of a nanofinger is a nanometer-sized hinge, that is, a structure which can repeatedly open and close. In this section, we will show examples of such nanofingers or hinges based on top-down microlithography and on bottom-up self-assembly of biological molecules.

3.2.1 Top-Down Nanotweezers

Lithographically defined tweezers have been realized as a so-called micro-electromechanical system (MEMS). Similar micron-sized mechanical resonators have already been applied in inkjet printers and in airbags of cars. MEMS resonators have also triggered major scientific interest, since the reduction in size of an electromechanical resonator down to the nanometer regime would allow researchers to study, for example, mechanical effects in the quantum realm [33–36] or the coupling of the eigenmode of such a nanoscale resonator to the electromagnetic field of a photon [37, 38]. Furthermore, the small mass of nanoelectromechanical devices makes them extremely sensitive adsorptive sensors, since the mass of adsorbed molecules can significantly change the eigenfrequency of a nanoscale mechanical device [39, 40].

For lifting and manipulation of objects on the nanoscale, it is necessary to use opposing forces to seize and hold a floating or a freely suspended nanostructure, such as a single protein in liquids or an individual nanowire within a three-dimensional electrical circuit. Kim and Lieber reported the first nanoscale tweezers in operation [42]. They attached two carbon nanotubes to metal electrodes on opposite sides of a micron-thick glass needle, in order to manipulate nanoscale clusters and nanowires with a size of about 500 nm. The tweezers were closed electrostatically using a voltage difference between the nanotubes. Generally, nanotweezers can act as manipulators, sensors and injectors. For instance, they can investigate the interaction between nanomaterials and they can measure the electrical conductance of nanostructures by employing the two probes of the tweezer as electrodes. The tweezers combine a mechanical degree of control of a two probe sensing instrument with the spatial resolution of a scanning probe microscope. To date, scanning tunneling microscopy and atomic force microscopy have evolved to a level where individual atoms on surfaces can be arranged into the shape of nanoscale letters [43] and clusters of atoms can be pushed into tiny junctions to make quantum devices [44]. In order to move a nanoscale object with a single probe, there needs to be sufficient adhesion to the probe to ensure that the object can be picked up and then transported to its new destination. To deposit the object at the new location, the adhesion at the final site needs to be larger than the adhesion to the single probe. One way to overcome the problem is to use tweezers – nanoelectromechanical probes, which are attached to the tips of an atomic force microscope [45].

In most cases nanotweezers are closed and opened electrostatically using a voltage difference between the two nanoprobes. The electrical field and the potential difference between the probes may cause problems, especially when biological samples are investigated. Here, mechanical manipulation of the nanotweezers is favorable. Recently, piezo-driven tweezers have been developed with a size of about 1 μm that can grab, hold and transport bacteria in addition to micron-sized particles in liquids [46]. The experiment demonstrates that lithographically defined tweezers are a promising alternative to optical tweezers commonly applied for the manipulation of biological samples [47]. The latter utilize the effect that a focused laser beam provides an attractive force for dielectric objects.

One of the smallest tweezers devices – with a gap of only 25 nm – was fabricated by a technique which combines conventional silicon microlithography with electron beam deposition of carbon [48]. How carbon nanofingers can be fabricated by this technique is shown schematically in Figure 3.5 [41]. The fabrication process follows a classical top-down approach by microstructuring silicon with lithographic techniques [49]. To this end, silicon is partially covered by a protective polymer layer (see Figure 3.3 for top-down lithography). Then, the unprotected areas are etched away [Figure 3.5(a)–(c)]. The process uses a heterostructure with an intermediate silicon oxide layer, which can be selectively etched by hydrofluoric acid. As depicted in Figure 3.5(d), the technique enables freely suspended silicon fingers to be defined. Finally, a carbon nanofinger is defined on top of the freely suspended structure by focusing an electron beam on the silicon nanostructure in the presence of residual gases [Figure 3.5(e)] [41]. As a result, the electron beam-deposited carbon builds up to form the carbon nanofingers [Figure 3.5(f)]. The functioning of the resulting device [Figure 3.6(a) and (b)] is very intuitive. It looks like tweezers known in everyday life, only orders of magnitude smaller. By applying a voltage to the inner two, freely suspended silicon electrodes, the gap between the carbon nanofingers can be opened and closed. Most importantly, the carbon nanofingers are nonconductive. No voltage difference is applied between the nanofinger tips, making the device ideal for application with such fragile structures as organic objects.

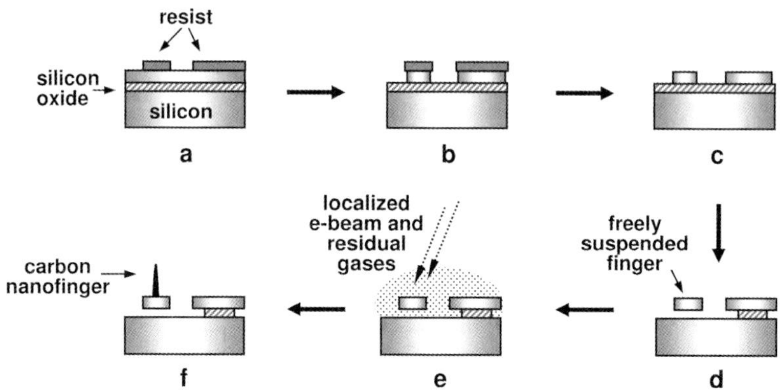

Figure 3.5 Fabrication of a freely suspended carbon nanofinger: (a) the photo- or e-beam resist is prepatterned by microlithographic steps on top of a multilayered silicon-on-insulator (SOI) wafer. The latter consists of a silicon top layer, an intermediate silicon oxide layer and a silicon substrate. (b), (c) In accordance with the pattern, the silicon top layer is removed either by wet-chemical or by reactive ion etching. (d) By the use of hydrofluoric acid, the silicon oxide layer is partially removed. Thereby, the smaller features of the prestructured silicon top layer become freely suspended. (e), (f) If an electron beam is focused on specific locations of the structure in the presence of residual gases, an electron beam-deposited carbon nanofinger can be formed [41].

As already mentioned, the release of objects picked up by the tweezers constitutes a severe conceptual problem. An object can only be released when it is deposited at a site to which its adhesion is higher than the adhesion to the nanofingers. This concept is based on hierarchical binding forces. Although this concept would allow for repeatedly picking up objects from one site A, transporting them to site B and releasing them there (in the case that $F_A < F_F < F_B$, where F_A, F_F and F_B represent the binding forces of the object to site A, to the nanofinger and to site B, respectively), it would fail in the reverse direction to transport objects from B to A (as F_B is larger than F_F, the holding force of the nanofinger would not be strong enough to detach the object from binding site B). Alternatively, the attachment of the object to the

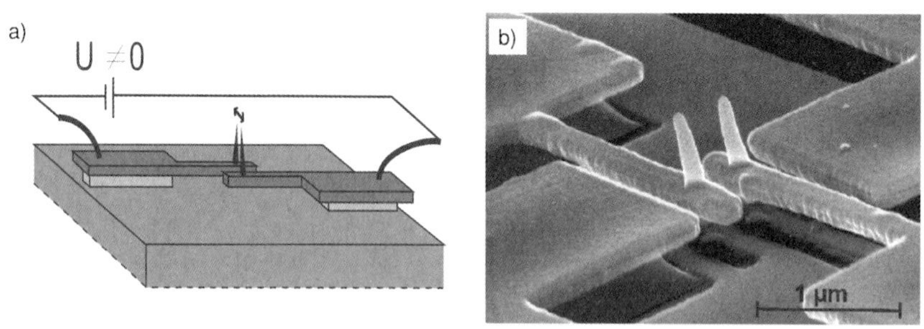

Figure 3.6 (a) By applying an electric voltage to the conductive silicon arms, the "fingers" move. (b) Electrically operated nanotweezers. The structure is made out of silicon with lithographic techniques. The tips on the silicon arms have subsequently been grown with electron beam deposition (EBD). The figure is based on the original work of Dr. Christine Meyer, Dr. Bert Lorenz and Professor Dr. K. Karrai and is reproduced with their permission [41].

nanofinger might be modulated by a biological molecule, for example, an antibody whose conformation can be modulated by light. In this way, the binding properties of the antibody could be remotely controlled by switching on or off the light and thus objects could be held and released. However, so far this concept has not been realized practically.

3.2.2
Bottom-Up Nanotweezers

In recent years, the unique biochemical and mechanical properties of DNA have been increasingly utilized for non-biological applications, for example, to realize tiny nanomechanical devices. Many of these devices exploit the different rigidity of single- and double-stranded DNA and can be switched back and forth between several structures by the addition or removal of DNA "fuel" strands. One of the prototype devices in this field are DNA nanotweezers [50]. Their operational principle is shown in Figure 3.7. In the open state, the DNA tweezers consist of three strands of DNA.

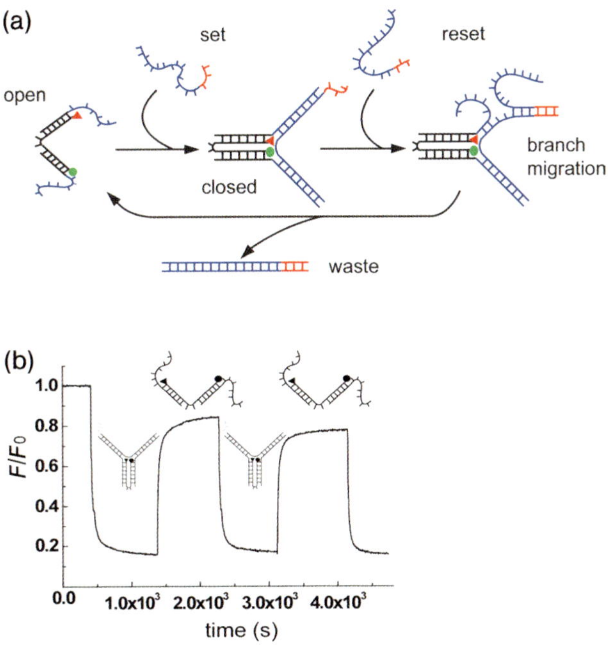

Figure 3.7 DNA nanotweezers. (a) Principle of operation. In the open state, three strands are hybridized together to form a molecular structure resembling tweezers. Due to hybridization with a "fuel" strand, the arms of the tweezers can be closed. Another DNA strand can be used to remove the fuel strand and restore the original open state via branch migration. (b) FRET experiment. The movement of the tweezers can be followed in fluorescence measurements. To this end, the arms of the tweezers are fluorescently labeled. One of the labels quenches the fluorescence of the other due to fluorescence resonance energy transfer. As this effect is strongly distance dependent, the motion of the arms is accompanied by a change in fluorescence intensity.

One central DNA strand is hybridized to two other strands in such a way that together they form two roughly 6-nm long, rigid double-stranded “arms” connected by a short, single-stranded flexible “hinge”. In the open state, each of the arms still has single-stranded extensions available for hybridization. The addition of a long “fuel” strand which is complementary to these extensions can then be used to close the tweezer structure, that is, the two arms are forced together by the hybridization with the fuel strand. The device can be switched back to its original configuration with a biochemical “trick”. In the closed state, a short single-stranded section of the device is deliberately left unhybridized. These nucleotides serve as an attachment point for an “anti-fuel” strand, which is exactly complementary to the fuel strand. A biochemical process known as “branch migration” unzips the structure, when fuel and anti-fuel try to bind with each other. After completion of this process, a waste “duplex” is ejected and the DNA tweezers have returned to their open configuration again. The device can be operated cyclically by the alternate addition of set and reset strands. The transition between the different states of the device can be characterized in fluorescence measurements, utilizing the distance dependent quenching of fluorophores due to fluorescence resonance energy transfer (FRET).

The same operation principle has since been used in many other DNA devices [51]. A simple variation of the DNA tweezers is the DNA actuator device [52]. Here, instead of two single-stranded extensions, the arms of the tweezers are connected by a single-stranded loop. Depending on the sequence of the fuel strands, the device can either be closed similarly to the tweezers or stretched into an elongated conformation. Whereas the intermediate configuration is a rather floppy structure (like the open tweezers), the closed and the stretched configurations are much more rigid. While these devices resemble macroscopic tweezers in shape, they cannot actually be used to “grab” nano-objects. However, such a function may be achieved by the incorporation of so-called “aptamers” into DNA devices [53]. These are special DNA or RNA sequences with a high binding affinity for other molecules, for example, proteins.

3.3 Second Example: Nanomotors

Another important mechanical element is the motor. A motor is a device that can generate periodic movements and carry a load with it. Similarly to the previous section, we will show two examples of nanomotors. The version using the top-down approach is again based on microlithography and follows the ideas of micromechanical engineering. The version using the bottom-up approach is based on functional organic molecules.

3.3.1 Top-Down Nanomotors

Nanomotors can also be created using refined lithographic techniques. As the strategies employed are somehow similar to those described in Section 3.2.1, we

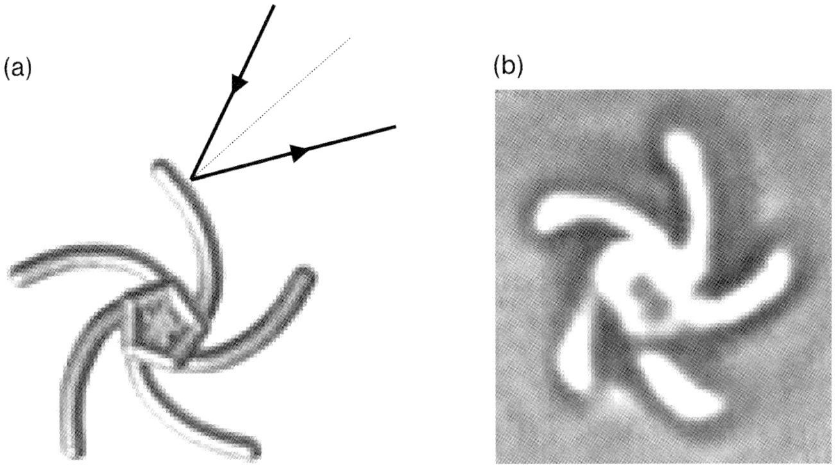

Figure 3.8 (a) A micrometer-sized rotor is driven by the momentum transferred due to reflection of an incident light beam. (b) Phase contrast image of the rotor in solution. Adopted from an paper by the group of Professor Dr. P. Ormos and is used with his permission. Details can be found in the original article [55, 56].

will only give two brief examples in this section. As with macroscopic motors, nanomotors also periodically convert input energy into mechanical work. The energy to drive the motor can originate from different sources, such as light, electric fields or chemical gradients.

A simple light-driven propeller has been demonstrated by Ormos's group. The propeller itself is created by selectively illuminating the parts of a light curing resin that resemble the shape of the propeller. The resin is cured at the illuminated regions due to photo-polymerization. Dissolution of the noncured parts of the resin finally results in a freestanding propeller (Figure 3.8) [54–57]. The propeller can be driven by incident light into rotation. When light is reflected at the arms of the propeller, momentum is transferred which causes the rotation. For a detailed description of the underlying physics we refer to the original article by Galajda and Ormos [55]. In order to achieve controlled rotation of the propeller, either a freestanding propeller is trapped in the focus of laser tweezers [55, 56] or a propeller bound to an axis on top of a substrate is driven by the light origination from an integrated waveguide [54]. So far the smallest propellers created in this way still have a size of a few micrometers.

An even smaller synthetic motor has been created by Zettl's group by scaling down MEMS technology to the nanometer scale [58]. The axis of this motor is formed by a carbon nanotube which is fixed between two anchor electrodes (Figure 3.9). A metal plate fixed to the carbon nanotube acts as a rotor. The outer shell of the nanotube beneath the rotor has been detached from the inner shells of the nanotube by shear forces so that a rotational bearing has been formed. The rotor is driven by an oscillating electric field between three additional electrodes. In practice, the rotor was

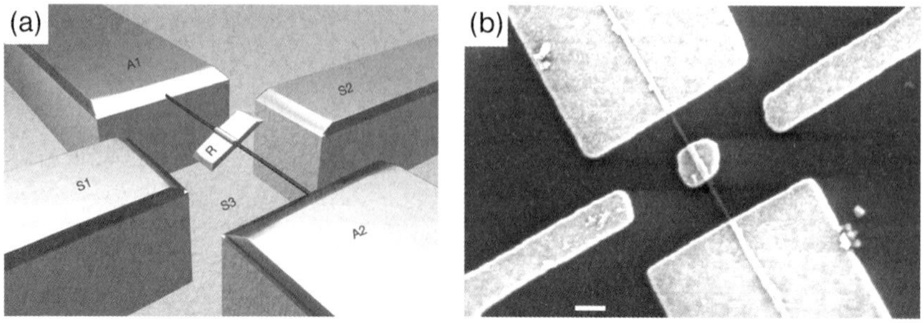

Figure 3.9 (a) Sketch of the nanomotor. A metal plate rotor (R) has been attached to a multi-walled carbon nanotube of which the outer shell beneath the metal plate can rotate freely around the inner shells and which is anchored on two electrodes (A1, A2). The motor is driven by electric fields applied from three stator electrodes, two on the SiO_2 surface (S1, S2) and one buried beneath the surface (S3). (b) Scanning electron microscope image of the nanomotor. Image reproduced with permission from the original article by Zettl's group [58].

realized by first depositing a multi-walled carbon nanotube on a silicon oxide substrate, followed by patterning the rotor and the electrodes using electron beam lithography and by etching down the silicon oxide below the rotor. More details can be found in the original breakthrough article by Zettl *et al.* [58].

3.3.2 Bottom-Up Nanomotors

Motors play an important role not only in our technical macroscopic life, but also on a molecular level in any biological organism. Molecular motors, for example, help bacteria to move and to transport cargo inside cells, and they are also the basis for contracting and expanding our muscles [59]. Since the field of nanotechnology emerged, scientists have thought about how to harness molecular motors for the construction of artificial machines. For example, the molecular motor kinesin has been used to transport colloidal quantum dots along the tracks of microtubules [60]. Such a concept might eventually lead to a scenario where building blocks could be transported along rails to their designated positions. Another example is the membrane-bound protein ATP synthase [61]. This rotary motor is driven by a proton gradient and is used in cells for the synthesis of ATP [61]. In the reverse direction the motor uses the energy of ATP hydrolysis to create a proton gradient. The rotation of the motor could be used to propel an actin filament that had been attached to the upper subunit of the motor with ATP as fuel [62]. Such a concept might eventually be used as a nanopropeller for moving small vesicles. While these concepts are based on using naturally existing molecular motors, synthetic molecular motors have also been chemically synthesized [63]. In this section we will briefly describe the concept and realization of such a motor that has been demonstrated by Gaub's group [64].

Molecules can often assume different structural arrangements called "conformations". Azobenzene, for example, can be reversibly switched upon illumination at two different excitation wavelengths between an extended *trans* and a shorter *cis* conformation [65]. In order to obtain longer effective length changes upon switching between the two different conformations, many azobenzene molecules can be linearly connected to one long polyazobenzene molecule. Single force spectroscopy techniques allow for attaching single molecules between the tip of an atomic force microscopy (AFM) cantilever and a substrate that is placed on top of a piezo-actuator [66]. In a periodic cycle, a single polyazobenzene molecule fixed between the tip and piezo of an AFM can now be elongated and shortened by illumination, which can be used to lift a load (as realized by the pulling force of the piezo-actuator) (Figure 3.10). In this way, in each cycle light energy is converted into mechanical energy [64]. Although the estimated efficiency of the motor is relatively low (total mechanical energy output/total optical energy input $\approx 10^{-4}$), it nevertheless demonstrates nicely how a light-driven artificial nanomotor can look.

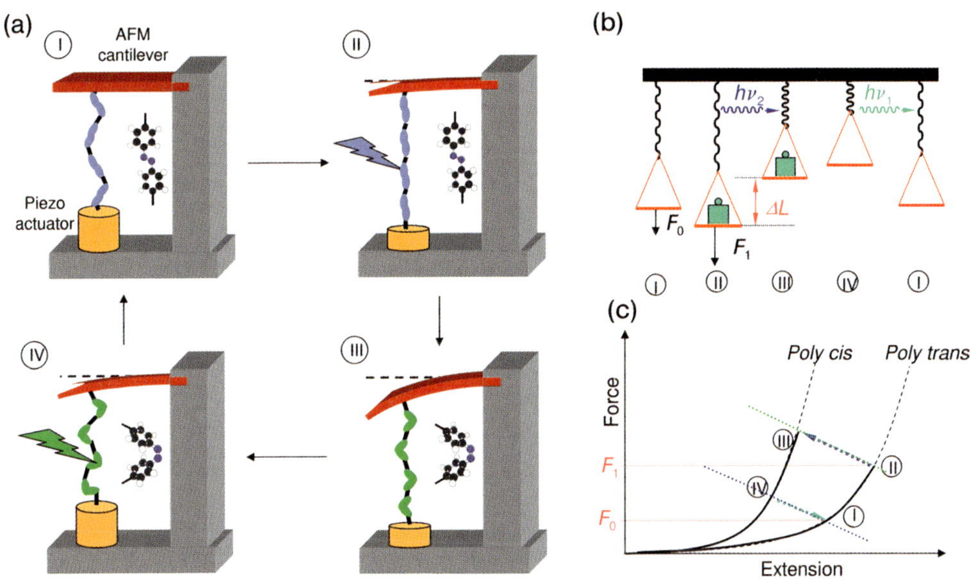

Figure 3.10 (a) A single polyazobenzene molecule is fixed between the tip of an AFM cantilever and a glass substrate mounted on top of a piezo-actuator. In a periodic cycle first the polyazobenzene is stretched by applying a load by retracting the piezo-actuator (I ⇒ II). Then, upon a light flash $h\nu_2$ the polyazobenzene is brought from the stretched *trans* to the shortened *cis* configuration (II ⇒ III). The load is now released by moving the piezo-cantilever upwards and thus the polyazobenzene can elongate (III ⇒ IV). With another light flash $h\nu_1$ the polyazobenzene is driven back to the extended *trans* configuration (IV ⇒ I). (b) Visualization of the load that is applied to the polyazobenzene by moving the piezo-actuator. (c) The cyclic process can be described in a force–extension diagram, where *F* is the force applied to the polyazobenzene by stretching it between the AFM cantilever and the glass substrate on top of the piezo and the extension describes the effective length of the polyazobenzene. Adapted from a presentation by Professor Dr. H. Gaub; details can be found in the original publication by Gaub's group [64].

3.4 Third Example: Patterning

One basic requirement for many technologies and applications is the ability to form controlled patterns on a surface. This can be exemplified with the components and techniques needed to fabricate an electronic circuit [67]. In order to make circuits, one essentially has to master three different steps. First, active elements such as transistors must be realized which are able to process information [68]. Second, the active elements have to be arranged into a functional geometry [28]. Third, the active elements have to be connected by wires [69]. Arranging elements on a surface is basically the same as forming a pattern. Another example is controlled cell attachment. If a surface is patterned partly with molecules that promote cell adhesion and partly with molecules that repel cells, cells only will grow on the desired parts of the surface. This is very important, for example, for bioelectronic interfaces, where cells have to be guided in such a way that they adhere on top of the active electronic elements, but not to other parts of the surface [70].

In this section we will briefly describe two examples for making small surface patterns. Soft lithography is a top-down approach, whereby surfaces can be structured with lithographically generated stamps. Following a self-assembly strategy, two-dimensional lattices of biological molecules, in particular DNA, can be formed in a bottom-up approach.

3.4.1 Soft Lithography

Soft lithography is an increasingly popular application of traditional lithographic techniques for the ordered deposition of "soft" materials such as small molecules, biomacromolecules or even live cells on surfaces. A variety of different techniques can be summarized under the term "soft lithography", for example, microcontact printing, replica molding and micromolding in capillaries [71, 72]. Most of these techniques utilize the soft, rubber-like material polydimethylsiloxane (PDMS). A schematic depiction of a typical soft lithographic process is shown in Figure 3.11. Usually the desired structure is first patterned into a "hard" material using conventional lithography. In Figure 3.11(a)–(c) a pattern defined on a mask is transferred into a negative resist on a silicon substrate. The structured resist film can be used as a template to transfer the patter into PDMS. To this end [Figure 3.11(d)–(f)], polymer precursors are poured over the template, followed by a curing step. After complete polymerization of the PDMS, the resulting "soft" structure can be peeled off from the substrate. There are many different applications for the structured PDMS material. In microcontact stamping, for example, the PDMS is used as a stamp with which molecules can be transferred on to a substrate in a pattern. Examples are shown in Figure 3.11(g) and (h), where fluorescently labeled proteins and also live cells have been patterned into an array of squares. PDMS-based structures are not only utilized for patterning, but have also been extensively used for the definition of microfluidic systems [73].

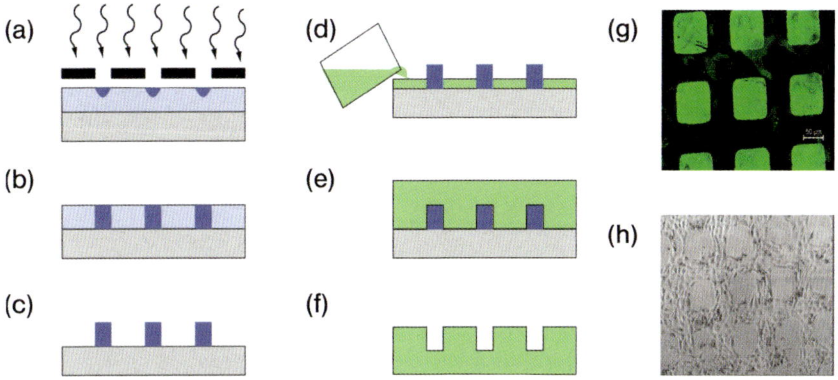

Figure 3.11 Soft lithography and microcontact printing of molecules. (a)–(c) Using conventional photolithography, a pattern is defined in a negative photoresist (blue) on a silicon wafer (gray). After exposure and development of the resist, a relief pattern remains on the surface. (d)–(f)This pattern can be used as a "master template" to create flexible microstamps from PDMS. Polymer precursors are poured over the master and cured. The resulting stamp can be peeled off the master and then used for soft lithographic processes such as stamping. (g) A square pattern of fluorescently labeled proteins stamped on a glass surface. (h) Live cells only reside outside of the square pattern which has been stamped with hydrophobic molecules which do not promote cell adhesion.

3.4.2
Two-Dimensional DNA Lattices

Whereas soft lithography can be mainly regarded as a top-down method to arrange molecules into a particular pattern, a different approach relies on molecular recognition events and self-assembly. As mentioned before, interactions between complementary strands of DNA can be utilized for the "programmable" assembly of molecular structures. Starting with the synthesis of a DNA molecule with the topology of a cube by Chen and Seeman in 1991 [74], DNA has been used to construct molecules with the structure of a truncated octahedron [75], DNA catenanes [76], tetrahedra and octahedra [77, 78] and other geometric objects. DNA has also been used to construct two-dimensional molecular lattices from DNA branched junctions [19, 79–81]. One of the central building blocks of such lattices is the so-called "double-crossover" (DX) construction. In a DX molecule, two double-stranded DNA molecules exchange strands twice, creating an entity which can be regarded as the molecular analogue of a LEGO building block [Figure 3.12(a)]. The rigid DX molecule has dimensions on the order of 4×15 nm. Depending on the sequence of single-stranded "sticky ends" at the corners of the DX molecules, they can arrange into large two-dimensional lattices [Figure 3.12(b) and (c)]. In fact, the assembly of such DX "tiles" closely resembles a computational process – regarding DNA self-assembly as the execution of a molecular "program" can actually be exploited to realize more complex molecular structures than with any other self-assembly technique [82]. When these networks fold back to themselves, they can also form DNA "nanotubes" [83–85]. Recently, such networks have even been utilized to arrange nanoparticles and proteins in two dimensions

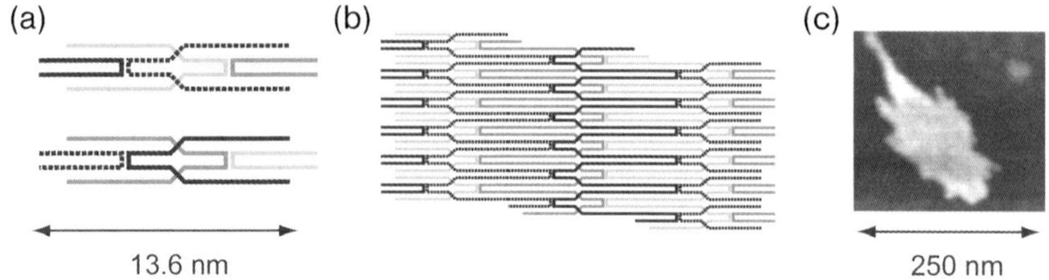

Figure 3.12 DX assembly. (a) The basic building blocks – DX molecules – are composed of two DNA double strands which exchange strands twice, creating a structure held together by two DNA crossovers. (b) Depending on the "sticky ends" at the corners of the DX tiles, a variety of two-dimensional lattices may be formed in a self-assembly process. (c) A small patch of a 2D DX lattice imaged with an atomic force microscope under air. Functionalization of these lattices allows for the arrangement of nanoparticles or proteins into sequence-programmed patterns.

[28, 86–89] and the impressively powerful "DNA origami" scheme [82] will soon allow the arrangement of these nanoparticles into arbitrary geometries and patterns.

3.5 Fourth Example: Quantum Dots[1)]

In Chapter 4, the physical principles of quantum dots are described. Quantum dots are arguably the ultimate examples of nanometer structures. In this section, we will show that we can manufacture quantum dots both with top-down (Section 3.5.2) and with bottom-up techniques (Section 3.5.4). Depending of the manufacturing process used, the properties of the quantum dots vary and we will explain this in particular with respect to their optical properties. Also the possible applications vary among the different types of quantum dots.

3.5.1 Different Methods for Making Quantum Dots

Here we will give a brief overview of the most popular methods used to fabricate quantum dots in practice. Lithographically defined quantum dots are the classical example of the application of a top-down method, whereby the quantum dot is created by locally etching away parts of the raw material. Colloidal quantum dots, on the other hand, are an example of a bottom-up approach, in which they are assembled from small building blocks (in this case surfactant-stabilized atoms).

The ultimate technique for the fabrication of quantum dots should be able to produce significant amounts of sample, with such high control of quantum dot size, shape and monodispersity that single-particle properties are not averaged out by sample inhomogeneity. So far, ensembles of quantum dots produced by the best

1) This paragraph has been adopted from a previous edition and the authors acknowledge their former coauthors Dr. Liberato Manna, Dr. Daniele Gerion and Prof. Dr. Paul Alivisatos 90.

available techniques still show a behavior deriving from a distribution of sizes, but this field is evolving very rapidly. In this section we give a short survey of the most popular fabrication approaches. Different techniques lead to different typologies of quantum dots. The confinement can be obtained in several different ways and in addition the quantum dot itself can have a peculiar geometry, it can be embedded into a matrix or grown on a substrate or it can be a “free” nanoparticle. Each of these cases is strictly related to the preparative approach chosen.

3.5.2 Lithographically Defined Quantum Dots

Lithographically defined quantum dots are formed by isolating a small region of a two-dimensional electron system (see Chapter 4) via tunneling barriers from its environment. Such two-dimensional electron systems (2DES) or 2D electron gases (2DEGs) can be found in metal–oxide–semiconductor field effect transistors (MOSFETs) or in the so-called semiconductor heterostructures [91, 92]. Heterostructures are composed of several thin layers of different semiconductor materials grown on top of each other, using molecular beam epitaxy (MBE). The layer sequence can be chosen in such a way that all free charge carriers are confined to a thin slice of the crystal, forming essentially a two-dimensional electron system. A superstructure derived from the periodic repetition of this sequence of layers is also called a “multiple quantum well”. One of the most widely investigated systems is, for instance, the aluminum gallium arsenide/gallium arsenide (AlGaAs/GaAs) quantum well. AlGaAs has the same lattice constant as GaAs but a wider bandgap, whose exact value depends on the aluminum content of the layer. Therefore, electrons in the GaAs layer are confined to this layer and form a two-dimensional electron gas.

Quantum dot systems can be generated in a lateral or in a vertical arrangement, as shown in Figure 3.13. In the lateral geometry, the 2DEG is locally electrostatically depleted by applying negative voltages to electrodes deposited on top of the crystal. We can understand this effect with the following argument. Let us assume that we apply a negative voltage to the metal electrodes above the two-dimensional electron gas. Due to electrostatic repulsions, electrons will be repelled by the electric field of the electrodes and the region of the 2DEG below the electrodes will be depleted of electrons. A charge-depleted region behaves like an insulator. Therefore, by applying an electric field with metal electrodes of an appropriate shape, it is possible to create an island of charges insulated from the rest of the 2DEG. If small enough, the island within the 2DEG behaves as a quantum dot. In the vertical geometry, a small pillar of the 2DEG is isolated by etching away the heterostructure around it. In such an arrangement charge carriers are again confined in all three dimensions.

Most of the electron transport measurements on quantum dots that have been performed to date have used the two types of quantum dots as samples that we have just described. The lateral arrangement offers a relatively high degree of freedom for the design of the structure, as this is determined by the choice of the electrode geometry. In addition, it is possible to fabricate and study “artificial molecules” [93–99] composed of several quantum dots linked together. With the

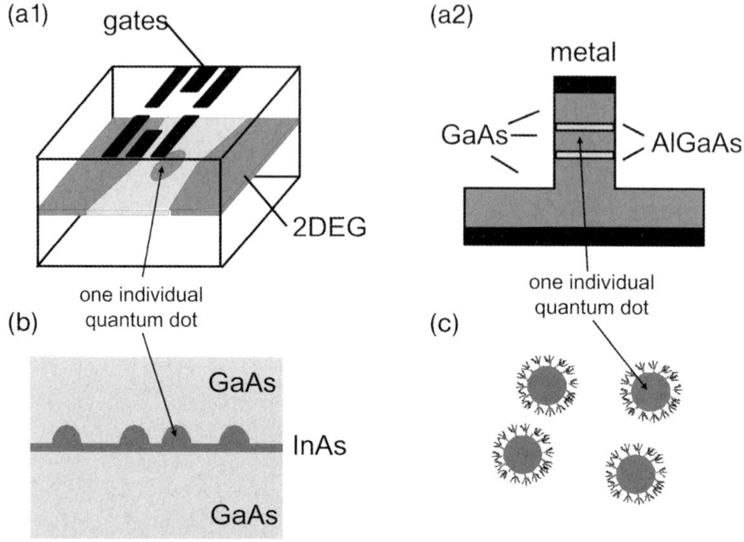

Figure 3.13 Three different types of quantum dots. (a1) A lithographically defined quantum dot in a lateral arrangement can be formed by electrostatic depletion of a two-dimensional electron gas (2DEG, shown in dark gray) via gate electrodes. The 2DEG is formed typically 20–100 nm below the surface of a semiconductor heterostructure (usually GaAs/AlGaAs). Application of negative voltages to metal gates on top of the heterostructure depletes the 2DEG below the gates (shown in light gray) and cuts out a small electron island from the 2DEG. Electrons can still tunnel on to and from the island. Electrical contact to the 2DEG is realized through ohmic contacts (not shown). (a2) A vertical quantum dot can be formed in a double barrier heterostructure. A narrow pillar is etched out of a GaAs/AlGaAs/GaAs/AlGaAs/GaAs heterostructure. The AlGaAs layers (light gray) form tunnel barriers that isolate the central GaAs region from the contact region. This central GaAs region behaves now as a quantum dot (shown in dark gray). Electrical contact is made via metal contacts (depicted in black) on top of the pillar and below the heterostructure. (b) Self-assembled quantum dots: molecular beam epitaxy (MBE) growth of InAs (dark gray) on GaAs (light gray) first leads to the formation of an extended layer of InAs (the wetting layer) and then to the formation of small InAs islands. Single electrons or electron–hole pairs (excitons) can be confined in these InAs quantum dots, either electrically or optically. (c) Colloidal quantum dots: these colloidal particles, having a diameter of only a few nanometers, are formed using wet chemistry and can be produced for most of the type II–VI, III–V, IV–VI and some type IV semiconductors. The surface of colloidal quantum dots is coated with a layer of surfactant molecules that prevents aggregation of the particles.

vertical arrangement, structures with very few electrons can be realized [100]. Recently, several research efforts have been focused on the investigation of many-body phenomena in these quantum dot systems. Relevant examples are, for instance, the study of the Kondo effect [101–104] and the design and control of coherent quantum states with the ultimate goal of quantum information processing [105, 106].

A remarkable advantage of lithographically defined quantum dots is that their electrical connection to the "macro-world" is straightforward. The manufacturing processes are similar to those used in chip fabrication and in principle such structures could be embedded within conventional electronic circuits. However, as the geometry of these quantum dots is determined lithographically, it is restricted by the usual size and resolution limits of lithographic techniques. Even by using electron beam lithography for the creation of the quantum dots, it is not possible to tailor their size with nanometer precision. Lithographically fabricated quantum dots are typically larger than 10 nm and so relatively low lateral confining energies can be achieved.

3.5.3 Epitaxially Self-Assembled Quantum Dots

A breakthrough in the field of epitaxially grown nanostructures was the discovery of epitaxial growth regimes that favored the formation of nanometer-sized islands of semiconductor materials on suitable substrates (Figure 3.13). These islands, exhibiting quantum dot behavior, are obtained naturally by epitaxially growing a thin layer of a low bandgap material over a higher bandgap material, using MBE or MOCVD techniques [107–110]. The respective crystal faces in contact must have a significant lattice mismatch (1–8%), as in the case of InAs on GaAs [111, 112] and Ge on Si [113]. During the growth, a strained film, called the "wetting layer", initially forms. The maximum thickness of this layer is related to the difference in the lattice constants between the two materials. Past this critical thickness, a 2D → 3D transition in the growth regime is observed, with the spontaneous formation of an array of nanometer-sized islands (Stranski–Krastanov regime), leading to partial release of the strain. If the growth is not interrupted at this step, misfit dislocations form because the energy of formation of these defects becomes smaller than the elastic energy accumulated in the strained film. The formation of dislocations in highly strained epilayers (when the lattice mismatch is of the order of 10% or more) before the formation of islands limits the range of possible substrate–island materials. The shape of the islands can be controlled by the growth conditions. Usually, the islands have a truncated pyramidal shape, but it is also possible to form, for example, ring-shaped quantum dots [114]. The final step consists in the growth, on the top of the islands, of several layers of the substrate material, so that the dots are completely buried and interfaces are passivated. The relative alignment of the bandgaps creates a confining potential for charge carriers that accumulate inside the quantum dots. In addition, strain fields in the proximity of the island–substrate interface, due to the lattice mismatch between the two materials, create potentials that modify the bandgap of the quantum dots at the bottom of the island. Holes are more likely to be localized in this region, as they are heavier than the electrons.

Self-assembled quantum dots can have a diameter as small as a few nanometers and so very pronounced quantum size effects can be observed in these systems. Self-assembled quantum dots have predominantly been characterized using optical or capacitance spectroscopy in a regime where they contain only a small number of charge carriers. Measurements on ensembles still suffer from inhomogeneous broadening of the spectroscopic features. However, in recent years it has been possible to look at only a few or even single self-assembled quantum dots at a time by reducing the number of quantum dots by mesa-etching [115] or by using confocal microscopy techniques [116]. Photoluminescence from single self-assembled quantum dots is a highly efficient process, characterized by several narrow emission lines, related to different exciton states in the dots, and is reminiscent of the emission from atoms. As mentioned already for lithographically defined quantum dots, many parallels can be drawn between atoms and quantum dots [115, 117–119]. For these reasons, quantum dots have gained the nickname "artificial atoms". Current research efforts are devoted to quantum dot ordering and positioning and also to the reduction of the quantum dot size distribution. In contrast to the case of lithographically defined quantum dots, it is challenging to make electrical contact to self-assembled quantum dots and therefore most of the possible applications can be found in optics. One of the major goals of research on self-assembled quantum dots is the fabrication of non-classical light sources from single dots. Another is to use them as light-addressable storage devices.

3.5.4
Colloidal Quantum Dots

Colloidal quantum dots are remarkably different from the quantum dot systems mentioned above as they are chemically synthesized using wet chemistry and are free-standing nanoparticles or nanocrystals grown in solution (Figure 3.13) [120]. In this case, colloidal quantum dots are just a subgroup of a broader class of materials that can be synthesized at the nanoscale using wet chemical methods. In the fabrication of colloidal nanocrystals, the reaction chamber is a reactor containing a liquid mixture of compounds that control the nucleation and the growth. In a general synthesis of quantum dots in solution, each of the atomic species that will form the nanocrystal building blocks is introduced in the reactor as a precursor. A precursor is a molecule or a complex containing one or more atomic species required for growing the nanocrystal. Once the precursors have been introduced into the reaction flask they decompose, forming new reactive species (the monomers) that will cause the nucleation and the growth of the nanocrystals. The energy required to decompose the precursors is provided by the liquid in the reactor, either by thermal collisions or by a chemical reaction between the liquid medium and the precursors or by a combination of the two mechanisms [121].

The key parameter in the controlled growth of colloidal nanocrystals is the presence of one or more molecular species in the reactor, broadly termed here "surfactants". A surfactant is a molecule that is dynamically adsorbed on the surface of the growing quantum dot under the reaction conditions. It must be mobile enough to provide access for the addition of monomer units, but stable enough to prevent the aggregation of nanocrystals. The choice of surfactants varies from case to case: a molecule that binds too strongly to the surface of the quantum dot is not suitable, as it would not allow the nanocrystal to grow. On the other hand, a weakly coordinating molecule would yield large particles or aggregates [122]. Some examples of suitable surfactants include alkanethiols, phosphines, phosphine oxides, phosphates, phosphonates, amides, amines, carboxylic acids and nitrogen-containing aromatics. If the growth of nanocrystals is carried out at high temperatures (e.g., at 200–400 °C) then the surfactant molecules must be stable under such conditions in order to be a suitable candidate for controlling the growth.

At low temperatures, or more generally when the growth is stopped, the surfactants are more strongly bound to the surface of the nanocrystals and provide their solubility in a wide range of solvents. This coating allows for great synthetic flexibility in that it can be exchanged with another coating of organic molecules having different functional groups or polarity. In addition, the surfactants can be temporarily removed and an epitaxial layer of another material with different electronic, optical or magnetic properties can be grown on the initial nanocrystal [123, 124].

By controlling the mixture of surfactant molecules that are present during the generation and the time of growth of the quantum dots, excellent control of their size and shape is possible [121, 125–127] (Figure 3.14). As described in Chapter 4, the wavelength of emission of quantum dots depends on their size. This can be nicely seen by observing the fluorescence light of solutions of colloidal quantum dots with different size [128, 129] (Figures 3.14 and 3.15).

In contrast to organic fluorophores, colloidal quantum dots have a continuous absorption spectrum for energies higher than their bandgap, a symmetric emission

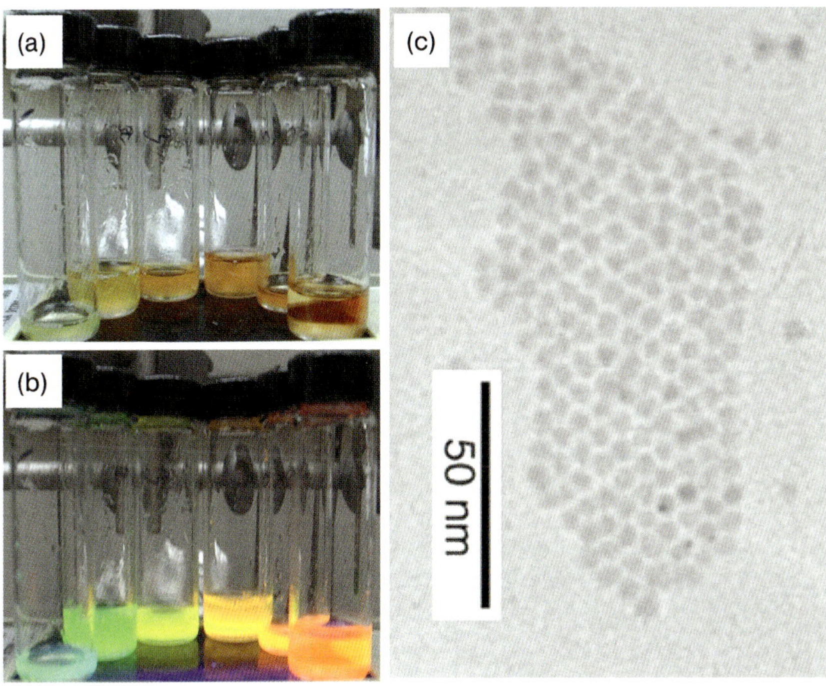

Figure 3.14 Colloidal CdSe quantum dots of different size dissolved in chloroform. The size of the quantum dots increases from the left to the right vial. (a) Photograph of the solutions. (b) Photograph of the solutions upon UV illumination from below. The different colors of fluorescence can be seen. (c) Each quantum dot corresponds to a spherical CdSe nanoparticle that is dispersed in the solvent. When a drop of this solution is put on a grid, the solvent evaporates and the quantum dots can be imaged with transmission electron microscopy (TEM). The authors are grateful to Stefan Kudera for recording the TEM image.

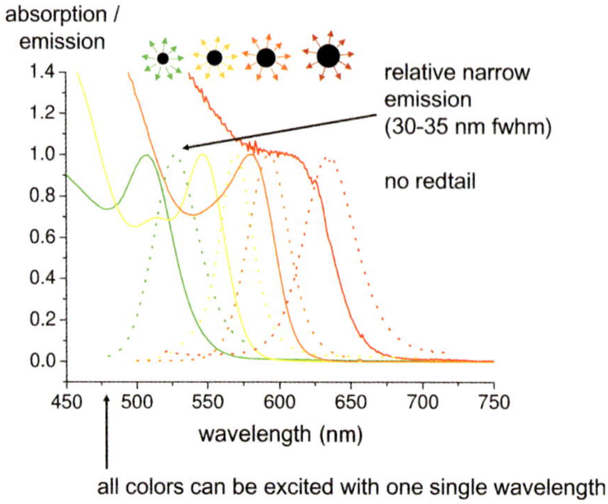

Figure 3.15 Absorption (solid lines) and emission spectra (dotted lines) of colloidal CdSe quantum dots of different sizes. The absorption peaks of green/yellow/orange/red fluorescent nanocrystals of 2.3/4.0/3.8/4.6 nm diameter are at 507/547/580/605 nm and the fluorescence peaks are at 528/57/592/637 nm.

spectrum without a red tail and, most importantly, reduced photobleaching [130, 131] (Figure 3.15).

Since colloidal nanocrystals are dispersed in solution, they are not bound to any solid support as is the case for the other two quantum dots systems described above. Therefore, they can be produced in large quantities in a reaction flask and later they can be transferred to any desired substrate or object. It is possible, for example, to coat their surface with biological molecules such as proteins or oligonucleotides. Many biological molecules perform tasks of molecular recognition with extremely high efficiency. This means that ligand molecules bind with very high specificity to certain receptor molecules, similarly to a key-and-lock system. If a colloidal quantum dot is tagged with ligand molecules, it specifically binds to all the positions where a receptor molecule is present. In this way it has been possible, for example, to make small groupings of colloidal quantum dots mediated by molecular recognition [22, 23, 132] and to label specific compartments of cell with different types of quantum dots [133–135].

Although colloidal quantum dots are rather difficult to connect electrically, a few electron transport experiments have been reported. In these experiments, nanocrystals were used as the active material in devices that behave as single electron transistors [68, 136].

3.6
Perspectives and Limits of Top-Down and Bottom-Up Approaches

In 1965, Gordon Moore, co-founder of Intel Corporation, predicted that the number of transistors on a computer chip would double about every 18 months [137]. The exponential law, also known as Moore's first law, has described the development of integrated circuits surprisingly well for decades. As the market for information technology continues to grow, the demand for computer hardware instigates more and more sophisticated top-down techniques to build more densely packed transistor circuits. Moore's second law states that the implementation of a next generation of integrated circuits at minimum cost will be exponentially more expensive. Until all the constraints finally limit the growth of the semiconductor top-down industry, scientists and engineers assume that nanotechnology will give answers to most of the technological challenges. For instance, as soon as the feature size of the semiconductor transistors reaches the level at which quantum phenomena are important, different concepts for the assembly need to be considered. One possibility is the bottom-up approach, which is based on molecular recognition and chemical self-assembly of molecules [13]. In combination with chemical synthesis techniques, the bottom-up approach allows the assembly of macromolecular complexes with new functionalities. Assuming Moore's laws apply [137], the final limit for optical top-down lithography is likely to be reached in less than a decade.

There are limitations also for the bottom-up assembly of complex nanostructures. We can illustrate this by the example of assembling nanoparticles with DNA to groupings of particles. Although building blocks exist in which each nanoparticle is modified with an exactly defined number of binding sites, still no controlled assemblies of more than around five particles exist. There are two fundamental technological problems: nonspecific adsorption and "floppyness" of biological molecules. Nonspecific adsorption causes particles to stick together, although they are not supposed to be connected. Many biological molecules that can be used as "glue" for the assembly of particle groupings, such as proteins and DNA, tend to adsorb nonspecifically on the surface of nanoparticles. Although covalent attachment of these molecules dominates over nonspecific adsorption and thus connection of the particles via their designated binding sites, there is a non-negligible amount of nonspecific interaction between the particles. For a larger particle grouping, just a single nonspecific connection can destroy the build-up of the whole grouping. One major task in this direction is to improve the surface chemistry of particles in order to obtain inert surfaces to which biological molecules do not adsorb nonspecifically. Biological molecules are intrinsically "soft" compared with inorganic materials. This implies that the connection between particles that are connected via biological molecules will always retain a certain degree of flexibility. In particular, the attachment of the "glue", that is, the biological molecule, to the particle surface is not rigid. Therefore, it will be almost impossible to form large, nonperiodic three-dimensional stiff structures with static geometry of particle groupings connected via biological molecules. Further, particle assemblies involving biological molecules as glue will be always limited in stability. Biological molecules are bound to their natural environment and cannot withstand many artificial conditions, such a high temperatures.

Although both top-down and bottom-up strategies have clear limitations, we are still far away from having reached them. In fact, the intrinsics may finally be overcome by combining both approaches. Still, Feynman's statement is true: "there is plenty of room at the bottom!"

References

1 Whitesides, G.M. (2005) Nanoscience, nanotechnology and chemistry. *Small*, **1**, 172–179.

2 Whitesides, G.M. (1998) Nanotechnology: art of the possible. *Technology Review*, **101**, 84–87.

3 Lehn, J.M. (2004) Supramolecular chemistry: from molecular information towards self-organization and complex matter. *Reports on Progress in Physics*, **67**, 249–265.

4 Binnig, G., *et al.* (1982) Surface studies by scanning tunneling microscopy. *Physical Review Letters*, **49**, 57–61.

5 Binnig, G., Quate, C.F. and Gerber, C. (1986) Atomic force microscope. *Physical Review Letters*, **56**, 930–933.

6 Gimzewski, J.K. and Joachim, C. (1999) Nanoscale science of single molecules using local probes. *Science*, **283**, 1683–1688.

7 Lieber, C.M., Liu, J. and Sheehan, P.E. (1996) Understanding and manipulating inorganic materials with scanning probe microscopes. *Angewandte Chemie International Edition in English*, **35**, 687–704.

8 Poggi, M.A., *et al.* (2004) Scanning probe microscopy. *Analytical Chemistry*, **76**, 3429–3443.

9 Wiesendanger, R. (1997) Scanning-probe-based science and technology. *Proceedings of the National Academy of Sciences of the United States of America*, **94**, 12749–12750.

10 Wiesendanger, R. (1994) Contributions of scanning probe microscopy and spectroscopy to the investigation and fabrication of nanometer-scale structures. *Journal of Vacuum Science & Technology B*, **12**, 515–529.

11 Friedbacher, G. and Fuchs, H. (1999) Classification of scanning probe microscopies (technical report). *Pure and Applied Chemistry*, **71**, 1337–1357.

12 Quate, C.F. (1997) Scanning probes as a lithography tool for nanostructures. *Surface Science*, **386**, 259–264.

13 Wilbur, J.L. and Whitesides, G.M. (1999) Self-assembly and self-assembled monolayers in micro- and nanofabrication, in *Nanotechnology*, (ed. G. Timp), Springer, New York.

14 Crommie, M.F., Lutz, C.P. and Eigler, D.M. (1993) Imaging standing waves in a 2-dimensional electron-gas. *Nature*, **363**, 524–527.

15 Irmer, B., *et al.* (1998) Josephson junctions defined by a nanoplough. *Applied Physics Letters*, **73**, 2051–2053.

16 Tennant, D.M. (1999) Limits of conventional lithography, in Nanotechnology, in *Nanotechnology*, (ed. G. Timp), Springer, New York.

17 Di Ventra, M., Evoy, S. and Heflin, J.R. (2004) *Introduction to Nanoscale Science and Technology*, Kluwer, Dordrecht.

18 Whitesides, G.M. and Grzybowski, B. (2002) Self-assembly at all scales. *Science*, **295**, 2418–2421.

19 Winfree, E., *et al.* (1998) Design and self-assembly of two-dimensional DNA crystals. *Nature*, **394**, 539–544.

20 Bowden, N., *et al.* (1997) Self-assembly of mesoscale objects in ordered two-dimensional arrays. *Science*, **276**, 233–234.

21 Alivisatos, A.P., *et al.* (1996) Organization of "nanocrystal molecules" using DNA. *Nature*, **382**, 609–611.

22 Mirkin, C.A., *et al.* (1996) A DNA-based method for rationally assembling nanoparticles into macroscopic materials. *Nature*, **382**, 607–609.

23 Zanchet, D., *et al.* (2002) Electrophoretic and structural studies of DNA-directed Au nanoparticle groupings. *Journal of Physical Chemistry B*, **106**, 11758–11763.

24 Sperling, R.A., *et al.* (2006) Electrophoretic separation of nanoparticles with a discrete number of functional groups. *Advanced Functional Materials*, **16**, 943–948.

25 Love, J.C., *et al.* (2005) Self-assembled monolayers of thiolates on metals as a

form of nanotechnology. *Chemical Reviews*, **105**, 1103–1169.

26 Reed, M.A., *et al.* (1997) Conductance of a molecular junction. *Science*, **278**, 252–254.

27 Lahav, M., Shipway, A.N. and Willner, I. (1999) Au-nanoparticle–bis-bipyridinium cyclophane superstructures: assembly, characterization and sensoric applications. *Journal of the Chemical Society-Perkin Transactions 2*, 1925–1931.

28 Zheng, J., *et al.* (2006) Two-dimensional nanoparticle arrays show the organizational power of robust DNA motifs. *Nano Letters*, **6**, 1502–1504.

29 Simmel, F.C. and Yurke, B. (2004) DNA-based nanodevices, in *Encyclopedia of Nanoscience and Nanotechnology*, (ed. H.S. Nalwa), American Scientific Publishers, Stevenson Ranch, 495–504.

30 Chan, V.Z-H., *et al.* (1999) Ordered bicontinuous nanoporous and nanorelief ceramic films from self assembling polymer precursors. *Science*, **286**, 1716–1719.

31 Smalley, R.E. (2001) Chemie, Liebe und dicke Finger. *Spektrum der Wissenschaft Spezial Nanotechnologie 2*, November 2001, 66–67.

32 Smalley, R.E. (September, 2001) Of chemistry, love and nanobots. *Scientific American*, 76–77.

33 Cleland, A.N. and Roukes, M.L. (1996) Fabrication of high frequency nanometer scale mechanical resonators from bulk Si crystals. *Applied Physics Letters*, **69**, 2653–2655.

34 Carr, D.W. and Craighead, H.G. (1997) Fabrication of nanoelectromechanical systems in single crystal silicon using silicon on insulator substrates and electron beam lithography. *Journal of Vacuum Science & Technology B*, **15**, 2760–2763.

35 Craighead, H.G. (2000) Nanoelectromechanical systems. *Science*, **290**, 1532–1535.

36 Knobel, R.G. and Cleland, A.N. (2003) Nanometre-scale displacement sensing using a single electron transistor. *Nature*, **424**, 291–293.

37 Karrai, K. (2006) Photonics – a cooling light breeze. *Nature*, **444**, 41–42.

38 Höhberger Metzger, C. and Karrai, K. (2004) Cavity cooling of a microlever. *Nature*, **432**, 1002–1005.

39 Ilic, B., Yang, Y. and Craighead, H.G. (2004) Virus detection using nanoelectromechanical devices. *Applied Physics Letters*, **85**, 2604–2606.

40 Ekinci, K.L., Huang, X.M.H. and Roukes, M.L. (2004) Ultrasensitive nanoelectromechanical mass detection. *Applied Physics Letters*, **84**, 4469–4471.

41 Meyer, C., Lorenz, H. and Karrai, K. (2003) Optical detection of quasi-static actuation of nanoelectromechanical systems. *Applied Physics Letters*, **83**, 2420–2422.

42 Kim, P. and Lieber, C.M. (1999) Nanotube nanotweezers. *Science*, **286**, 2148–2150.

43 Eigler, D.M. and Schweizer, E.K. (1990) Positioning single atoms with a scanning tunneling microscope. *Nature*, **344**, 524–526.

44 Carlsson, S.B., *et al.* (1999) Mechanical tuning of tunnel gaps for the assembly of single-electron transistors. *Applied Physics Letters*, **75**, 1461–1463.

45 Akita, S., *et al.* (2001) Nanotweezers consisting of carbon nanotubes operating in an atomic force microscope. *Applied Physics Letters*, **79**, 1691–1693.

46 Jericho, S.K., *et al.* (2004) Micro-electro-mechanical systems microtweezers for the manipulation of bacteria and small particles. *Review of Scientific Instruments*, **75**, 1280–1282.

47 Ashkin, A., *et al.* (1986) Observation of a single-beam gradient force optical trap for dielectric particles. *Optics Letters*, **11**, 288–290.

48 Boggild, P., *et al.* (2001) Fabrication and actuation of customized nanotweezers with a 25 nm gap. *Nanotechnology*, **12**, 331–335.

49 Blick, R.H., *et al.* (2002) Nanostructured silicon for studying fundamental aspects

of nanomechanics. *Journal of Physics-Condensed Matter*, **14**, R905–R945.
50 Yurke, B., *et al.* (2000) A DNA-fuelled molecular machine made of DNA. *Nature*, **406**, 605–608.
51 Simmel, F.C. and Dittmer, W.U. (2005) DNA nanodevices. *Small*, **1**, 284–299.
52 Simmel, F.C. and Yurke, B. (2002) A DNA-based molecular device switchable between three distinct mechanical states. *Applied Physics Letters*, **80**, 883–885.
53 Dittmer, W.U., Reuter, A. and Simmel, F.C. (2004) A DNA-based machine that can cyclically bind and release thrombin. *Angewandte Chemie-International Edition*, **43**, 3550–3553.
54 Kelemen, L., Valkai, S. and Ormos, P. (2006) Integrated optical motor. *Applied Optics*, **45**, 2777–2780.
55 Galajda, P. and Ormos, P. (2002) Rotors produced and driven in laser tweezers with reversed direction of rotation. *Applied Physics Letters*, **80**, 4653–4655.
56 Galajda, P. and Ormos, P. (2002) Rotation of microscopic propellers in laser tweezers. *Journal of Optics B: Quantum and Semiclassical Optics*, **4**, S78–S81.
57 Galajda, P. and Ormos, P. (2001) Complex micromachines produced and driven by light. *Applied Physics Letters*, **78**, 249–251.
58 Fennimore, A.M., *et al.* (2003) Rotational actuators based on carbon nanotubes. *Nature*, **424**, 408–410.
59 Schliwa M. (ed.) (2002) *Molecular Motors*, Wiley-VCH, Weinheim.
60 Muthukrishnan, G., *et al.* (2006) Transport of semiconductor nanocrystals by kinesin molecular motors. *Small*, **2**, 626–630.
61 Stock, D., Leslie, A.G.W. and Walker, J.E. (1999) Molecular architecture of the rotary motor in ATP synthase. *Science*, **286**, 1700–1705.
62 Sambongi, Y., *et al.* (1999) Mechanical rotation of the c subunit oligomer in ATP synthase (F0F1): direct observation. *Science*, **286**, 1722–1724.
63 Fletcher, S.P., *et al.* (2005) A reversible, unidirectional molecular rotary motor driven by chemical energy. *Science*, **310**, 80–82.
64 Hugel, T., *et al.* (2002) Single-molecule optomechanical cycle. *Science*, **296**, 1103–1106.
65 Hartley, G.S. (1937) The *cis*-form of azobenzene. *Nature*, **140**, 281–281.
66 Grandbois, M., *et al.* (1999) How strong is a covalent bond? *Science*, **283**, 1727–1730.
67 Eichen, Y., *et al.* (1998) Self-assembly of nanoelectronic components and circuits using biological templates. *ACTA Polymerica*, **49**, 663–670.
68 Klein, D.L., *et al.* (1997) A single-electron transistor made from a cadmium selenide nanocrystal. *Nature*, **389**, 699–701.
69 Braun, E., *et al.* (1998) DNA-templated assembly and electrode attachement of a conducting silver wire. *Nature*, **391**, 775–778.
70 Offenhäusser, A., Rühe, J. and Knoll, W. (1995) Neuronal cells cultured on modified microelectronic device surfaces. *Journal of Vacuum Science & Technology A-Vacuum Surfaces and, Films* **13**, 2606–2612.
71 Xia, Y.N. and Whitesides, G.M. (1998) Soft lithography. *Annual Review of Materials Science*, **28**, 153.
72 Quake, S.R. and Scherer, A. (2000) From micro- to nanofabrication with soft materials. *Science*, **290**, 1536–1540.
73 Unger, M.A., *et al.* (2000) Monolithic microfabricated valves and pumps by multilayer soft lithography. *Science*, **288**, 113–116.
74 Chen, J.H. and Seeman, N.C. (1991) Synthesis from DNA of a molecule with the connectivity of a cube. *Nature*, **350**, 631–633.
75 Zhang, Y.W. and Seeman, N.C. (1994) Construction of a DNA-truncated octahedron. *Journal of the American Chemical Society*, **116**, 1661–1669.
76 Mao, C., Sun, W. and Seeman, N.C. 1997 Assembly of borromean rings from DNA. *Nature* **386** 137–138.
77 Goodman, R.P., *et al.* (2005) Rapid chiral assembly of rigid DNA building blocks for

molecular nanofabrication. *Science*, **310**, 1661–1665.

78 Shih, W.M., Quispe, J.D. and Joyce, G.F. (2004) A 1.7-kilobase single-stranded DNA that folds into a nanoscale octahedron. *Nature*, **427**, 618–621.

79 LaBean, T.H., *et al.* (2000) Construction, analysis, ligation and self-assembly of DNA triple crossover complexes. *Journal of the American Chemical Society*, **122**, 1848–1860.

80 Liu, D., *et al.* (2004) Tensegrity: construction of rigid DNA triangles with flexible four-arm DNA junctions. *Journal of the American Chemical Society*, **126**, 2324–2325.

81 Liu, H.P., *et al.* (2006) Approaching the limit: can one DNA oligonucleotide assemble into large nanostructures? *Angewandte Chemie-International Edition*, **45**. 1942,

82 Rothemund, P.W.K. (2006) Folding DNA to create nanoscale shapes and patterns. *Nature*, **440**, 297–302.

83 Liu, D., *et al.* (2004) DNA nanotubes self-assembled from triple-crossover tiles as templates for conductive nanowires. *Proceedings of the National Academy of Sciences of the United States of America*, **101**, 717–722.

84 Mitchell, J.C., *et al.* (2004) Self-assembly of chiral DNA nanotubes. *Journal of the American Chemical Society*, **126**, 16342–16343.

85 Rothemund, P.W.K., *et al.* (2004) Design and characterization of programmable DNA nanotubes. *Journal of the American Chemical Society*, **126**, 16344–16352.

86 Le, J.D., *et al.* (2004) DNA-templated self-assembly of metallic nanocomponent arrays on a surface. *Nano Letters*, **4**, 2343–2347.

87 Pinto, Y.Y., *et al.* (2005) Sequence-encoded self-assembly of multiple-nanocomponent arrays by 2D DNA scaffolding. *Nano Letters*, **5**, 2399–2402.

88 Malo, J., *et al.* (2005) Engineering a 2D protein-DNA crystal. *Angewandte Chemie-International Edition*, **44**, 3057–3061.

89 Liu, Y., *et al.* (2005) Protein nanoarrays – aptamer-directed self-assembly of protein arrays on a DNA nanostructure. *Angewandte Chemie-International Edition*, **44**, 4333–4338.

90 Parak, W.J. *et al.* (2004) Quantum dots, in *Nanoparticles – From Theory to Application*, (ed. G. Schmid), Wiley-VCH, Weinheim, pp. 4–49.

91 Davies, J.H. (1998) *The Physics of Low-dimensional Semiconductors*, Cambridge University Press, Cambridge.

92 Ando, T., Fowler, A.B. and Stern, F. (1982) Electronic properties of two-dimensional systems. *Reviews of Modern Physics*, **54**, 437–672.

93 Fuhrer, A., *et al.* (2001) Energy spectra of quantum rings. *Nature*, **413**, 822–825.

94 Blick, R.H., *et al.* (1998) Complex broadband millimeter wave response of a double quantum dot: Rabi oscillations in an artificial molecule. *Physical Review Letters*, **81**, 689–692.

95 Kemerink, M. and Molenkamp, L.W. (1994) Stochastic Coulomb blockade in a double quantum dot. *Applied Physics Letters*, **65**, 1012–1014.

96 Waugh, F.R., *et al.* (1995) Single-electron charging in double and triple quantum dots with tunable coupling. *Physical Review Letters*, **75**, 705–708.

97 Hofmann, F. and Wharam, D.A. (1995) Investigation of the Coulomb blockade in a parallel quantum dot geometry. *Adv. Solid State Phys.*, **35**, 197–214 (e.g. at http://www.springerlink.com.contents/92863331-121165wl).

98 Blick, R.H., *et al.* (1996) Single-electron tunneling through a double quantum dot: the artificial molecule. *Physical Review B-Condensed Matter*, **53**, 7899–7902.

99 Bayer, M., *et al.* (1998) Optical modes in photonic molecules. *Physical Review Letters*, **81**, 2582–2585.

100 Tarucha, S., *et al.* (1996) Shell filling and spin effects in a few electron quantum dot. *Physical Review Letters*, **77**, 3613–3616.

101 Goldhaber-Gordon, D., *et al.* (1998) Kondo effect in a single-electron transistor. *Nature*, **391**, 156–159.

102 Cronenwett, S.M., Oosterkamp, T.H. and Kouwenhoven, L.P. (1998) *Science*, **281**, 540.

103 Simmel, F., *et al.* (1999) Anomalous Kondo effect in a quantum dot at nonzero bias. *Physical Review Letters*, **83**, 804–807.

104 Schmid, J., *et al.* (2000) Absence of odd–even parity behavior for Kondo resonances in quantum dots. *Physical Review Letters*, **84**, 5824–5827.

105 Petta, J.R., *et al.* (2005) Coherent manipulation of coupled electron spins in semiconductor quantum dots. *Science*, **309**, 2180–2184.

106 Koppens, F.H.L., *et al.* (2006) Driven coherent oscillations of a single electron spin in a quantum dot. *Nature*, **442**, 766–771.

107 Petroff, P.M., Lorke, A. and Imamoglu, A. (May, 2001) Epitaxially self-assembled quantum dots. *Physics Today, Physics Today*, **54**, *5*, 46–52.

108 Cho, A.Y. (1999) How molecular beam epitaxy (MBE) began and its projections into the future. *Journal of Crystal Growth*, **202**, 1–7.

109 Fafard, S., *et al.* (1994) 0-Dimensional-induced optical properties in self-assembled quantum dots. *Superlattices and Microstructures*, **16**, 303–309.

110 Petroff, P.M. and DenBaars, S.P. (1994) MBE and MOCVD growth and properties of self-assembling quantum dot arrays in III–V semiconductor structures. *Superlattices and Microstructures*, **15**, 15–21.

111 Leon, R., *et al.* (1995) Spatially resolved visible luminescence of self-assembled semiconductor. *Science*, **267**, 1966–1968.

112 Luyken, R.J., *et al.* (1999) The dynamics of tunneling into self-assembled InAs dots. *Applied Physics Letters*, **74**, 2486–2488.

113 Paul, D.J. (1999) Silicon–germanium strained layer materials in microelectronics. *Advanced Materials*, **11**, 191–204.

114 Garcia, J.M., *et al.* (1997) Intermixing and shape changes during the formation of InAs self-assembled quantum dots. *Applied Physics Letters*, **71**, 2014–2016.

115 Bayer, M., *et al.* (2000) Hidden symmetries in the energy levels of excitonic “artificial atoms”. *Nature*, **405**, 923–926.

116 Warburton, R.J., *et al.* (2000) Optical emission from a charge-tunable quantum ring. *Nature*, **405**, 926–929.

117 Tarucha, S. (1998) Transport in quantum dots: observation of atomlike properties. *MRS Bulletin*, **23**, 49–53.

118 Lorke, A. and Luyken, R.J. (1998) Many-particle ground states and excitations in nanometer-size quantum structures. *Journal of Physics B: Condensed Matter*, **256–258**, 424–430.

119 Gammon, D. (2000) Semiconductor physics: electrons in artificial atoms. *Nature*, **405**, 899–900.

120 Alivisatos, A.P. (1996) Semiconductor clusters, nanocrystals and quantum dots. *Science*, **271**, 933–937.

121 Murray, C.B., Norris, D.J. and Bawendi, M.G. (1993) Synthesis and characterization of nearly monodisperse CdE (E = S, Se, Te) semiconductor nanocrystallites. *Journal of the American Chemical Society*, **115**, 8706–8715.

122 Peng, X., Wickham, J. and Alivisatos, A.P. (1998) Kinetics of II–VI and III–V colloidal semiconductor nanocrystal growth: “focusing” of size distributions. *Journal of the American Chemical Society*, **120**, 5343–5344.

123 Dabbousi, B.O., *et al.* (1997) (CdSe)ZnS core–shell quantum dots: synthesis and characterization of a size series of highly luminescent nanocrystallites. *Journal of Physical Chemistry B*, **101**, 9463–9475.

124 Peng, X., *et al.* (1997) Epitaxial growth of highly luminescent CdSe/CdS core/shell nanocrystals with photostability and electronic accessibility. *Journal of the American Chemical Society*, **119**, 7019–7029.

125 Peng, X., *et al.* (2000) Shape control of CdSe nanocrystals. *Nature*, **404**, 59–61.

126 Puntes, V.F., Krishnan, K. and Alivisatos, A.P. (2002) Synthesis of colloidal cobalt nanoparticles with controlled size and shapes. *Topics in Catalysis*, **19**, 145–148.

127 Kudera, S., *et al.* (2006) Synthesis and perspectives of complex crystaline nano-structures. *Phys. Status Solidi C*, **203**, 1329–1336.

128 Alivisatos, A.P.(August, 1995) Semiconductor nanocrystals. *MRS Bulletin*, 23–32.

129 Alivisatos, A.P. (1996) Perspectives on the physical chemistry of semiconductor nanocrystals. *Journal of Physical Chemistry A*, **100**, 13226–13239.

130 Wu, M.X., *et al.* (2003) Immunofluorescent labeling of cancer marker Her2 and other cellular targets with semiconductor quantum dots – corrigenda. *Nature Biotechnology*, **21**, 452.

131 Bruchez, M.P. (2005) Turning all the lights on: quantum dots in cellular assays. *Current Opinion in Chemical Biology*, **9**, 533–537.

132 Loweth, C.J., *et al.* (1999) DNA-based assembly of gold nanocrystals. *Angewandte Chemie-International Edition*, **38**, 1808–1812.

133 Bruchez, M.J., *et al.* (1998) Semiconductor nanocrystals as fluorescent biological labels. *Science*, **281**, 2013–2016.

134 Chan, W.C.W. and Nie, S. (1998) Quantum dot bioconjugates for ultrasensitive nonisotopic detection. *Science*, **281**, 2016–2018.

135 Dubertret, B., *et al.* (2002) *In vivo* imaging of quantum dots encapsulated in phospholipid micelles. *Science*, **298**, 1759–1762.

136 Klein, D.L., *et al.* (1996) An approach to electrical studies of single nano-crystals. *Applied Physics Letters*, **68**, 2574–2576.

137 Moore, G.E. (1965) Cramming more components onto integrated circuits. *Electronics*, **38**, 8.

4
Fundamental Principles of Quantum Dots[1)]

Wolfgang J. Parak, Liberato Manna, and Thomas Nann

4.1
Introduction and Outline

4.1.1
Nanoscale Science and Technology

In the last decade new directions of modern research, broadly defined as "nanoscale science and technology" have emerged [2, 3]. These new trends involve the ability to fabricate, characterize and manipulate artificial structures, whose features are controlled at the lower nanometer scale. They embrace areas of research as diverse as engineering, physics, chemistry, materials science and molecular biology. Research in this direction has been triggered by the recent availability of revolutionary instruments and approaches that allow the investigation of material properties with a resolution close to the atomic level. Strongly connected to such technological advances are pioneering studies that have revealed new physical properties of matter at a level which is intermediate between the atomic and molecular level and bulk.

Materials science and technology is a rapidly evolving field and is currently making the most significant contributions to research in nanoscale science. It is driven by the desire to fabricate materials with novel or improved properties. Such properties can be, for instance, strength, electrical and thermal conductivity, optical response, elasticity and wear resistance. Research is also evolving towards materials that are designed to perform more complex and efficient tasks. Examples include materials with a higher rate of decomposition of pollutants, a selective and sensitive response towards a given biomolecule, an improved conversion of light into current and more efficient energy storage. For such and more complex tasks to be realized, novel materials have to be based on several components whose spatial organization is engineered at the molecular level. This class of materials can be defined as

1) This chapter has been partly adapted from a previous version which included contributions also from Dr. Daniele Gerion, Dr. Friedrich Simmel and Professor Dr. Paul Alivisatos [1].

Nanotechnology. Volume 1: Principles and Fundamentals. Edited by Günter Schmid

ISBN: 978-3-527-31732-5

"nano-composites". They are made of assembled nanosized objects or molecules. Their macroscopic behavior arises from the combination of the novel properties of the individual building blocks and their mutual interaction.

In electronics, the design and assembly of functional materials and devices based on nanoscale building blocks can be seen as the natural, inevitable evolution of the trend towards miniaturization. The microelectronics industry, for instance, is fabricating integrated circuits and storage media whose basic units are approaching the size of few tens of nanometers. For computers, "smaller" goes along with higher computational power at lower cost and with greater portability. However, this race towards higher performance is driving current silicon-based electronics to the limits of its capability [4–7]. The design of each new generation of smaller and faster devices involves more sophisticated and expensive processing steps and requires the solution of new sets of problems, such as heat dissipation and device failure. If the trend towards further miniaturization persists, silicon technology will soon reach limits at which these problems become insurmountable. In addition, scientists have found that device characteristics in very small components are strongly altered by quantum mechanical effects. In many cases, these effects will undermine the classical principles on which most of today's electronic components are based. For these reasons, alternative materials and approaches are currently being explored for novel electronic components, in which the laws of quantum mechanics regulate their functioning in a predictable way. Perhaps in the near future a new generation of computers will rely on fundamental processing units that are made of only a few atoms.

Fortunately, the advent of new methods for the controlled production of nanoscale materials has provided new tools that can be adapted for this purpose. New terms such as nanotubes, nanowires and quantum dots are now common jargon in scientific publications. These objects are among the smallest, man-made units that display physical and chemical properties which make them promising candidates as fundamental building blocks of novel transistors. The advantages envisaged here are higher device versatility, faster switching speed, lower power dissipation and the possibility of packing many more transistors on a single chip. Prototypes of these new single nanotransistors are nowadays fabricated and studied in research laboratories and are still far from commercialization. How millions of such components could be arranged and interconnected in complex architectures and at low cost still remains a formidable task to solve.

With a completely different objective, the pharmaceutical and biomedical industries try to synthesize large supramolecular assemblies and artificial devices that mimic the complex mechanisms of nature or that can potentially be used for more efficient diagnoses and better cures for diseases. Examples in this direction are nanocapsules such as liposomes, embodying drugs that can be selectively released in living organs or bioconjugate assemblies of biomolecules and magnetic (or fluorescent) nanoparticles that might provide faster and more selective analysis of biotissues in addition to less invasive cures for several types of diseases. These prototype systems might one day evolve into more complex nanomachines, with highly sophisticated functional features, able to carry out complicated tasks at the cellular level into a living body.

Nanoscience and nanotechnology will definitely have a strong impact on many aspects of future society. The scientific community envisages that nanotechnology will strongly permeate key areas such as information technology and telecommunications, medicine, energy production and storage and transportation. This chapter, however, is not meant as a survey of the present state and future developments of nanoscale science and technology and the list of examples mentioned above is far from complete. Here, we simply want to stress the point that any development in nanoscience must necessarily follow the understanding of the physical laws that govern matter at the nanoscale and how the interplay of the various physical properties of a nanoscopic system translates into a novel behavior or into a new physical property. In this sense, this chapter will serve as an overview of basic physical rules governing nanoscale materials, with particular emphasis on quantum dots, including their various physical realizations and their possible applications. Quantum dots are the ultimate example of a solid in which all dimensions shrink down to a few nanometers. Moreover, semiconductor quantum dots are probably the most studied nanoscale systems.

The outline of this chapter is as follows. In Section 4.2 we will try to explain with a few examples why the behavior of nanoscale materials can be very different from that of their bulk and from their atomic counterparts and how quantum mechanics can help us in rationalizing this. Following this discussion, we will give a definition of "quantum dot". In Section 4.3, we follow a bottom-up approach and give a simplified picture of a solid as being a very large molecule, where the energy levels of each individual atomic component have merged to form bands. The electronic structure of a quantum dot, being intermediate between the two extreme cases of single atoms and the bulk, will then be an easier concept to grasp. In Section 4.4, we will use the model of a free electron gas and the concept of quantum confinement to explain what happens to a solid when its dimensions shrink one by one. This will lead us to a more accurate definition of quantum wells, quantum wires and quantum dots. In Section 4.5, we will examine in more detail the electronic structure of quantum dots, although we will try to keep the level of the discussion relatively simple.

4.2 Nanoscale Materials and Quantum Mechanics

4.2.1 Nanoscale Materials are Intermediates Between Atomic and Bulk Matter

Nanoscale materials frequently show a behavior which is intermediate between that of a macroscopic solid and that of an atomic or molecular system. Consider, for instance, the case of an inorganic crystal composed of very few atoms. Its properties will be different from those of a single atom, but we cannot imagine that they will be the same as those of a bulk solid. The number of atoms on its surface, for instance, is a significant fraction of the total number of atoms and therefore will have a large influence on the overall properties of the crystal. We can easily imagine that this crystal might have higher chemical reactivity than the corresponding bulk solid and that it will probably melt at lower temperatures. Consider now the example of a carbon nanotube, which can be thought of as a sheet of graphite wrapped in such a way that the carbon atoms on one edge of the sheet are covalently bound to the atoms on the opposite edge of the sheet. Unlike its individual components, a carbon nanotube is chemically extremely stable because the valences of all its carbon atoms are saturated. Moreover, we guess that carbon nanotubes can be good conductors because electrons can move freely along these tiny, wire-like structures. Once again, we see that such nanoscopic objects can have properties which do not belong to the realm of their larger (bulk) or smaller (atoms) counterparts. However, there are many additional properties specific to such systems which cannot be understood by such a simple reasoning. These properties are related to the sometimes counterintuitive behavior that charge carriers (electrons and holes) can exhibit when they are forced to dwell in such structures. These properties can only be explained by the laws of quantum mechanics.

4.2.2 Quantum Mechanics

A fundamental aspect of quantum mechanics is the particle–wave duality, introduced by de Broglie, according to which any particle can be associated with a matter wave whose wavelength is inversely proportional to the particle's linear momentum. Whenever the size of a physical system becomes comparable to the wavelength of the particles that interact with such a system, the behavior of the particles is best described by the rules of quantum mechanics [8]. All the information we need about the particle is obtained by solving its Schrödinger equation. The solutions of this equation represent the possible physical states in which the system can be found. Fortunately, quantum mechanics is not required to describe the movement of objects in the macroscopic world. The wavelength associated with a macroscopic object is in fact much smaller than the object's size and therefore the trajectory of such an object can be derived with the principles of classical mechanics. Things change, for instance, in the case of electrons orbiting around a nucleus, since their associated wavelength is of the same order of magnitude as the electron–nucleus distance.

We can use the concept of particle–wave duality to give a simple explanation of the behavior of carriers in a semiconductor nanocrystal. In a bulk inorganic semiconductor, conduction band electrons (and valence band holes) are free to move throughout the crystal and their motion can be described satisfactorily by a linear combination of plane waves whose wavelength is generally of the order of nanometers. This means that, whenever the size of a semiconductor solid becomes comparable to these wavelengths, a free carrier confined in this structure will behave as a particle in a potential box [9]. The solutions of the Schrödinger equation are standing waves confined in the potential well and the energies associated with two distinct wavefunctions are, in general, different and discontinuous. This means that the particle energies cannot take on any arbitrary value and the system exhibits a discrete energy level spectrum. Transitions between any two levels are seen, for instance, as discrete peaks in the optical spectra. The system is then also referred to as "quantum confined". If all the dimensions of a semiconductor crystal shrink down to a few nanometers, the resulting system is called a "quantum dot" and will be the subject of our discussion throughout this chapter. The main point here is that in order to rationalize (or predict) the physical properties of nanoscale materials, such as their electrical and thermal conductivity or their absorption and emission spectra, we need first to determine their energy level structure.

For quantum confined systems such as quantum dots, the calculation of the energy structure is traditionally carried out using two alternative approaches. One approach was just outlined above. We take a bulk solid and we study the evolution of its band structure as its dimensions shrink down to few nanometers. This method will be described in more detail later (Section 4.4). Alternatively, we can start from the individual electronic states of single isolated atoms as shown in Section 4.3, and then study how the energy levels evolve as atoms come closer and start to interact with each other.

4.3 From Atoms to Molecules and Quantum Dots

The chemical approach towards quantum dots resembles the "bottom-up" strategy by which molecules are formed by chemical reactions between individual atoms or smaller molecules. Molecules are stable assemblies of an exactly defined finite number of atoms, whereas solids are stable assemblies of atoms or molecules with quasi-infinite dimensions. If the assembly of units within a solid does not follow translational symmetry, the solid is called amorphous. If, on the other hand, the units are repeated regularly, the solid is called crystalline. Since quantum dots have finite dimensions and are regular assemblies of atoms, such nano-objects are regarded as large molecules from a chemist's point of view, whereas physicists see them usually as small crystals.

The electronic properties of quantum dots can now be described and calculated by linear combinations of atomic orbitals (LCAO method or Hückel theory) and other approximations [10]. Here, the starting point is an atom, whereas the physical approach in Section 4.4 starts with an infinite wavefunction. It will be shown that the results of both approaches are basically the same.

The fundamental idea of quantum theory was developed at the beginning of the last century and affirms that particles have wave-like properties and vice versa. In about 1923, de Broglie suggested his famous momentum–wavelength relationship (4.3)[2] by combining Einstein's relativistic energy (4.1) with the energy of a photon (4.2):

2) (1) = (2) $\Rightarrow mc^2 = h\nu$ (h = Plank's constant, c = speed of light).
Momentum of a photon $p = mc \Rightarrow pc = h\nu$.
Wavelength of a photon $\lambda = c/\nu \Rightarrow p = h/\lambda$ $\Rightarrow \lambda = h/p$.

$$E = mc^2 \tag{4.1}$$

$$E = h\nu \tag{4.2}$$

$$\lambda = h/p \tag{4.3}$$

The left-hand term (wavelength) in Equation 4.3 represents the wave nature of a particle, whereas the right-hand term (momentum) represents the particle-nature of a wave. Equation 4.3 can be written as[3)]

$$\lambda = h/p = h/\sqrt{2m(E-V)} \tag{4.4}$$

where m is the mass of the particle and E and V are its total and potential energy, respectively. Combining Equation 4.4 with the classical three-dimensional wave equation:

$$\nabla^2\psi(x,y,z) = -(2\pi/\lambda)^2\,\psi(x,y,z) \tag{4.5}$$

where Ψ is the wavefunction results in

$$\nabla^2\psi = -\left[2\pi\sqrt{2m(E-V)}/h\right]^2\psi = -8\pi^2 m/h^2(E-V)\psi \tag{4.6}$$

Some trivial rearrangements and insertion of $\hbar = h/2\pi$ result in Schrödinger's equation:

$$\nabla^2\psi + \frac{2m}{\hbar^2}(E-V)\psi = 0 \tag{4.7}$$

or

$$\hat{H}\cdot\psi = E\cdot\psi \tag{4.8}$$

using the Hamiltonian operator $\hat{H} = H\left(x,\ldots,\frac{\hbar}{i}\frac{\partial}{\partial x},\ldots\right)$; ψ is called eigenfunction of the operator and E is the eigenvalue, which represents the energy.

The first step to calculate electronic properties of matter is to apply Schrödinger's equation to the hydrogen atom. Therefore, we look at the wavefunction of the electron in the potential field of the nucleus, which is basically the coulomb attraction between electron and nucleus:

$$V = -\frac{e^2}{4\pi\varepsilon_0 r} \tag{4.9}$$

Furthermore, the wavefunction has to be separated into equations in terms of spherical coordinates:

$$\psi(r,\vartheta,\varphi) = R(r)\cdot\Theta(\vartheta)\cdot\Phi(\varphi) \tag{4.10}$$

The solution of Schrödinger's equation with the separated variables leads to three quantum numbers associated with the three functions and the hydrogen atom's

3) $E = p^2/2m + V$ (total energy E = kinetic energy $p^2/2m$ + potential energy V) $\Rightarrow p^2 = 2m(E-V) \Rightarrow (3)\lambda = h[2m(E-V)]^{-1/2}$

energy levels. Furthermore, each one-electron wavefunction can exist in two forms called "spin states".[4] The function ψ^2 ($\psi\psi^*$, respectively) gives the probability density of finding an electron at a given point. The different eigenfunctions ψ for different sets of quantum numbers are called "atomic orbitals" (AOs) with corresponding energies (eigenvalues) E [10, 11].

Schrödinger's equation for multi-electron atoms and molecules becomes increasingly complicated, since all of the interactions between different electrons and nuclei contribute to the potential energy. In the Born–Oppenheimer approximation, the terms that describe movement of the nuclei are decoupled from those that describe the movement of the electrons, which move much faster compared with the nuclei. Thus the Hamiltonian for a molecule with N atoms and K electrons reads

$$\hat{H} = \left(-\frac{\hbar^2}{2m} \sum_{i=1,K} \nabla^2 - \sum_{j=1,N} \sum_{i=1,K} Z_j/r_{ij} \right) + \sum_{i=1,K-1} \sum_{l=i+1,K} 1/r_{il} + \sum_{j=1,N-1} \sum_{m=j+1,N} Z_j Z_m/r_{jm} \tag{4.11}$$

where Z_i are charges of the nuclei and r_{ij} distances between two charges. Schrödinger equations for molecules cannot be solved analytically. Therefore, approximation methods have to be used. These methods can be categorized as either "*ab initio*" or "semiempirical". *Ab initio* methods use only natural constants to solve Schrödinger's equation. The most prominent *ab initio* method is the Hartree–Fock method. Semiempirical methods use measured values (usually spectroscopic data) for the same purpose.

The numerical solution of the Hartree–Fock equation is only valid for atoms and linear molecules. Therefore, another approximation is needed: orbitals are written as the product of so-called basis functions (usually exponential functions) and the wavefunctions of hydrogen. These functions are called basis sets and vary for different approaches. *Ab initio* methods have high computing requirements, so that they are restricted to the calculation of some dozens of atoms.

In order to calculate electronic states of larger molecules, one has to introduce further simplifications. Semiempirical methods make use of experimental or fitted data to replace variables or functions in the Hartree–Fock matrices. The general procedure for calculating solutions for Schrödinger's equation remains the same as described above, but due to the simplifications, semiempirical methods allow for the calculation of molecules with several hundred atoms, including nanocrystals.

Figure 4.1 displays schematically the "transition" from atomic orbitals (s, p or sp^3) over molecular orbitals (σ, σ^*) to quantum dots and semiconductor energy bands. Since electrons populate the orbitals with lowest energy first, there is a highest occupied (molecular) or "binding" orbital (the "valence band" in semiconductors) and a lowest unoccupied (molecular) or "antibinding" orbital (the "conduction band"). The energy gap between the highest occupied (molecular) orbital (HOMO)

4) Readers who are interested in the details of this solution can find it in every common physical phemistry textbook [10, 11].

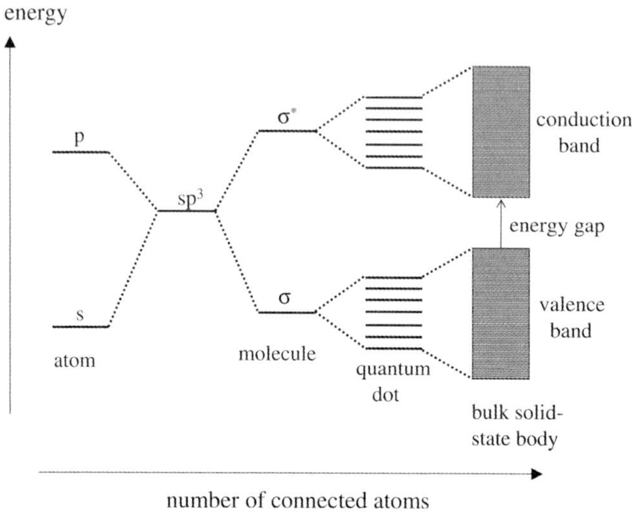

Figure 4.1 Electronic energy levels depending on the number of bound atoms. By binding more and more atoms together, the discrete energy levels of the atomic orbitals merge into energy bands (here shown for a semiconducting material) [12]. Therefore, semiconducting nanocrystals (quantum dots) can be regarded as a hybrid between small molecules and bulk material.

and lowest unoccupied (molecular) orbital (LUMO) is characteristic for luminophores, quantum dots and semiconductors. The energy gap decreases with increasing number of atoms in the transition region between molecules and bulk solids, as indicated in Figure 4.1. The exact calculation of this (most interesting) transition region follows the mathematical scheme described above. Alternatively, one can view the problem similar to a "particle-in-a-box" approach, which is outlined in Section 4.4.

First calculations of the energy of the first excited state in semiconductor quantum dots were carried out in the early 1980s by Brus [13, 14]. Brus did not solve Schrödinger's equation for the quantum dot, but for the exciton within a semiconductor nanocrystal by means of a variational method [effective-mass approximation (EMA)]. This approach thus resembles the particle-in-a-box method (see Section 4.4). The first semiempirical calculation was published in 1989 by Lippens and Lannoo [15]. They used the tight-binding approach to model CdS and ZnS quantum dots. As depicted in Figure 4.1 and calculated firstly by Brus, they found an increasing energy gap between HOMO and LUMO with decreasing nanocrystal size. Moreover, their results fit much better with experimental data than those obtained with the effective mass approximation (EMA).

Further refinements include the linear combination of atomic orbitals (LCAO) [16], the semiempirical pseudopotential calculation [17] and the *kp* method [18]. All of these methods provide estimates for the size-dependent bandgap of quantum dots. Even though the agreement of the calculations with the experimental data differs slightly, the general result is clear: the bandgap of the quantum dots increases with decreasing size of the nanocrystals. These results were as expected and thus not very useful for the experimental scientist so far. However, they paved the way for more sophisticated calculations including the inclusion of defects in nanocrystals and the presence of surface ligands. Very few examples of such calculations have been published so far. One example is the calculation of the size-dependent behavior of the quantum dot–ligand bond [19].

4.4 Shrinking Bulk Material to a Quantum Dot

In this section, we will go back to the concept of quantum confinement of carriers in a solid from a physicist's point of view and will use it to derive a more detailed description of the electronic band structure in a low-dimensional solid. This description will catch the general physics of a solid when its dimensions shrink one by one down to few nanometers. We will start first with an elementary model of the behavior of electrons in a bulk solid. This model will then be adapted to the case of confined carriers.

4.4.1 Three-Dimensional Systems (Bulk Material)

We now consider the case of a three-dimensional solid with size d_x, d_y, d_z containing N free electrons. "Free" means that those electrons are delocalized and thus not bound to individual atoms. Furthermore, we will make the assumption that the interactions between the electrons, and also between the electrons and the crystal potential, can be neglected as a first approximation. Such a model system is called "free electron gas" [20, 21]. Astonishingly, this oversimplified model still captures many of the physical aspects of real systems. From more complicated theories, it has been learnt that many of the expressions and conclusions from the free electron model remain valid as a first approximation even when one takes electron–crystal and electron–electron interactions into account. In many cases it is sufficient to replace the free electron mass m by an "effective" mass m^* which implicitly contains the corrections for the interactions. To keep the story simple, we proceed with the free electron picture. In the free electron model, each electron in the solid moves with a velocity $\vec{v} = (v_x, v_y, v_z)$. The energy of an individual electron is then just its kinetic energy:[5)]

$$E = \frac{1}{2} m \vec{v}^2 = \frac{1}{2} m (v_x^2 + v_y^2 + v_z^2) \tag{4.12}$$

According to Pauli's exclusion principle, each electron must be in a unique quantum state. Since electrons can have two spin orientations ($m_s = +^1/_2$ and

5) The total energy E is the sum of the kinetic energy $p^2/2m$ and the potential energy V. For free particles (free electron gas) there is no potential energy and therefore their total energy is equal to their kinetic energy.

$m_s = -^1/_2$), only two electrons with opposite spins can have the same velocity $\vec{v}$. This case is analogous to the Bohr model of atoms, in which each orbital can be occupied by two electrons at maximum. In solid-state physics, the wavevector $\vec{k} = (k_x, k_y, k_z)$ of a particle is more frequently used instead of its velocity to describe the particle's state. Its absolute value $k = |\vec{k}|$ is the wavenumber. The wavevector $\vec{k}$ is directly proportional to the linear momentum $\vec{p}$ and thus also to the velocity $\vec{v}$ of the electron:

$$\vec{p} = m\,\vec{v} = \frac{h}{2\pi}\,\vec{k} \qquad (4.13)$$

The scaling constant is Planck's constant h and the wavenumber is related to the wavelength λ associated with the electron through the de Broglie relation [20, 21] (Figure 4.2):

$$\pm k = |\vec{k}| = \pm\frac{2\pi}{\lambda} \qquad (4.14)$$

The calculation of the energy states for a bulk crystal is based on the assumption of periodic boundary conditions (Figure 4.2). Periodic boundary conditions are a mathematical trick to "simulate" an infinite ($d \to \infty$) solid. This assumption

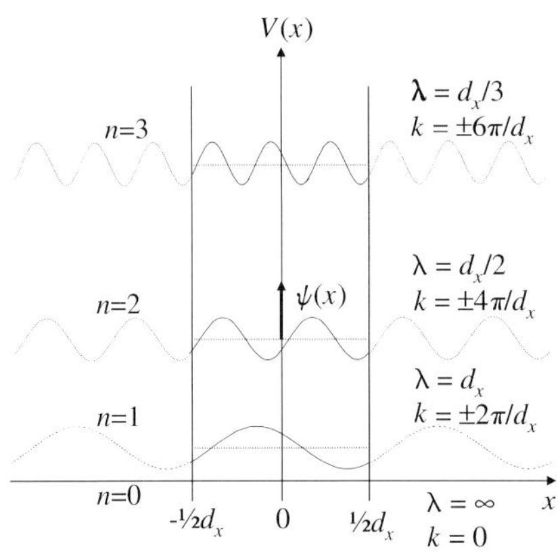

Figure 4.2 Periodic boundary conditions (only drawn for the *x*-dimension) for a free electron gas in a solid with thickness *d*. The idea of periodic boundary conditions is to "simulate" mathematically an infinite solid. Infinite extension is similar to an object without any borders. This means that a particle close to the "border" must not be affected by the border, but "behaves" exactly as in the bulk. This can be realized by using a wavefunction $\psi(x)$ that is periodic within the thickness *d* of the solid. Any electron that leaves the solid from its right boundary would re-enter under exactly the same conditions on its left side. For the electron the borders are quasi-nonexistent. The probability density $|\psi(x)|^2$ is the probability that an electron is at the position *x* in the solid. Different states for the electrons ($n = 0, 1, 2, \ldots$) have different wavefunctions. λ is the de Broglie wavelength of the electrons and *k* is their corresponding wavenumber. A "real" bulk solid can be approximated by an infinite solid ($d \to \infty$) and its electronic states in *k*-space are quasi-continuously distributed: $\Delta k = 2\pi/d_x \to 0$.

implies that the conditions at opposite borders of the solid are identical. In this way, an electron that is close to the border does not really “feel” the border. In other words, the electrons at the borders “behave” exactly as if they were in the bulk. This condition can be realized mathematically by imposing the following condition on the electron wavefunctions: $\psi(x, y, z) = \psi(x + d_x, y, z)$, $\psi(x, y, z) = \psi(x, y + d_y, z)$ and $\psi(x, y, z) = \psi(x, y, z + d_z)$. In other words, the wavefunctions must be periodic with a period equal to the whole extension of the solid number [21, 22]. The solution of the stationary Schrödinger equation under such boundary conditions can be factorized into the product of three independent functions $\psi(x, y, z) = \psi(x)\psi(y)\psi(z) = A\exp(ik_x x)\exp(ik_y y)\exp(ik_z z)$. Each function describes a free electron moving along one Cartesian coordinate. In the argument of the functions $k_{x,y,z}$ is equal to $\pm n\Delta k = \pm n2\pi/d_{x,y,z}$ and n is an integer [20–22]. These solutions are waves that propagate along the negative and positive directions for $k_{x,y,z} > 0$ and $k_{x,y,z} < 0$, respectively. An important consequence of the periodic boundary conditions is that all the possible electronic states in the $\vec{k}$ space are equally distributed. There is an easy way of visualizing this distribution in the ideal case of a one-dimensional free electron gas: there are two electrons ($m_s = \pm^1/_2$) in the state $k_x = 0$ ($v_x = 0$), two electrons in the state $k_x = +\Delta k$ ($v_x = +\Delta v$), two electrons in the state $k_x = -\Delta k$ ($v_x = -\Delta v$), two electrons in the state $k_x = +2\Delta k$ ($v_x = +2\Delta v$) and so on.

For a three-dimensional bulk material we can follow an analogous scheme. Two electrons ($m_s = \pm^1/_2$) can occupy each of the states $(k_x, k_y, k_z) = (\pm n_x\Delta k, \pm n_y\Delta k, \pm n_z\Delta k)$, again with $n_{x,y,z}$ being an integer. A sketch of this distribution is shown in Figure 4.3. We can easily visualize the occupied states in $\vec{k}$-space because all these states are included in a sphere whose radius is the wavenumber associated with the highest energy electrons. At the ground state, at 0 K, the radius of the sphere is the Fermi wavenumber k_F (Fermi velocity v_F). The Fermi energy $E_F \propto k_F^2$ is the energy of the last occupied electronic state. All electronic states with an energy $E \le E_F$ are occupied, whereas all electronic states with higher energy $E > E_F$ are empty. In a solid, the allowed wavenumbers are separated by $\Delta k = \pm n2\pi/d_{x,y,z}$. In a bulk material $d_{x,y,z}$ is large and so Δk is very small. Then the sphere of states is filled quasi-continuously [21].

We need now to introduce the useful concept of the density of states $D_{3d}(k)$, which is the number of states per unit interval of wavenumbers. From this definition $D_{3d}(k)\Delta k$ is the number of electrons in the solid with a wavenumber between k and $k + \Delta k$. If we know the density of states in a solid we can calculate, for instance, the total number of electrons having wavenumbers less than a given k_{max}, which we will call $N(k_{max})$. Obviously, $N(k_{max})$ is equal to $\int_0^{k_{max}} D_{3d}(k)dk$. In the ground state of the solid all electrons have wavenumbers $k \le k_F$, where k_F is the Fermi wavenumber. Since in a bulk solid the states are homogeneously distributed in $\vec{k}$-space, we know that the number of states between k and $k + \Delta k$ is proportional to $k^2\Delta k$ (Figure 4.3). This can be visualized in the following way. The volume in three-dimensional $\vec{k}$-space scales with k^3. If we only want to count the number of states with a wavenumber between k and $k + \Delta k$, we need to determine the volume of a spherical shell with radius k and thickness Δk. This volume is proportional to product of the surface of the sphere (which scales as k^2)

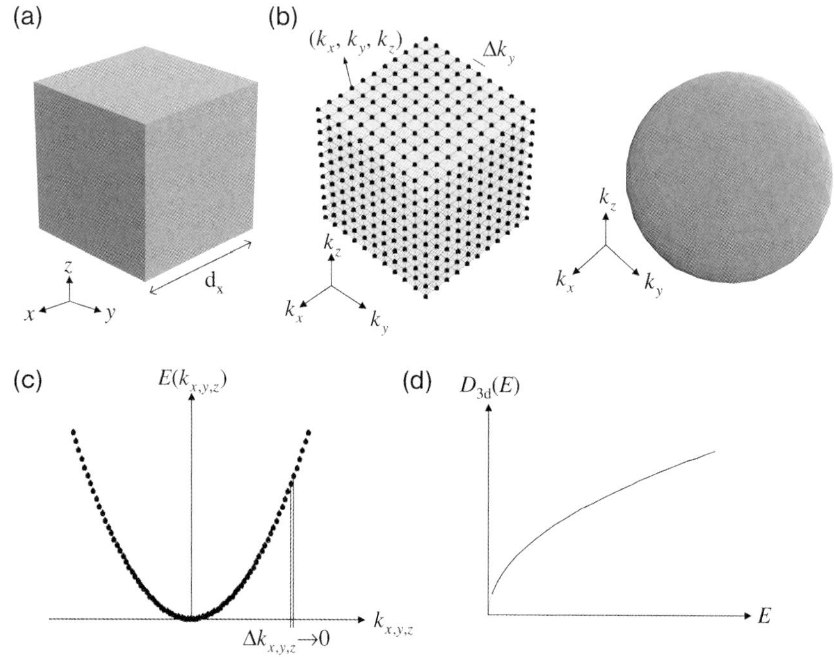

Figure 4.3 Electrons in a three-dimensional bulk solid [21]. (a) Such solid can be modeled as an infinite crystal along all three dimensions *x*, *y*, *z*. (b) The assumption of periodic boundary conditions yields standing waves as solutions for the Schrödinger equation for free electrons. The associated wavenumbers (k_x, k_y, k_z) are periodically distributed in the reciprocal *k*-space number [22]. Each of the dots shown in the figure represents a possible electronic state (k_x, k_y, k_z). Each state in *k*-space can be only occupied by two electrons. In a large solid the spacing $\Delta k_{x,y,z}$ between individual electron states is very small and therefore the *k*-space is quasi-continuously filled with states. A sphere with radius k_F includes all states with $k = (k_x^2 + k_y^2 + k_z^2)^{1/2} < k_F$. In the ground state, at 0 K, all states with $k < k_F$ are occupied with two electrons and the other states are empty. Since the *k*-space is homogeneously filled with states, the number of states within a certain volume scales with k^3. (c) Dispersion relation for free electrons in a three-dimensional solid. The energy of free electrons scales with the square of the wavenumber and its dependence on *k* is described by a parabola. For a bulk solid the allowed states are quasi-continuously distributed and the distance between two adjacent states (here shown as points) in *k*-space is very small. (d) Density of states D_{3d} for free electrons in a three-dimensional system. The allowed energies are quasi-continuous and their density scales with the square root of the energy $E^{1/2}$.

with the thickness of the shell (which is Δk). $D_{3d}(k)\Delta k$ is thus proportional to $k^2\Delta k$ and, in the limit when Δk approaches zero, we can write

$$D_{3d}(k) = \frac{dN(k)}{dk} \propto k^2 \tag{4.15}$$

Instead of knowing the density of states in a given interval of wavenumbers, it is more useful to know the number of electrons that have energies between E and E + ΔE. From Equations 4.12 and 4.13 we know that $E(k)$ is proportional to k^2 and thus $k \propto \sqrt{E}$. Consequently, $dk/dE \propto 1/\sqrt{E}$. By using Equation 4.15, we obtain for the density of states for a three-dimensional electron gas [22]

$$D_{3\mathrm{d}}(E) = \frac{\mathrm{d}N(E)}{\mathrm{d}E} = \frac{\mathrm{d}N(k)}{\mathrm{d}k}\frac{\mathrm{d}k}{\mathrm{d}E} \propto E \times 1/\sqrt{E} \propto \sqrt{E} \tag{4.16}$$

This can be seen schematically in Figure 4.3. With Equation 4.16 we conclude our simple description of a bulk material. The possible states in which an electron can be found are quasi-continuous. The density of states scales with the square root of the energy. More details about the free electron gas model and more refined descriptions of electrons in solids can be found in any solid-state physics textbook [20].

4.4.2 Two-Dimensional Systems

We now consider a solid that is fully extended along the x-and y-directions, but whose thickness along the z-direction (d_z) is only a few nanometers (Figure 4.5). Free electrons can still move freely in the x–y plane. However, movement in the z-direction is now restricted. Such a system is called a two-dimensional electron gas (2DEG) [23]. As mentioned in Section 4.2, when one or more dimensions of a solid become smaller than the de Broglie wavelength associated with the free charge carriers, an additional contribution of energy is required to confine the component of the motion of the carriers along this dimension. In addition, the movement of electrons along such a direction becomes quantized. This situation is shown in Figure 4.4. No electron can leave the solid and electrons that move in the z-direction are trapped in a "box". Mathematically this is described by infinitely high potential wells at the border $z = \pm ^1/2 d_z$.

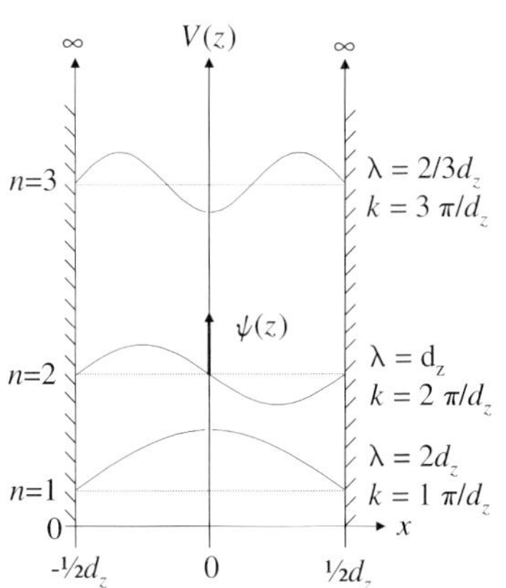

Figure 4.4 Particle in a box model for a free electron moving along in the z-axis. The movement of electrons in the z-direction is limited to a "box" with thickness d. Since electrons cannot "leave" the solid (the box), their potential energy V(x) is zero within the solid, but is infinite at its borders. The probability density $|\psi(z)|^2$ is the probability that an electron is located at position x in the solid. Different states for the electrons ($n = 1, 2, \ldots$) differ in their wavefunction.

The solutions for the particle in a box situation can be obtained by solving the one-dimensional Schrödinger equation for an electron in a potential $V(z)$, which is zero within the box but infinite at the borders. As can be seen in Figure 4.4, the solutions are stationarywaves with energies[6] $E_{nz} = \nabla^2 k_z^2/2m = h^2 k_z^2/8\pi^2 m = h^2 n_z^2/8md_z^2$, $n_z = 1, 2, \ldots$ [10, 22]. This is similar to states $k_z = n_z \Delta k_z$ with $\Delta k_z = \pi/d_z$. Again, each of these states can be occupied at the maximum by two electrons.

Let us compare the states in the k-space for three- and two-dimensional materials (Figures 4.3 and 4.5). For a two-dimensional solid that is extended in the x–y plane there are only discrete values allowed for k_z. The thinner the solid in the z-directions, the larger is the spacing Δk_z between those allowed states. On the other hand, the distribution of states in the $k_x - k_y$ plane remains quasi-continuous. Therefore, one can describe the possible states in the k-space as planes parallel to the k_x- and k_y-axes, with a separation Δk_z between the planes in the k_z-direction. We can number the individual planes with n_z. Since within one plane the number of states is quasi-continuous, the number of states is proportional to the area of the plane. This means that the number of states is proportional to $k^2 = k_x^2 + k_y^2$. The number of states in a ring with radius k and thickness Δk is therefore proportional to $k\Delta k$. Integration over all rings yields the total area of the plane in k-space. Here,

6) The particle-in-a-box approach (Figure 4.4) looks similar to the case of the periodic boundary conditions (Figure 4.2). There are indeed important differences between the two cases. Periodic boundary conditions "emulate" an infinite solid. A quantum mechanical treatment of this problem yields propagating waves that are periodic within the solid. Such waves can be seen as superposition of plane waves. For an idealized one-dimensional solid, with boundaries fixed at $x = \pm d/2$, a combination of plane waves can be for instance $\psi(x) = A\exp(ikx) + B\exp(-ikx)$ with $k = n \times 2\pi/d$. Written in another way, the solutions are of the type $\exp(ikx)$, with $k = \pm n \times 2\pi/d$. The solutions for $k = +n \times 2\pi/d$ and $k = -n \times 2\pi/d$ are linearly independent. The waves $\exp(+in \times 2\pi x/d)$ propagate to the right, the waves $\exp(-in \times 2\pi x/d)$ to the left side of the solid. Neither wave feels the boundaries. Since $\exp(ikx) = \cos(kx) + i\sin(kx)$ and $\exp(-ikx) = \cos(kx) - i\sin(kx)$, we also can write $\psi(x) = C\sin(kx) + D\cos(kx)$ with $k = n \times 2\pi/d$ as solutions. The only constraint here is that the wavefunction must be periodic throughout the solid. The state with wavenumber $k = 0$ is a solution, since $C\sin(0) + D\cos(0) = D \neq 0$. Therefore, the state with the lowest kinetic energy is $E \propto k^2 = 0$ for $k = 0$. The individual states in k-space are very close to each other because $\Delta k = 2\pi/d$ tends to 0 when d increases. On the other hand, the particle in a box model describes the case in which the motion of the electrons is confined along one or more directions. Outside the box the probability of finding an electron is zero. For a one-dimensional problem the solutions are standing waves of the type $\psi(x) = A\sin(kx)$ with $k = n\pi/d$. There is only one solution of this type. The function $\psi(x) = B\sin(-kx)$ can be written as $\psi(x) = -B\sin(kx)$ and therefore is still of the type $\psi(x) = A\sin(kx)$. Because of the boundary conditions $\psi(x = \pm d/2) = 0$ there is no solution of the type $\psi(x) = B\cos(kx)$. Since the standing wave is confined into the box, there is only the solution $k = +n\pi/d > 0$. For a small box the energy states are far apart from each other in k-space and the distribution of states and energies is discrete. An important difference with respect to the extended solid is the occurrence of a finite zero-point energy . There is no solution for $k = 0$, since $\psi(0) = A\sin(0) = 0$. Therefore, the energy of the lowest possible state ($n = 1$) is equal to $E = h^2/8md^2$, that is $k = \pi/d$. This energy is called zero-point energy and is a purely quantum mechanical effect. It can be understood as the energy that is required to "confine" the electron inside the box. For a large box the zero-point energy tends to zero. However, for small boxes this energy becomes significant as it scales with the square of the reciprocal of the box size d^2.

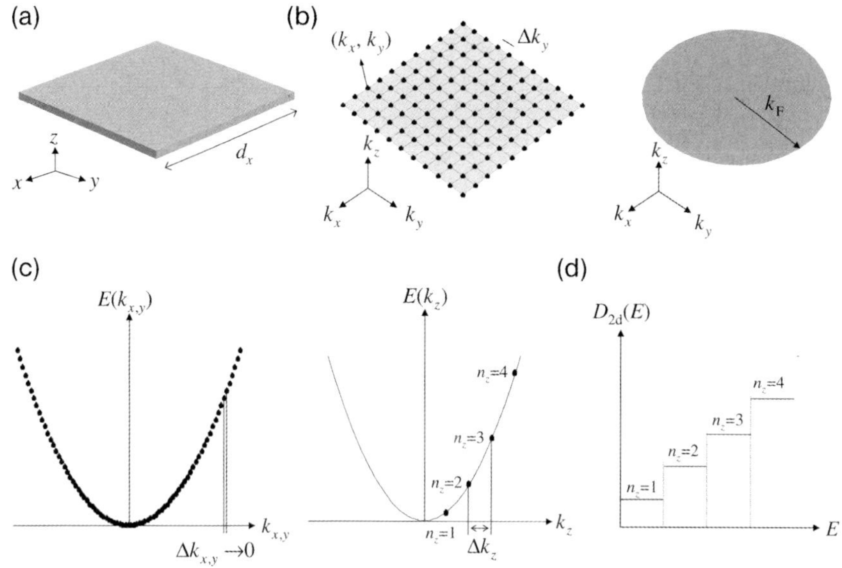

Figure 4.5 Electrons in a two-dimensional system. (a) A two-dimensional solid is (almost) infinitely extended in two dimensions (here x, y), but is very thin along the third dimension (here denoted z), which is comparable to the de Broglie wavelength of a free electron ($d_z \to \lambda$). (b) Electrons can still move freely along the x- and y-directions. The wavefunctions along such directions can be found again by assuming periodic boundary conditions. k_x and k_y states are quasi-continuously distributed in k-space. The movement of electrons in the z direction is restricted and electrons are confined to a "box". Only certain quantized states are allowed along this direction. For a discrete k_z state, the distribution of states in three-dimensional k-space can be described as a series of planes parallel to the k_x- and k_y-axes. For each discrete k_z state, there is a separate plane parallel to the k_x- and k_y-axes. Here only one of those planes is shown. The k_x and k_y states within one plane are quasi-continuous, since $\Delta k_{x,y} = 2\pi/d_{x,y} \to 0$. The distance between two planes for two separate k_z states is large, since $\Delta k_z = \pi/d_z \gg 0$. For each k_z value the k_x and k_y states are homogeneously distributed on the $k_x - k_y$ plane [22]. The number of states within this plane is therefore proportional to the area of a disk around $k_x = k_y = 0$. This means that the number of states for a certain wavenumber scales with k^2. In the ground state all states with $k \leq k_F$ are occupied with two electrons, while the remaining states are empty. (c) Free electrons have a parabolic dispersion relation [$E(k)$ (k^2]. The energy levels $E(k_x)$ and $E(k_y)$ for the electron motion along the x- and y-directions are quasi-continuous (they are shown here as circles). The wavefunction $\psi(z)$ at the border of a small "box" must be zero, leading to standing waves inside the box. This constraint causes discrete energy levels $E(k_z)$ for the motion along the z-direction. Electrons can only occupy such discrete states (n_{z1}, n_{z2}, …, shown here as circles). The position of the energy levels now changes with the thickness of the solid in the z-direction or in other words with the size of the "box". (d) Density of states for a two-dimensional electron gas. If electrons are confined in one direction (z) but can move freely in the other two directions (x, y), the density of states for a given k_z state ($n_z = 1, 2, \ldots$) does not depend on the energy E.

in contrast to the case of a three-dimensional solid, the density of states scales linearly with k:

$$D_{2d}(k) = \frac{\mathrm{d}N(k)}{\mathrm{d}k} \propto k \tag{4.17}$$

In the ground state, all states with $k \leq k_F$ are occupied with two electrons. We now want to know how many states exist for electrons that have energies between E and $E + \Delta E$. From Equations 4.12 and 4.13 we know the relation between k and E: $E(k)$ (k^2 and thus $k \propto \sqrt{E}$ and $dk/dE \propto 1/\sqrt{E}$. By using Equation 4.17 we obtain the density of states for a two-dimensional electron gas; see also Figure 4.5 [22].

$$D_{2d}(E) = \frac{dN(E)}{dE} = \frac{dN(k)}{dk}\frac{dk}{dE} \propto \sqrt{E} \times 1/\sqrt{E} \propto 1 \tag{4.18}$$

The density of electronic states in a two-dimensional solid is therefore remarkably different from the three-dimensional case. The spacing between the allowed energy levels in the bands increases, because fewer levels are now present. As soon as one dimension is reduced to nanometer size, dramatic changes due to quantum confinement occur, as for example the non-negligible zero-point energy. In two-dimensional materials the energy spectrum is still quasi-continuous, but the density of states is now a step function [22, 24].

The quantum mechanical behavior of electrons in a two-dimensional solid is the origin of many important physical effects. With recent progress in nanoscience and -technology, the fabrication of two-dimensional structures has become routine. Two-dimensional systems are usually formed at interfaces between different materials or in layered systems in which some of the layers may be only a few nanometers thick. Structures such as this can be grown, for example, by successive deposition of the individual layers with molecular beam epitaxy. In such geometry, charge carriers (electrons and holes) can move freely parallel to the semiconductor layer, but their movement perpendicular to the interface is restricted. The study of these nanostructures led to the discovery of remarkable two-dimensional quantized effects, such as the integer and the fractional quantum Hall effect [25–28].

4.4.3 One-Dimensional Systems (Quantum Wires)

Let us now consider the case in which the solid also shrinks along a second (y) dimension. Now electrons can only move freely in the x-direction and their motion along the y- and z-axes is restricted by the borders of the solid (Figure 4.6). Such a system is called "quantum wire" and – when electrons are the charge carriers – a one-dimensional electron system (1DES). The charge carriers and excitations now can move only in one dimension and occupy quantized states in the other two dimensions.

The states of a one-dimensional solid can now be obtained by methods that are analogous to those described for the three- and two-dimensional materials. In the x-direction electrons can move freely and again we can apply the concept of periodic boundary conditions. This gives a quasi-continuous distribution of states parallel to the k_x-axis and for the corresponding energy levels. Electrons are confined along the remaining directions and their states can be derived from the Schrödinger equation for a "particle in a box" potential. Again, this yields discrete k_y and k_z states. We can now visualize all possible states as lines parallel to the k_x-axis. The lines are separated by discrete intervals along k_y and k_z, but within one line the distribution of k_x states is

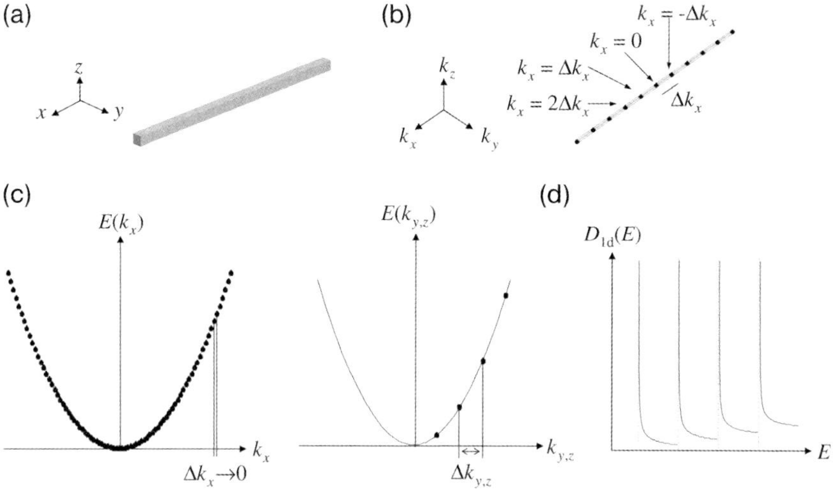

Figure 4.6 (a) One-dimensional solid. (b) The allowed (k_x, k_y, k_z) states can be visualized as lines parallel to the k_x-axes in the three-dimensional k-space. In this figure only one line is shows as an example. Within each line, the distribution of states is quasi-continuous, since $\Delta k_x \rightarrow 0$. The arrangement of the individual lines is discrete, since only certain discrete k_y and k_z states are allowed. (c) This can also be seen in the dispersion relations. Along the k_x-axes the energy band $E(k_x, k_y, k_z)$ is quasi-continuous, but along the k_y- and k_z-axes only certain energies exist. (d) The density of states within one line along the k_x-axes is proportional to $E^{-1/2}$. Each of the hyperbolas shown in the D_{1d} diagram corresponds to an individual (k_y, k_z) state.

quasi-continuous (Figure 4.6). We can count the number of states along one line by measuring the length of the line. The number of states is therefore proportional to $k = k_x$. Hence the number of states with wavenumbers in the interval between k and $k + \Delta k$ is proportional to Δk:

$$D_{1d}(k) = \frac{dN(k)}{dk} \propto 1 \tag{4.19}$$

In the ground state, all states with $k \leq k_F$ are occupied with two electrons. From Equations 4.12 and 4.13, we know the relation between k and E for free electrons: $E(k)$ (k^2, and thus $k \propto \sqrt{E}$ and $dk/dE \propto 1/\sqrt{E}$. By using Equation 4.19, we obtain the density of states for a one-dimensional electron gas:

$$D_{1d}(E) = \frac{dN(E)}{dE} = \frac{dN(k)}{dk}\frac{dk}{dE} \propto 1 \times 1/\sqrt{E} \propto 1/\sqrt{E} \tag{4.20}$$

The density of states is depicted in Figure 4.6. In one-dimensional systems the density of states has an $E^{-1/2}$ dependence and thus exhibits singularities near the band edges [22]. Each of the hyperbolas contains a continuous distribution of k_x states, but only one discrete k_y and k_z state.

The quantization of states in two dimensions has important consequences for the transport of charges. Electrons can only flow freely along the x-axes but are limited to discrete states in the y- and z-directions. Therefore, they are only transported in discrete "conductivity channels". This may be of considerable importance for the microelectronics industry. If the size of electronic circuits is reduced more and more, at one point the diameter of wires will become comparable to the de Broglie wavelength of the electrons. The wire will then exhibit the behavior of a quantum wire. Quantum aspects of 1D transport were first observed in so-called quantum point contacts which were lithographically defined in semiconductor heterostructures [29, 30]. More recent examples for such 1D wires include short organic semiconducting molecules [31–36], inorganic semiconductor and metallic nanowires [37–42] and break junctions [43–45]. A particular role is played by carbon nanotubes [32, 46–52]. Carbon nanotubes have been extensively studied both as model systems for one-dimensional confinement and for potential applications, such as electron emitters [53].

4.4.4 Zero-Dimensional Systems (Quantum Dots)

When charge carriers and excitations are confined in all three dimensions, the system is called "quantum dot". The division is somewhat arbitrary since, for instance, clusters made of very few atoms are not necessarily considered as quantum dots. Although clusters are smaller than the de Broglie wavelength, their properties depend critically on their exact number of atoms. Larger clusters have a well-defined lattice and their properties no longer depend critically on their exact number of atoms. We shall then refer to such systems with the term "quantum dots" [54–64] (Figure 4.7).

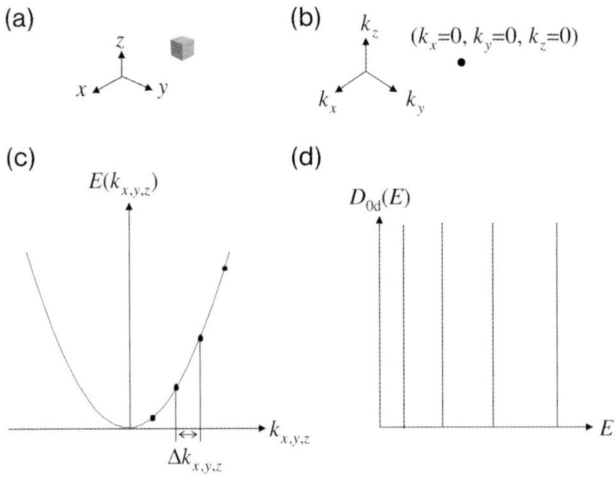

Figure 4.7 A zero-dimensional solid. (a) The solid is shrunk in all three dimensions to a thickness that is comparable to the de Broglie wavelength of its charge carriers. (b) Because of such confinement, all states (k_x, k_y, k_z) are discrete points in the three-dimensional k-space. (c) Only discrete energy levels are allowed. (d) The one-dimensional density of states $D_{0d}(E)$ contains delta peaks, that correspond to the individual states. Electrons can occupy only states with these discrete energies.

In a quantum dot, the movement of electrons is confined in all three dimensions and there are only discrete (k_x, k_y, k_z) states in the k-space. Each individual state in k-space can be represented by a point. The final consequence is that only discrete energy levels are allowed, which can be seen as delta peaks in the distribution $D_{0d}(E)$. As we can see, the energy bands converge to atom-like energy states, with the oscillator strength compressed into few transitions. This change is most dramatic at the edges of the bands and influences semiconductors more than metals. In semiconductors, the electronic properties are in fact strongly related to the transitions between the edges of the valence band and the conduction band. In addition to the discreteness of the energy levels, we want to stress again the occurrence of a finite zero-point energy. In a dot, even in the ground state electrons have energies higher than bulk electrons at the conduction band edge. These points will be discussed in more detail in the next section.

4.5 Energy Levels of a (Semiconductor) Quantum Dot

In this section, we will describe in more detail a zero-dimensional solid. Since many quantum effects are more pronounced in semiconductors than metals, we will focus on the case of a semiconducting material. In Section 4.4 we described how the properties of a free electron gas change when the dimensions of the solid are reduced. The model of the free electron gas does not include the "nature" of the solid. However, from a macroscopic point of view we distinguish between metals, semiconductors and insulators [20]. The model of a free electron gas describes relatively well the case of electrons in the conduction band of metals. On the other hand, electrons in an insulating material are only poorly described by the free electron model. In order to extend the model of free electrons for semiconducting materials, the concept of a new charge carrier, the hole, was introduced [21]. If one electron from the valence band is excited to the conduction band, the "empty" electronic state in the valence band is called a hole. Some basic properties of semiconducting materials can be described by the model of free electrons and free holes. The energy bands for electrons and holes are separated by a bandgap [20, 21]. The dispersion relations for the energy of electrons and holes in a semiconductor are parabolic to a first approximation. This approximation holds true only for electrons (holes) occupying the levels that lie at the bottom (top) of the conduction (valence) band. Each parabola represents a quasi-continuous set of electron (hole) states along a given direction in k-space. The lowest unoccupied energy band and the highest occupied energy band are separated by an energy gap E_g(bulk), as shown in Figure 4.8. The bandgap for a bulk semiconductor can range from a fraction of an electronvolt up to a few electronvolts.

We could expect that the energy dispersion relations would still be parabolic in a quantum dot. However, since only discrete energy levels can exist in a dot, each of the original parabolic bands of the bulk case is now fragmented into an ensemble of points. The energy levels of a quantum dot can be estimated with the particle-in-a-box

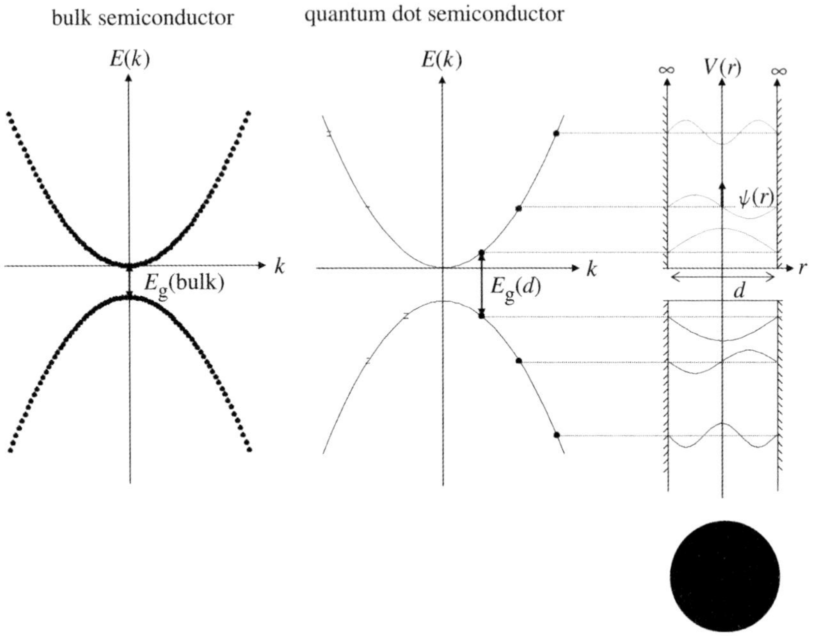

Figure 4.8 Free charge carriers in a solid have a parabolic dispersion relation [$E(k) \propto k^2$]. In a semiconductor the energy bands for free electrons and holes are separated by an energy gap E_g. In a bulk semiconductor, the states are quasi-continuous and each point in the energy bands represents an individual state. In a quantum dot the charges are confined to a small volume. This situation can be described as a charge carrier confined in an infinite potential well of width d. Here, the width d of the potential well corresponds to the diameter of the quantum dot. The only allowed states are those whose wavefunctions vanish at the borders of the well 8. This leads to discrete energy levels [8, 11]. The energy gap between the lowest possible energy level for electrons and holes $E_g(d)$ is larger than that of a bulk material E_g(bulk).

model. As described in the previous section (Figure 4.4), the lowest energy for an electron in a one-dimensional potential well is

$$E_{\text{well,1d}} = (1/8)h^2/md^2 \tag{4.21}$$

where d is the width of the well. In a quantum dot, the charge carriers are confined in all three dimensions and this system can be described as an infinite three-dimensional potential well. The potential energy is zero everywhere inside the well but is infinite on its walls. We can also call this well a "box". The simplest shapes for a three-dimensional box can be, for instance, a sphere or a cube. If the shape is cubic, the Schrödinger equation can be solved independently for each of the three translational degrees of freedom and the overall zero-point energy is simply the sum of the individual zero point energies for each degree of freedom [10, 65]:

$$E_{\text{well,3d(cube)}} = 3E_{\text{well,1d}} = (3/8)h^2/md^2 \tag{4.22}$$

If the box is a sphere of diameter d, the Schrödinger equation can be solved by introducing spherical coordinates and by separating the equation in a radial part and in a part that contains the angular momentum [66, 67]. The lowest energy level (with angular momentum = 0) is then

$$E_{\text{well},3\text{d(sphere)}} = (1/2)h^2/md^2 \tag{4.23}$$

The effect of quantum confinement is again remarkable. More confined charge carriers lead to a larger separation between the individual energy levels, and also to a greater zero-point energy. If carriers are confined into a sphere of diameter d, the zero-point energy is higher than that for charges that are confined to a cube whose edge length is equal to d [$E_{\text{well},3\text{d(sphere)}} > E_{\text{well},3\text{d(cube)}}$]. This is because such a sphere simply has a smaller volume [$(\pi/6)d^3$] than the cube (d^3).

An electron–hole pair can be generated in the quantum dot, for instance by a photoinduced process or by charge injection. The minimum energy E_g required for creating an electron–hole pair in a quantum dot is made up of several contributions. One contribution is the bulk band gap energy, E_g(bulk). Another important contribution is the confinement energy for the carriers, which we call $E_{\text{well}} = E_{\text{well}}(e^-) + E_{\text{well}}(h^+)$. For large particles (bulk: $d \to \infty$), E_{well} tends to zero. We can estimate the overall confinement energy for an electron–hole pair in a spherical quantum dot. It is the zero point energy of the potential well or in other words the energy of the state of a potential box with the lowest energy. This can be written as

$$E_{\text{well}} = h^2/2m^*d^2 \tag{4.24}$$

where m^* is the reduced mass of the exciton and is given by [68]

$$1/m^* = 1/m_e + 1/m_h \tag{4.25}$$

were m_e and m_h are the effective masses for electrons and holes, respectively. In order to calculate the energy required to create an electron–hole pair, another term (E_{Coul}) has to be considered. The Coulomb interaction E_{Coul} takes into account the mutual attraction between the electron and the hole, multiplied by a coefficient that describes the screening by the crystal. In contrast to E_{well}, the physical content of this term can be understood within the framework of classical electrodynamics. However, an estimate of such a term is only possible if the wavefunctions for the electron and the hole are known. The strength of the screening coefficient depends on the dielectric constant ε of the semiconductor. An estimate of the coulomb term yields

$$E_{\text{Coul}} = -1.8e^2/2\pi\varepsilon\varepsilon_0 d \tag{4.26}$$

This term can be fairly significant because the average distance between an electron and a hole in a quantum dot can be small [13, 14, 55, 69, 70]. We can now estimate the size-dependent energy gap of a spherical semiconductor quantum dot, which is given by the following expression [13, 14, 55, 68–70]:

$$E_g(\text{dot}) = E_g(\text{bulk}) + E_{\text{well}} + E_{\text{Coul}} \tag{4.27}$$

Then, by inserting Equations 4.24 and 4.26 into Equation 4.27, we obtain

$$E_g(d) = E_g(\text{bulk}) + h^2/2m^*d^2 - 1.8e^2/2\pi\varepsilon\varepsilon_0 d \qquad (4.28)$$

Here we have emphasized the size dependence in each term. Equation 4.28 is only a first approximation. Many effects, such as crystal anisotropy and spin–orbit coupling, have to be considered in a more sophisticated calculation. The basic approximation for the bandgap of a quantum dot comprises two size-dependent terms: the confinement energy, which scales as $1/d^2$, and the Coulomb attraction, which scales as $1/d$. The confinement energy is always a positive term and thus the energy of the lowest possible state is always raised with respect to the bulk situation. On the other hand, the Coulomb interaction is always attractive for an electron–hole pair system and therefore lowers the energy. Because of the $1/d^2$ dependence, the quantum confinement effect becomes the predominant term for very small quantum dot sizes (Figure 4.9).

The size-dependent energy gap can be a useful tool for designing materials with well-controlled optical properties. A much more detailed analysis on this topic can be found in, for example, a paper by Efros and Rosen [61].

In this chapter, we have shown how the dependence of the energy gap of semiconductors on the size of the material can be explained by either shrinking down the material from bulk to nanometer dimensions or by assembling the material atom by atom. Both views, one in a top-down and the other in a bottom-up approach, ultimately lead to the same physics. In Chapter 3, a more general description of these two types of approaches is given.

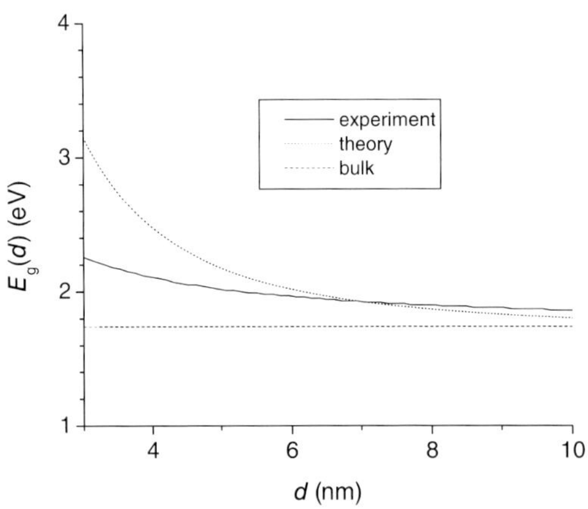

Figure 4.9 Size dependence of the energy gap $E_g(d)$ for colloidal CdSe quantum dots with diameter d. The bulk value for the energy gap is $E_g(\text{bulk}) = 1.74$ eV [68]. The theoretical curve was obtained using Equation 4.28 with the following parameters: effective mass of electrons/holes $m_e = 0.13m_0$, $m_h = 0.4m_0$, m_0 = mass of free electrons ($m = 9.1095 \times 10^{-31}$ kg) $\Rightarrow m^* = 0.098m$ [68]; dielectric constant $\varepsilon_{CdSe} = 5.8$ [71], permittivity constant $\varepsilon_0 = 8.854 \times 10^{-12}$ C^2 N^{-1} m^{-2}, Planck's constant $h = 6.63 \times 10^{-34}$ J s, 1 eV = 1.602×10^{-19} J. The experimental values were obtained by recording the absorption spectra of CdSe quantum dots of different sizes and determining the size of the quantum dots by transmission electron microscopy (TEM) [72].

References

1 Parak, W.J., Manna, L., Simmel, F.C., Gerion, D. and Alivisatos, P. (2004) *Nanoparticles – from Theory to Application* (ed. G. Schmid), 1st edn., Wiley-VCH, Weinheim, 4.

2 Lane, N. (2001) *Journal of Nanoparticle Research*, **3**, 95.

3 Service, R.F. (2000) *Science*, **290**, 1526.

4 Kingon, A.I., Maria, J.-P. and Streiffer, S.K. (2000) *Nature*, **406**, 1032.

5 Lloyd, S. (2000) *Nature*, **406**, 1047.

6 Ito, T. and Okazaki, S. (2000) *Nature*, **406**, 1027.

7 Peercy, P.S. (2000) *Nature*, **406**, 1023.

8 Cohen-Tannoudji, C., Diu, B. and Laloe, F. (1997) *Quantum Mechanics*, 1st edn., Wiley, New York.

9 Yoffe, A.D. (2001) *Advances in Physics*, **50**, 1.

10 Atkins, P.W. (1986) *Physical Chemistry*, 4th edn., W. H. Freeman, New York.

11 Karplus, M. and Porter, R.N. (1970) *Atoms and Molecules*, 1st edn., W. A. Benjamin, New York.

12 Alivisatos, A.P. (1997) *Endeavour*, **21**, 56.

13 Brus, L.E. (1983) *Journal of Chemical Physics*, **79**, 5566.

14 Brus, L.E. (1984) *Journal of Chemical Physics*, **80**, 4403.

15 Lippens, P.E. and Lannoo, M. (1989) *Physical Review B-Condensed Matter*, **39**, 10935.

16 Delerue, C., Allan, G. and Lannoo, M. (1993) *Physical Review B-Condensed Matter*, **48**, 11024.

17 Wang, L.-W. and Zunger, A. (1996) *Physical Review B-Condensed Matter*, **53**, 9579.

18 Fu, H., Wang, L.-W. and Zunger, A. (1998) *Physical Review B-Condensed Matter*, **57**, 9971.

19 Schrier, J. and Wang, L.-W. (2006) *Journal of Physical Chemistry B*, **110**, 11982.

20 Kittel, C. (1989) *Einführung in die Festkörperphysik*, 8th edn., R. Oldenbourg Verlag, Munich.

21 Ashcroft, N.W. and Mermin, N.D. (1976) *Solid State Physics*, Saunders College, Phiadelphia, PA.

22 Davies, J.H. (1998) *The Physics of Low-dimensional Semiconductors*, Cambridge University Press, Cambridge.

23 Ando, T., Fowler, A.B. and Stern, F. (1982) *Reviews of Modern Physics*, **54**, 437.

24 Moriarty, P. (2001) *Reports on Progress in Physics*, **64**, 297.

25 Zhitenev, N.B., Fulton, T.A., Yacoby, A., Hess, H.F., Pfeiffer, L.N. and West, K.W. (2000) *Nature*, **404**, 473.

26 Suen, Y.W., Engel, L.W., Santos, M.B., Shayegan, M. and Tsui, D.C. (1992) *Physical Review Letters*, **68**, 1379.

27 Stormer, H.L. (1998) *Solid State Communications*, **107**, 617.

28 Stormer, H.L., Du, R.R., Kang, W., Tsui, D.C., Pfeiffer, L.N., Baldwin, K.W. and West, K.W. (1994) *Semiconductor Science and Technology*, **9**, 1853.

29 Wharam, D.A., Thornton, T.J., Newbury, R., Pepper, M., Ahmed, H., Frost, J.E.F., Hasko, D.G., Peacock, D.C., Ritchie, D.A. and Jones, G.A.C. (1988) *Journal of Physics C: Solid State Physics*, **21**, L209.

30 van Wees, B.J., van Houten, H., Beenakker, C.W.J., Williams, J.G., Kouwenhoven, L.P., van der Marel, D. and Foxon, C.T. (1988) *Physical Review Letters*, **60**, 848.

31 Bumm, L.A., Arnold, J.J., Cygan, M.T., Dunbar, T.D., Burgin, T.P., L. Jones, II Allara, D.L., Tour, J.M. and Weiss, P.S. (1996) *Science*, **271**, 1705.

32 Anantram, M.P., Datta, S. and Xue, Y.Q. (2000) *Physical Review B-Condensed Matter*, **61**, 14219.

33 Cobden, D.H. (2001) *Nature*, **409**, 32.

34 Freemantle, M. (2001) *Chemical & Engineering News*, 5 March, 38.

35 Cui, X.D., Primak, A., Zarate, X., Tomfohr, J., Sankey, O.F., Moore, A.L., Moore, T.A., Gust, D., Harris, G. and Lindsay, S.M. (2001) *Science*, **294**, 571.

36 Reed, M.A. (2001) *MRS Bulletin*, 113.

37 Hu, J.T., Odom, T.W. and Lieber, C.M. (1999) *Accounts of Chemical Research*, **32**, 435.

38 Cui, Y., Duan, X., Hu, J. and Lieber, C.M. (2000) *Journal of Physical Chemistry B*, **104**, 5213.

39 Rodrigues, V., Fuhrer, T. and Ugarte, D. (2000) *Physical Review Letters*, **85**, 4124.

40 Rao, C.N.R., Kulkarni, G.U., Govindaraj, A., Satishkumar, B.C. and Thomas, P.J. (2000) *Pure and Applied Chemistry*, **72**, 21.

41 Häkkinen, H., Barnett, R.N., Scherbakov, A.G. and Landman, U. (2000) *Journal of Physical Chemistry B*, **104**, 9063.

42 Cui, Y. and Lieber, C.M. (2001) *Science*, **291**, 851.

43 Reed, M.A., Zhou, C., Muller, C.J., Burgin, T.P. and Tour, J.M. (1997) *Science*, **278**, 252.

44 van den Brom, H.E., Yanson, A.I. and Ruitenbeek, J.M. (1998) *Physica B*, **252**, 69.

45 Xe, H.X., Li, C.Z. and Tao, N.J. (2001) *Applied Physics Letters*, **78**, 811.

46 Tans, S.J., Devoret, M.H., Dai, H., Thess, A., Smalley, R.E., Geerligs, L.J. and Dekker, C. (1997) *Nature*, **386**, 474.

47 Saito, S. (1997) *Science*, **278**, 77.

48 McEuen, P.L., Bockrath, M., Cobden, D.H. and Lu, J.G. (1999) *Microelectronic Engineering*, **47**, 417.

49 Yao, Z., Postma, H.W.C., Balents, L. and Dekker, C. (1999) *Nature*, **402**, 273.

50 Odom, T.W., Huang, J.-L., Kim, P. and Lieber, C.M. (2000) *Journal of Physical Chemistry B*, **104**. 2794.

51 McEuen, P.L. (2000) *Physics World*, June 2000, 31.

52 Jacoby, M. (2001) *Chemical & Engineering News*, 30 April 13.

53 de Heer, W.A., Chatelain, A. and Ugarte, D. (1995) *Science*, **270**, 1179.

54 Bastard, G. and Brum, J.A. (1986) *IEEE Journal of Quantum Electronics*, **QE22**, 1625.

55 Bawendi, M.G., Steigerwald, M.L. and Brus, L.E. (1990) *Annual Review of Physical Chemistry*, **41**, 477.

56 Alivisatos, A.P. (1996) *Science*, **271**, 933.

57 Alivisatos, A.P. (1998) *MRS Bulletin*, **23**, 18.

58 Kouwenhoven, L.P., Marcus, C.M., McEuen, P.L., Tarucha, S., Westervelt, R.M. and Wingreen, N.S. (1997) *Mesoscopic Electron Transport, NATO ASI Series E*, (ed. L.P.K.L.L. Sohn), Kluwer, Dordrecht.

59 Warburton, R.J., Miller, B.T., Dürr, C.S., Bördefeld, C., Kotthaus, J.P., Medeiros-Riberio, G., Petroff, P.M. and Huant, S. (1998) *Physical Review B-Condensed Matter*, **58**, 16221.

60 Alivisatos, P. (2000) *Pure and Applied Chemistry*, **72**, 3.

61 Efros, A.L. and Rosen, M. (2000) *Annual Review of Materials Science*, **30**, 475.

62 Soloviev, V.N., Eichhofer, A., Fenske, D. and Banin, U. (2000) *Journal of the American Chemical Society*, **122**, 2673.

63 Zrenner, A. (2000) *Journal of Chemical Physics*, **112**, 7790.

64 Petroff, P.M., Lorke, A. and Imamoglu, A. (2001) *Physics Today*, May, 46.

65 Landau, L.D. and Lifschitz, E.M. (1979) *Quantenmechnik*, 9th edn., Vol. 3, Akademie-Verlag, Berlin.

66 Schwabl, F. (1990) *Quantenmechanik*, 2nd ed. Springer, Berlin.

67 Messiah, A. (1976) *Quantenmechanik*, Band 1 Walter de Gruyter, Berlin.

68 Trindade, T., O'Brien, P. and Pickett, N.L. (2001) *Chemistry of Materials*, **13**, 3843.

69 Brus, L. (1986) *Journal of Physical Chemistry*, **90**, 2555.

70 Steigerwald, M.L. and Brus, L.E. (1990) *Accounts of Chemical Research*, **23**, 183.

71 Gorska, M. and Nazarewicz, W. (1974) *Physica Status Solidi B-Basic Research*, **65**, 193.

72 Yu, W.W., Qu, L., Guo, W. and Peng, X. (2003) *Chemistry of Materials*, **15**, 2854.

5
Fundamentals and Functionality of Inorganic Wires, Rods and Tubes

Jörg J. Schneider, Alexander Popp, and Jörg Engstler

5.1
Introduction

Nanostructured one-dimensional inorganic tubes, wires and rods are known for a variety of single elements and combinations thereof. The number of studies towards their synthesis and properties is now vast. For nanowires – anisotropic nanocrystals with large aspect ratio (length to diameter) – around 5000 papers have been published during the last 2 years. An excellent comprehensive monograph and timely reviews presenting the current state of the art up to 2005 exist [1].

In this chapter, physical properties of 1D inorganic structures will be discussed first, followed by a section devoted to general techniques for the synthesis of inorganic wires, rods and tubes. The chapter then highlights some of the material developments made over the last few years in the very active area of nanostructured inorganic rods, wires and tubes with respect to their specific materials functionality. [The field of carbon nanotubes (CNTs) is probably still the fastest growing of all 1D materials. CNTs will only be touched upon in this chapter as far as sensing and nano–micro integration in functional devices are concerned. For further reading on this topic, the reader is referred to numerous excellent monographs in the field.] Due to the breadth of the field, the selection of materials and applications is somewhat subjective and reflects what the authors personally feel are "hot" topics. Where could the materials discussed impact on future technological developments? Fields of sensing and micro/nanoelectronics integration will be selectively addressed here. It is the intention of this chapter to introduce the reader to these fields and the currently ongoing rapid developments in these promising future fields of functional 1D nanomaterials.

A drastic change in materials properties is often connected with the nanoscale range (1–100 nm). In addition to an understanding of fundamental size-related electronic effects [quantum size effects (QSE)], which are connected with this "miniaturization" of matter (for an intriguing description of how quantum phenomena arise in 0D and 1D nanostructured matter, see Chapter 4), interest in nanostructured materials often arises from the fact that the small size connected with nanoscaled matter creates new chemistry. For example, the extremely high number of interfaces connected with

Nanotechnology. Volume 1: Principles and Fundamentals. Edited by Günter Schmid

ISBN: 978-3-527-31732-5

small-scale matter, be it in 0D (particles), 1D (wires, rods, tubes) or 2D (films) dimensions create high chemical reactivity. Interfaces control important material properties such as catalytic activity or analytical sensing behavior in addition to electronic properties of nanomaterials, which are highly dependent on such interfacial contacts of individual nano-building blocks and also on the individual QSE of the nano-building blocks. Besides such effects connected with the nanoscale regime, morphological properties of assembled nanomaterials such as habit (size, shape) and surface structure are also important for new and desired materials properties arising from the sequential build-up of larger structures from nano-building blocks.

The morphology of mesostructures can be tailored by synthetic techniques, for example, self-assembly, which arranges individual small building blocks such as molecules or even nanoparticles into larger mesostructured objects by employing secondary interactions such as hydrogen bonding, van der Waals, capillary or hydrophobic forces. Therefore, the shaping of materials, be it on the molecular or the nanoscale, presents a major task to the experimentalist since it is the key to organizing matter on the mesoscopic scale (Greek *mesos*, in between); the dimension between the pure nanostructured regime in which QSE reign nearly every type of material property and the macroscopic world in which solid-state physics is the key to describing and understanding material properties.

A very successful and intriguing example of controlled 1D assembly from the chemist's workbench can be found in the technique of supramolecular assembly of compact or porous 1D structures. Such structures can be built up sequentially starting first from defined molecular precursors, followed by self-assembly of 0D (nano)particle aggregates via controlled condensation. In this approach, the individual building blocks arrange in hierarchical order (from the molecular to the nanoscale and finally to the mesoscale) by supramolecular organization (Figure 5.1) [2].

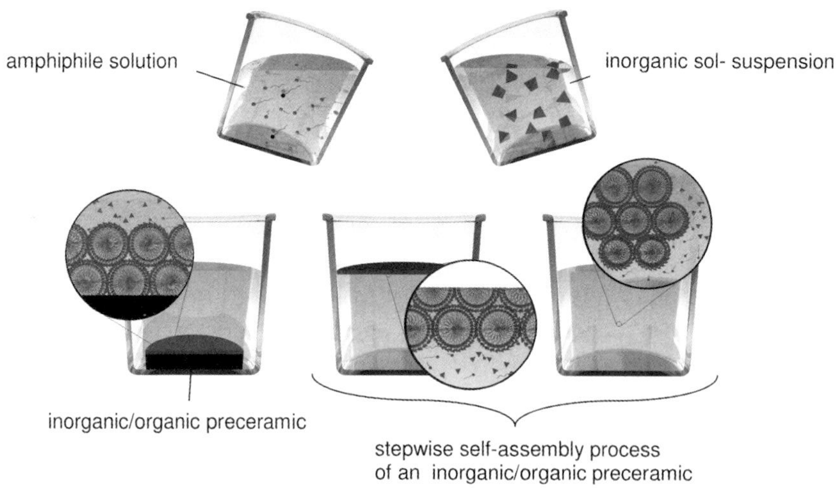

Figure 5.1 Schematic of the process of self assembly of inorganic sol particles on a polymeric micellar amphiphile assembly leading to an organic/inorganic composite. The composite structure is formed via self assembly. Adapted from Ref. [2f].

A polymeric macromolecule (called a template) steers, in addition to temperature, solvent and concentration of all reactants, the complex arrangement of individual building blocks into the final macro-sized structure.

5.2
Physical Properties of 1D Structures

Starting from a three-dimensional solid, the confinement of electrons into a one-dimensional structure leads to the quantization of electronic states in two directions (lets us say x and y). In the z-direction electrons can move freely and give rise to a quasi-continuous distribution of energy states along this dimension. Along x and y they are confined and only one discrete state is possible. Once the diameter of the 1D system is comparable to the de Broglie wavelength of the electrons, the 1D structure will become a quantum wire. Probably the most important application-related aspect of this dimensionality reduction in 1D materials is the restricted flow of charge carriers in only a single direction, the "conductivity channel" (see Section 5.4.3.1). An intriguing example can be found in single crystalline silicon nanowires smaller than

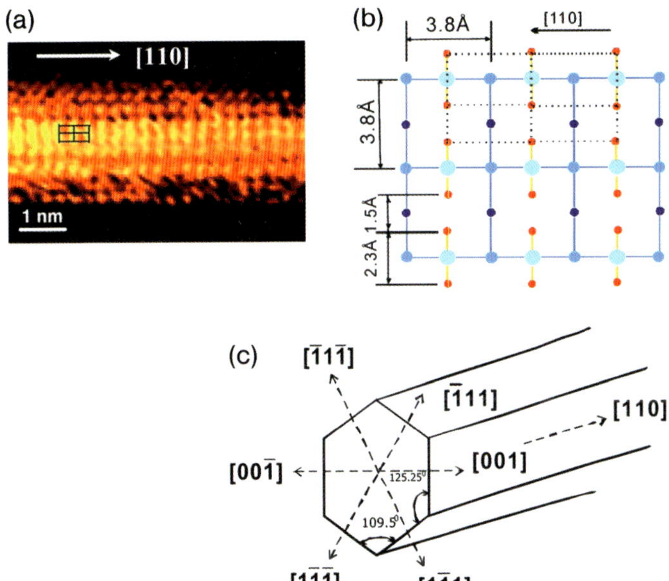

Figure 5.2 Scanning tunneling microscopic (STM) image and schematic view of an Si nanowire with an Si (0 0 1) facet. (a) Constant-current STM image of an Si nanowire on a HOPG substrate. The wire's axis is along the [110] direction. (c) Schematic view of an Si nanowire bounded by four (1 1)-type facets and two (0 0 1)-type facets. The wire's axis is along the [110] direction. Reprinted with permission from Ref. [3].

5 nm in diameter (5 nm is the exciton size for Si). Via scanning tunneling microscopy (STM), the electronic states and the bandgaps of such Si wires have been probed for wires with different diameters <10 nm. Gaps ranging from 1.1 eV (7 nm Si wire diameter) up to 3.5 eV (1.3 nm Si wire diameter) have been determined, demonstrating the extreme quantum confinement effect in such structures (Figure 5.2) [3].

A variety of direction-dependent electronic effects, for example, polarizability of light, are also different for anisotropic 1D materials compared with 0D materials [4].

Although electron conductivity in 1D wires and rods is favored along the preferred direction, phonon transport is greatly impeded, in thin 1D nanostructures, due to boundary scattering in the confined directions. Electrons may suffer elastic scattering events during their journey along the wire. This has important implications on the heat conductivity of 1D nanowire structures and for application of such structures in wiring or circuiting next-generation semiconductor devices. On the other hand, poor heat transport in confined nanowires may be used for development of thermoelectric materials. The thermoelectric effect (Seebeck effect), which is responsible for this property, describes the enhancement of the thermal electronic conductivity through a material as phonon transport in the structure worsens (due to the 1D confinement effect). Theory has predicted a significant increase over bulk values of this thermoelectric effect depending on the diameter, composition and charge carrier concentration of the 1D material of choice [5]. Nevertheless, research in that technologically important area is still in its infancy. An example is discussed in Section 5.3.1. For a detailed discussion of mesoscopic transport phenomena, for example, boundary scattering in 1D confined structures, the reader is referred to Ref. [6].

5.3 Synthetic Methods for 1D Structures

Synthetic methods for obtaining 1D nanostructures are numerous and have been reviewed by various authors recently [7]. They can be divided into top-down and bottom-up techniques; in the former a desired nanostructure is formed by sophisticated physical techniques, for example, electron beam structuring or laser structuring of a bulk material. Such techniques, however, are always restricted to the wavelength of the structuring beam, thus nanomaterials with dimensions well below 10 nm are still beyond the scope of these methods.

Bottom-up techniques, however, show a rich diversity towards desirable chemical and materials compositions and also structure (amorphous, crystalline), morphology (0D, 1D and 2D) and size. Several bottom-up synthetic techniques leading to the formation of 1D structures which seem to have a more general impact can be identified: the template method, self-assembly techniques, vapor to liquid–solid synthesis of 1D nanostructures and electrochemical techniques. These techniques show an enormous breadth since a variety of elemental compositions for different structures and morphologies and also sizes are accessible. Often combinations of individual techniques are employed, broadening further the scope for the experimentalist.

5.3.1
The Template Approach

This is probably the most versatile method for the synthesis of 1D structures. A host structure with a pore morphology – the template – is filled with a compact, more or less dense material or as film. The former produces compact wires or rods and the latter results in the formation of tubes.

Filling of the pores is possible either via solution (simply by capillary filling), by electrochemical deposition techniques or via the gas phase.

Oxidation of valve metals such as aluminum, titanium, tantalum and hafnium leads to porous metal oxide films on metal surfaces [8]. For aluminum this is a useful and outstanding technique to prepare both surface-attached and free-standing porous 2D alumina films (after detachment from the metal surface) with varying pore diameters. The pore size of these films is strongly dependent on the experimental conditions and can be varied between 10 and several hundred nm [9]. Especially in the case of aluminum, these films can be fairly thick (up to several tens of μm) or thin (down to several hundred nm), but still self-supporting, free-standing and handable. The diameter of the pores (D_p) and the cells (D_c) of a porous alumina membrane are dependent on the anodization potential applied in the electrolytic process [9]. Additional experimental parameters governing the pore size are temperature and current density. The latter is influenced by the concentration and type of electrolyte used in the electrolysis process. Nearly perfectly ordered pores are accessible by prestructuring the metallic surface [9a] (Figure 5.3).

Consequently, porous alumina membranes have been widely used to prepare various types of mesoscale materials within their pores. Their enormous synthetic impact in the area of mesostructured 1D materials comes from the ability to combine well-established wet chemical synthesis techniques (e.g., sol–gel chemistry) with the straightforward synthesis and subsequent filling of porous alumina templates.

Preparation of active sol precursors, filling and aging within the mesopores of alumina, followed by final calcination steps, have led to a huge variety of different mesostructured 1D materials, for example, in the ceramics field [1e,11]. The pores can even be used to arrange silica in a columnar or circular arrangement with a defined internal mesostructure. Further entrapment of a variety of 1D nanostructures into the mesopores of alumina such as metallic rods (Pt, Au, Pd), semiconductor rods and carbon nanostructures have been reported recently [1e,11].

Using a combination of sol–gel processing on the surface of (1 0 0)-oriented silicon templates, a porous structure with an ordered arrangement of pores is formed. These studies show that a topographic structure (here Si) can be used to engineer pores on this solid surface, allowing a high degree of freedom. The filling of such pores with cobalt has been demonstrated [12].

Another template technique giving porous, free-standing membranes, but of polymeric materials, uses heavy ion track etching of polymer foils [13] to fill the statistically formed ion tracks within the polymer with nanoscale materials in order to form rod-like structures [14, 15] (Figure 5.4). This led to bunches of randomly

Figure 5.7 SEM image of a nanorod bundle stapled via attractive forces between disk-shaped magnetic inner-rod sections. Reproduced with permission from Ref. [17].

stabilizing ligand allows control of the growth process and of the overall morphology of the rod (tip or whole body growth) [18]. Combining this technique with the template method should lead to 2D-arranged hybrid metallic rods.

Devising methods for aligning 1D materials with a uniform growth front is of general importance for optimizing device performance, for example, in thermoelectric materials such as bismuth telluride. For example, overgrowth of rods when the template pores are already filled has to be avoided in order not to cover the pores of the template with an active material, which may lead to short-circuiting. This has been achieved by applying a pulsed potential deposition technique and has yielded a uniform growth front of Bi_2Te_3 nanowire arrays in porous alumina [19]. Using nanorods as sacrificial templates to generate polymeric or ceramic nanorod structures is another method that uses porous alumina as the initial template and shows the high versatility of this porous template structure. Electrodeposition of, for example, nickel rods into porous alumina, followed by dissolution of the oxide template and coating of the metallic wires with either organic polymers or inorganic polyelectrolyte ceramics, finally gives the corresponding polymeric or ceramic (after subsequent calcinations) 1D structures. The complete method uses a hybrid technique of first templating followed by a layer-by-layer deposition technique [20].

Although the nano-templating approach using porous alumina (Figure 5.8) works well for template diameters down to about 20 nm, reports on 1D nanomaterials with smaller diameters prepared with this so-called "hard template technique" [1e] are still scarce so far [21].

5.3.2 Electrochemical Techniques

5.3.2.1 Electrospinning

This technique can be used to create nano- to microscale fibers of mainly polymeric materials [22]. Recent advances in the technique of electrospinning have brought this

Figure 5.8 Scanning electron microscopic (SEM) view of a porous alumina template membrane with pores in the range 15–17 nm (J.J. Schneider, J. Engstler, M. Naumann, TU Darmstadt).

already well-established method again in the focus of the experimentalists as a valuable method for synthesizing large quantities of, for example, ceramic nanofibers of various compositions with dense or hollow morphology [22–24].

In the technique, a precursor solution is held in the hollow tip of a needle-like capillary by its own surface tension. This capillary is subjected to an electric field and induces an electric field on the surface of the liquid precursor. On reaching a critical value, the repulsive electric field overcomes the surface tension and a charged jet of the solution is ejected from the capillary. Once the jet has started, its trajectory can be controlled by the electric field. These charged fibers are then deposited on a grounded collector. The process depends on a number of parameters such as solution properties (viscosity, conductivity and surface tension) and also process parameters such as hydrostatic pressure and electric potential at the capillary tip, distance between the tip and the collecting metal plate and general parameters such as temperature and humidity [23]. This technique gives access to isolated fibers [24–26], fiber agglomerates (mats) and dense or hollow fibers of different composition (e.g., single phase or composite) [27]. However, experimental electrospinning conditions are critical for the morphology of the resulting material; nevertheless, the morphology of the material can be tuned from 0D particles up to 1D filaments (hollow or dense morphology) (Figure 5.9).

5.3.2.2 Electrophoretic Deposition

This technique is widely used in the deposition of both mesoporous and dense thin films from colloidal suspensions. The electrophoretic technique can be ideally combined with the sol–gel synthetic approach since in the latter charged species are the active components (colloids or charged stabilized sols). These can be moved in an external electric field [28–30]. In general, reduction or oxidation of the particles occurs at the electrodes, the initial deposition surface for the preceramic material. Combining sol–gel and electrophoretic synthesis techniques basically uses the (field) oriented motion of the charged sol colloids (electrophoretic motion), (Figure 5.10) [31]. To produce, for example, a 1D ceramic material, a structure-directing template is necessary. Porous alumina is an ideal host for this purpose and a number of 1D ceramics obtained by this technique have been reported (see Chapter 4).

As the deposition of the particles is from suspension, an additional post-deposition compaction or annealing step has to be performed in order to obtain, for example, a

Figure 5.9 (a) Electrospun polymer fibers derived from Eudragit L100-55 (methacrylic acid–ethyl acrylate (1 : 1) copolymer. (b) Hollow spheres obtained via elctrospinning of aluminum *sec*-butylate in light petroleum (J.J. Schneider, J. Engstler, TU Darmstadt).

dense mesostructured ceramic. Since the as-prepared structure is already built from nanosized particles, this step can often be performed under smoother conditions as typically used in conventional ceramic processing [32].

5.3.3
Vapor–Liquid–Solid (VLS) and Related Synthesis Techniques

For the synthesis of 1D structures from the gas phase, vapor–liquid–solid (VLS) and vapor–solid (VS) processes are the typical growth mechanisms which are accepted to explain the 1D growth of mesoscopic structures [33]. In this growth process, a catalyst particle first melts, becomes saturated with a gaseous precursor and, when oversaturated, either an elemental wire or a compound wire depending on the precursor extrudes from this catalyst droplet to form a single-crystal nanowire. This is essentially the method proposed for whisker growth from the vapor [33]. Nanowires

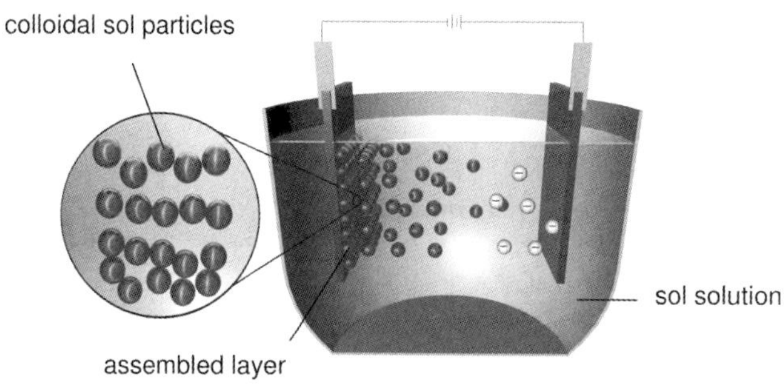

Figure 5.10 Schematic drawing of an experimental setup used for synthesis of mesostructured films by electrophoretic deposition. Particles are deposited from a colloid suspension to form a film on the surface of the electrodes. If the deposition is done in a porous template nanostructured rods are accessible. The compaction of the as prepared rods or films into a ceramic material follows in a post process.

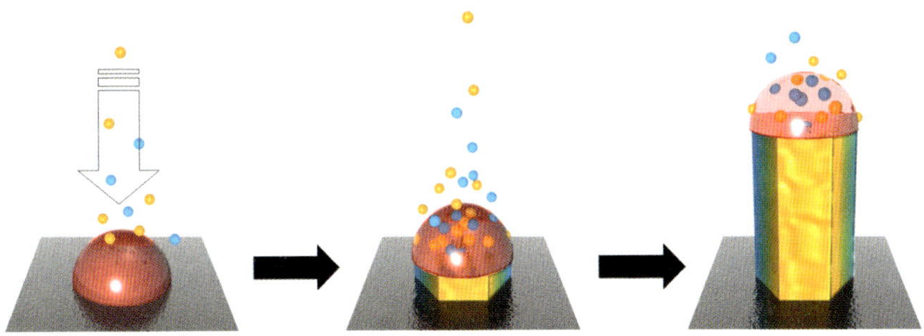

Figure 5.11 Growth mechanism for a pseudo-1D crystalline morphology. $V_{apor}L_{iquid}S_{olid}$ mechanism proposed by Wagner and Ellis for growth under CVD conditions.

grow as long as active catalyst is supplied and the growth temperature is maintained (Figure 5.11).

Recent model studies on the influence of various metal catalysts on the growth, structure morphology and size of inorganic nanowires have shown that different catalyst metals with different crystal morphologies are able to generate different wire morphologies based on the VLS formation process [34, 35]. For the growth of Si nanowires with Au nanocluster catalysts, it was found that the lowest energy surface, which is a {1 1 1} plane for Si, controls nucleation and growth (Figure 5.12) [34]. The results point towards the importance of an additional effect of oxygen in the gold-catalyzed growth kinetics of Si wires. The presence of oxygen can suppress Au catalyst

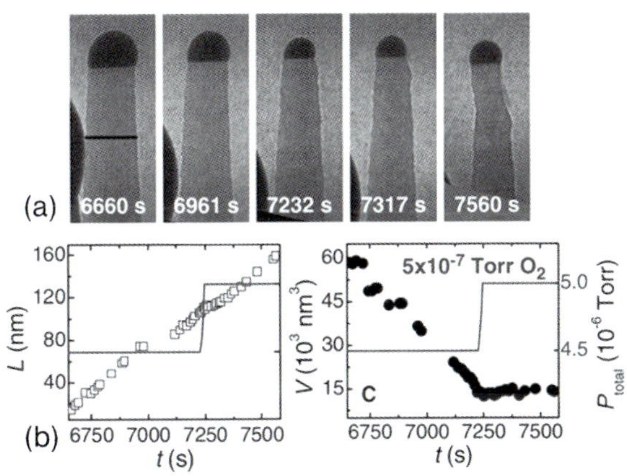

Figure 5.12 Effect of increasing oxygen pressure on Si nanowire growth kinetics. (a) Series of images extracted from a video sequence showing the effect of introducing oxygen to an Si wire that was previously growing in disilane (Si_2H_6). The growth was carried out in 4.5×10^{-6} torr disilane at 610 °C for 111 min; 5×10^{-7} torr oxygen was introduced while maintaining the disilane pressure constant. Scale bar is 50 nm. (b) Length L of the Si wire as a function of time t. (c) Volume V of the droplet as a function of time t. Reprinted with permission from Ref. [36].

migration and influences the wire diameter and the overall wire morphology [36]. The presence of oxygen, be it in the form of gas-phase oxygen or surface-bound oxygen, could be an important experimental parameter to modulate nanowire morphology further in a more general way during VLS growth [36].

For ZnO nanowires, it has been shown recently that control of the partial O_2 pressure under growth conditions is crucial for reproducible large-scale wire formation (Figure 5.13). It is very likely that the gas-phase conditions play a major role in the VLS growth of this and speculatively also other 1D materials made by this technique. Studies towards understanding the influence of reactive gas-phase species are central to deducing how individual morphologies of 1D materials depend on catalyst composition, shape and crystallinity in addition to the overall reaction conditions [37].

The current understanding of growth control of nanowires via the VLS technique is that (a) interplay given by the phase diagram of temperature–pressure and (b) the composition ofthe precursor elements and the catalyst particle under consideration are crucial. Careful control of these conditions can give rise to a variety of wire morphologies which are accessible at will once the conditions are thoroughly adjusted.

As shown, indicative for the VLS mechanism are catalyst tips at the faceted ends of the nanowires. Silicides are interesting semiconductors, with promising thermoelectric properties. Even though silicide wires (MSi_2, M = Fe, Co, Cr) were grown in the presence of nickel or iron [38], no catalyst metal particles are detected on the faceted nanowire ends as usually observed for the VLS growth mode. Obviously, in the chemical vapor transport technique (CVT) which is used for the synthesis of silicide wires a different gas-phase mechanism than in the most widely found catalyst-driven VLS growth may operate.

A promising technique related to the VLS Au seeded nanowire growth method has been reported. It allows oriented low-temperature gas-phase growth of pure Si nanowires [39]. Plasma growth at temperatures as low as 300 °C enables an additional degree of synthetic control over nanowire orientation (Figure 5.14). The crucial step in the formation process seems to be Si incorporation at the vapor–liquid interface.

Additional techniques also related to the traditional VLS-type growth mechanism are catalyst-driven growth from the solution/liquid phase into the solid state (SLS) and also from a supercritical liquid (fluid) solution (SLFS) into the solid state. The SLFS technique is synthetically complementary to the long-established technique of seed-mediated growth from solution, which yields one-dimensional colloidal metallic nanostructures [40]. In the former, a molecular precursor and in the latter a nanoparticle is decomposed under controlled conditions in solution or under supercritical conditions and then serves as a source for the metal catalyst particles from which the crystalline 1D structure grows (Figure 5.15) [41]. For this growth process from solution, low-melting metals as catalyst particles are essential (e.g., In, Bi, Sn). For metals with higher melting points (e.g., Au, Ge), supercritical solvent conditions have been successfully employed [41–43]. This synthetic technique allowed for the first time access to colloidal Au quantum wires which show a spectroscopic QSE.

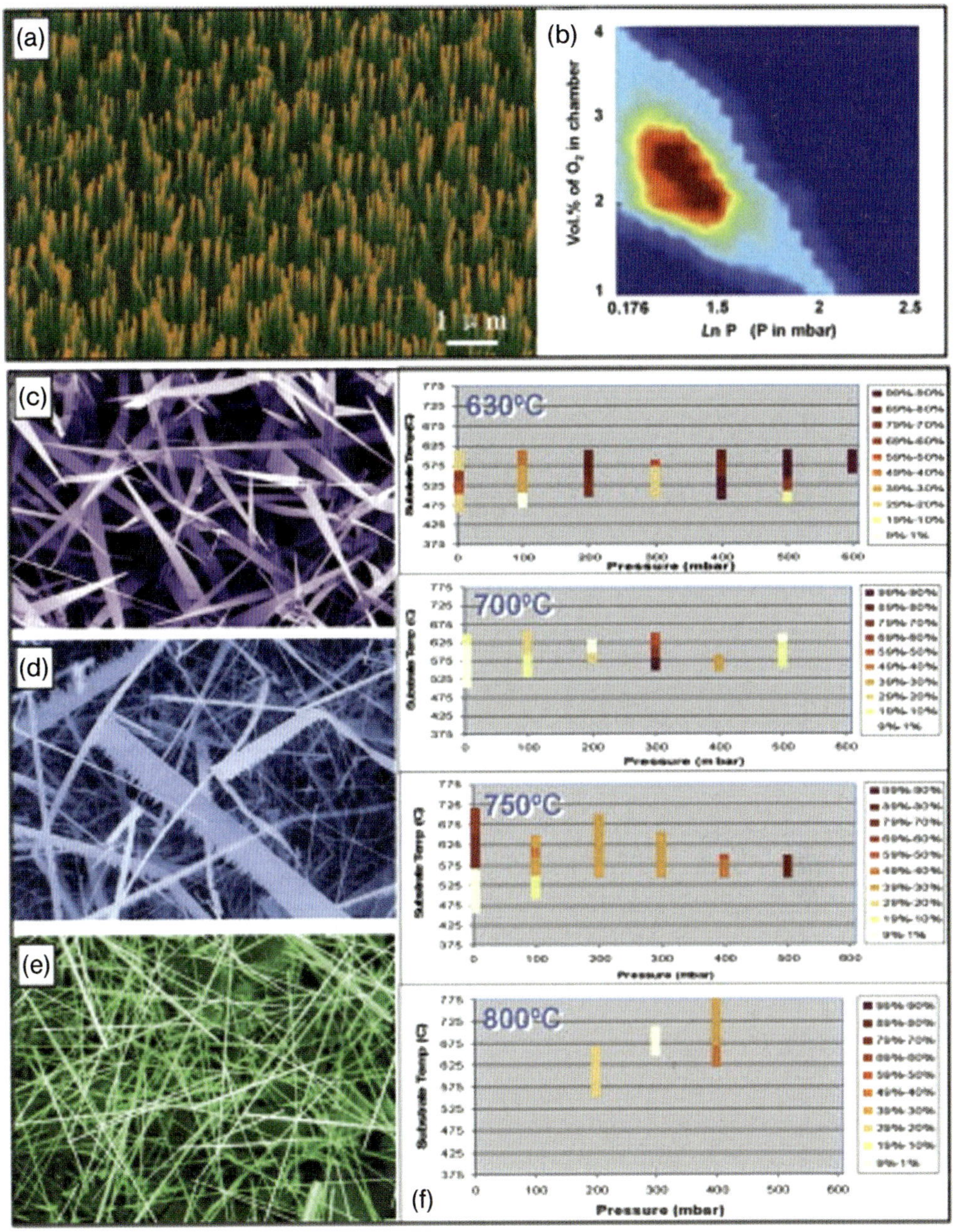

Figure 5.13 Aligned ZnO nanowires grown on a single-crystal alumina substrate with a honeycomb pattern which defined by the catalyst mask (a); (b) diagram showing the effect of oxygen partial pressure and total pressure in the growth chamber on the growth of ZnO NWs. Diversity of the morphology of possible ZnO nanomaterials: (c) ZnO nanobelts; (d) ZnO nanosaws; (e) nanowires of ZnO. (f) Yield of ZnO nanosaws as a function of reaction chamber pressure at four different furnace temperatures. The strength of the color indicates the percentage yield of the products. Reprinted with permission from Ref. [37b].

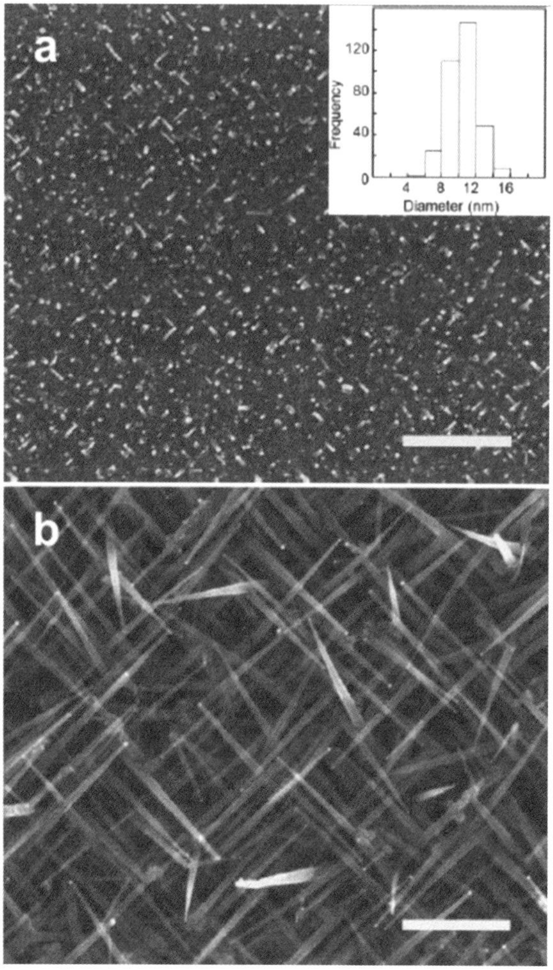

Figure 5.14 SEM images of Si nanowires grown on Si (1 0 0) substrates at 350 °C, 0.5 torr growth pressure for 15 min. (a) Thermal; (b) r.f. plasma enhanced conditions. All scale bars are 500 nm. Inset shows Au seed nanodot diameter distribution. Reprinted with permission from Ref. [39].

5.4 Contacting the Outer World: Nanowires and Nanotubes as Building Blocks in Nano/Micro/Macro-Integration

The key for applying 1D structures in future nanotechnological applications will be first assembly of nanostructures and then integration of such assembled nanostructures into existing micro-building blocks and their subsequent packaging into larger structures to allow for micro and macro manipulation of these integrated

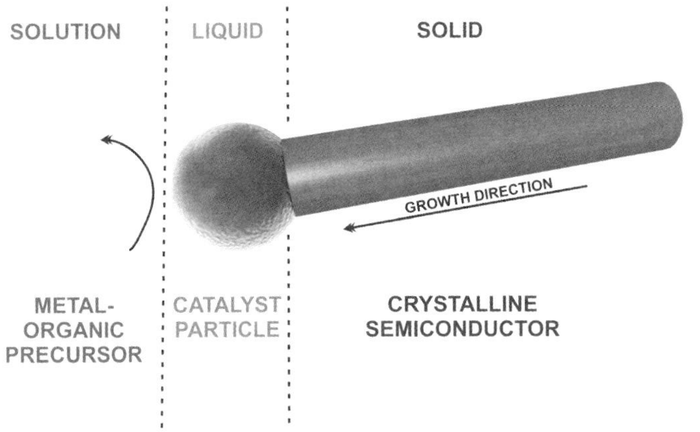

Figure 5.15 $S_{olution}L_{iquid}S_{olid}$ mechanism for nanowire growth from solution. Adapted from Ref. [41].

assemblies. This is especially challenging since a priori pertinent and incompatible length scales, nano vs. micro vs. macro, have to be bridged. Here the so called pitch problem is an important issue. It is related to the controlled formation of any nanowire structure into an array of individual wires packed into a defined order with definite distances between the individual objects.

For this purpose, techniques such as field-assisted assembly, fluid flow alignment and Langmuir–Blodgett techniques for placement and assembly of nanowires have been successfully studied and progress in these areas has been impressive in recent years. Precise positioning which allows the use of a huge number of 1D structures is no longer fiction [44–46]. Furthermore, the ability to develop even more powerful methods which have the potential to bridge that gap is still a major goal, especially for large-scale integration into electronic and sensing devices.

In 2003 it was still pointed out that "precise positioning of single-walled CNTs is beyond the capability of current growth and assembly technology and presents a major hurdle for CNT based electronic applications" [47]. In the following section selected examples of assembly and integration of 1D structures from different material classes will be highlighted to show current progress in that field.

5.4.1
Nanowire and Nanotube Sensors

Sensors based on a one-dimensional morphology (wire, tube) offer selective recognition for biological and chemical species of interest [48, 49]. This is based on their unique electronic and optical properties, which have been studied in detail either for isolated objects or in an unordered fashion for bundles of them. However, the control and use of well-arranged, aligned 1D structures have also made enormous progress within the last few years and have an impact on the applicability of such structures in sensor devices. The basis for this is the field effect transistor (FET) geometry, which

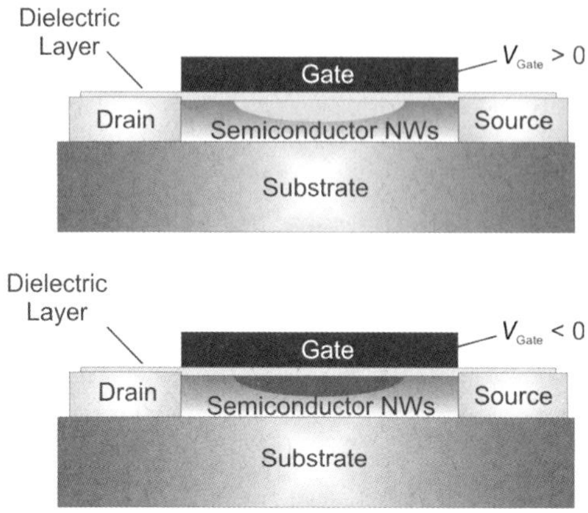

Figure 5.16 Schematics of field effect transistor (FET) geometry with nanowires as active channel material and different channel sensitivity. With a depletion of charge carriers conductance increases (upper FET schematic) When charge carriers are accumulated the conductance increases. For a p-type nanowire FET device, generation of positive charges on the nanowire's surface leads to a decrease in conductance. This can be induced, for example, by binding a protein to the wire's surface, which has a net positive charge in aqueous solution.

has been extensively explored as a device bridging the micro- and the nanoworld. In a FET device the contact between the source and drain electrode consists of a semiconductor. Its conductivity can be modulated by a third electrode, the gate coupled to the semiconductor by a dielectric layer through which charge is injected into the semiconductor. Since the binding of charged or polar molecules to the 1D semiconductor structure alters the channel electrode characteristics, this may lead to an accumulation or depletion of charge carriers and thus an increase or decrease in device conductance (Figure 5.16).

The idea was already put forth in the 1980s, but only for planar devices, which however, show only limited applicability for this sensor principle [50–52]. In a planar 2D film, only the near-surface region is altered by this effect, whereas in a single-crystalline rod surface, binding of an analyte has a strong impact on the depletion or accumulation of charge carriers in the overall structure of the nanoscale 1D object. The size of this effect depends, of course, on the size of the wire (which needs to be in the region of 2–5 nm) and its hybridization with a biomolecule (e.g., DNA). This can create a high charge density on the nanowire surface, which produces an electrostatic gating effect. This so-called field effect reduces the charge carrier concentration and results in an increase in resistance ($V_{gate} > 0$) (Figure 5.16).

With respect to CNT structures, recently the separation of metallic and semiconducting single-walled CNTs, albeit still in low quantities, has been achieved. Dielectrophoresis is the key to the nanoscale manipulation and separation of CNTs. An AC field induces polarization in a nanoscale object and this results in a force which can be used to manipulate and assemble nano-objects [53].

Dielectrophoresis in physiologically relevant saline solution has been performed and it has been shown that this technique is useful in manipulating nanowires across electrode gaps in saline solutions. Individual carbon nanowires can be switched between bridging and unbridging events in microelectrode gaps. This may be important for, for example, contacting biomolecules, DNA recognition or protein binding, all of which typically operate in highly conducting aqueous saline solutions [54]. Also intriguing is the arrangement of semiconducting or metallic CNTs using electrophoresis techniques into highly integrated structures [55]. Altogether these findings these findings may pave the way to integrated sensor devices since they allow to use the whole individual tube length of the CNT as active element.

When dispersing commercial single-walled CNTs into an appropriate solvent with the aid of a dispensing polymer, printable and conductive inks have been obtained which allow the fabrication of robust, flexible, transparent (85%) and conductive (100 kΩ) single-walled CNT–polymer composite films on plastic substrates via inkjet printing. Their use as sensors for alcohol vapors under static and dynamic flow conditions has been reported [56]. Obviously no discrimination between metallic and semiconducting CNT can be made here. However, the approach's beauty lies in its simplicity of handling, storage and fabrication of mechanically stable, flexible and sensitive CNT sensor devices.

The precise alignment of single Si wires (diameter 20–30 nm) has been achieved by a combination of electrostatic positioning and conventional lithography (Figure 5.17) [57]. With this technique, the individual Si nanowire resistance and their contact resistance has been determined. It can be envisaged that this technique is capable of arranging 1D structures up to a high precision and opens up new avenues into sensing of single particles with individually arranged ensembles of nanowires.

Directed assembly of Si nanowires between two electrodes has been achieved by electrical field assembly out of a nanowire suspension. The aligned nanowires remained bound to the electrodes by van der Waals forces. So far arrays of 18 planar electrode pairs with micrometer spacings have been studied for nanowire alignment [58].

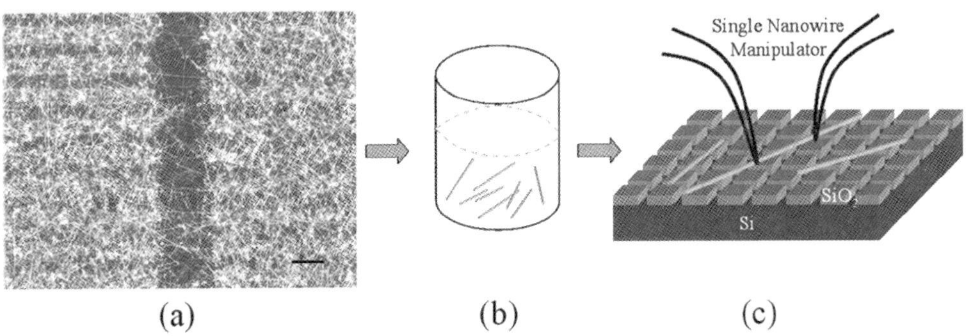

Figure 5.17 (a) SEM image of Si nanowires grown from patterned Au catalyst on a silicon wafer (VLS growth). Nanowires were removed into a suspension of deionized water. (b) A drop of the nanowire solution was dispersed on a template substrate and evaporated under vacuum. (c) Schematic of manipulator tips picking up a nanowire from a template substrate. The patterned structure of the template minimizes the adhesion forces of the wire.

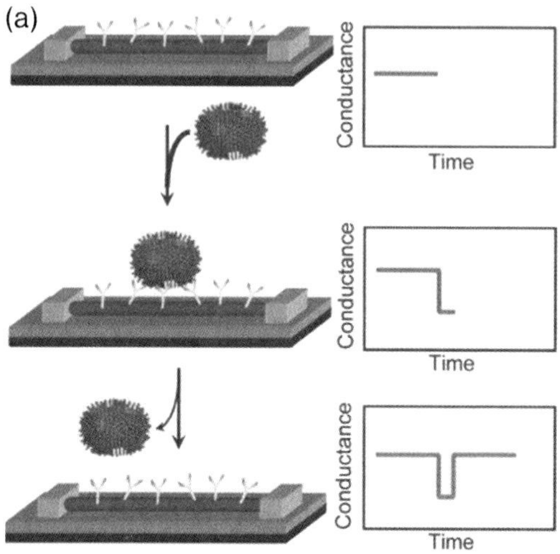

Figure 5.18 Schematic of a single virus binding and unbinding event at the surface of an Si nanowire device modified with antibody receptors and the corresponding time-dependent change in conductance. Reprinted with permission from Ref. [62].

The sensing mechanism in nanowires is due to a change in charge density on the wire's surface and probably deep into the "bulk" of the nanowire structure. This has been proposed to lead to ultrasensitive biosensors [59] and indeed was realized a few years later [60]. The upper sensing limitation of viruses as disease carriers and potential warfare agents with single-wire devices has been the subject of recent studies [61]. The working principle relies on the modification of a single wire, modified with antibody receptors (Figure 5.18) [62]. The corresponding time-dependent change in conductance is monitored. A real binding event monitored with an influenza A virus has been unraveled by measuring the change in the time-dependent conductance (Figure 5.19) [62].

With the use of the recently developed electrostatic positioning technique [57] for individual nanowires (see Figure 5.17), an integrated device configuration seems conceivable. Other successful approaches for assembling arrays of wires for multi-detection of analytes have been described [63]. Intriguing with respect to multiplexing methods for the detection of analytes is a transfer technique for arranging arrays of doped silicon wires on flexible plastic substrates. First, on to a prepatterned p-type doped Si substrate, highly aligned Si nanowires were prepared via anisotropic etching. Transfer on to plastic substrates was made possible by a polydimethylsiloxane (PDMS) stamping technique. Via selective chemical functionalization techniques of the Si nanowires using silanes, a sensor library with different end-group functionalization of the wires could be formed (Figure 5.20) [64]. This array finally worked as a highly sensitive integrated electronic nose for different vapors (Figure 5.21) [64].

The arrangement and contacting of up to 1 million cm^{-2} CNTs into individually contacted nanotube devices was achieved by dielectrophoretic deposition of metallic

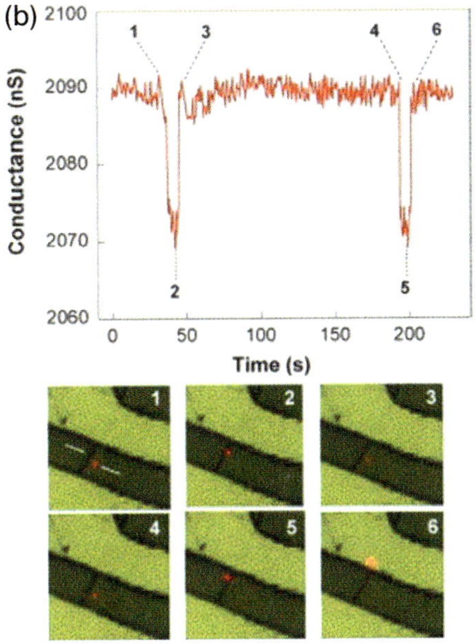

Figure 5.19 Simultaneous conductance and optical data recorded for an Si nanowire device after introduction of a influenza A virus solution. The images correspond to the two binding/unbinding events highlighted by time points 1–3 and 4–6 in the conductance data, with the virus appearing as a red dot in the images. Reprinted with permission from Ref. [62].

and semiconducting CNTs. The technique uses a field change occurring during nanotube deposition between individual electrodes of a preformed microstructured device (Figure 5.22) [55, 65]. Although not fully understood yet, redistribution of the electric field around the CNT in the gap seems to be important for the organization, rather than short-circuiting of the gap electrodes by the entrapped object [65].

In this approach, the overall dimensions of an individually contacted CNT device are only limited by the dimensions of the contacting electrodes rather than the

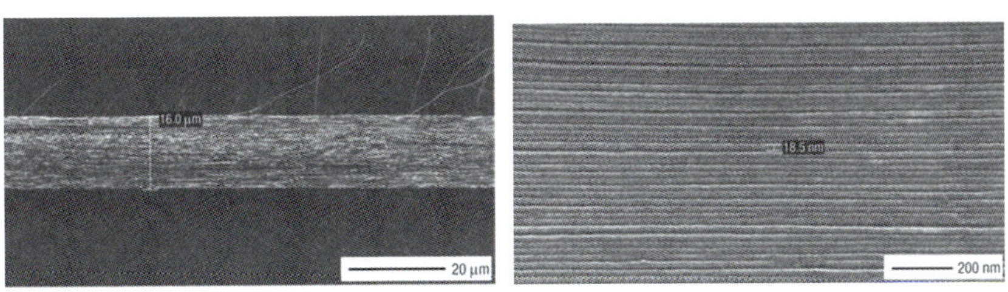

Figure 5.20 Superlattice nanowire pattern transfer film consisting of about 400 nanowires. High-magnification SEM image. The diameter of a typical nanowire is about 18.5 nm. Reprinted with permission from Ref. [64].

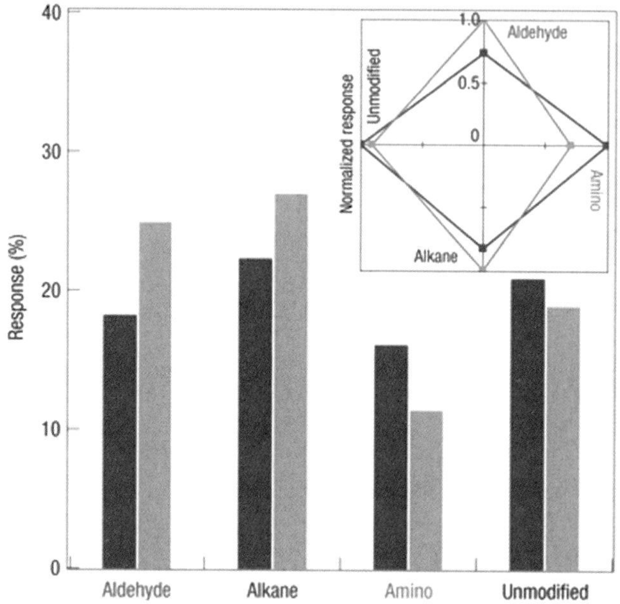

Figure 5.21 Sensor characteristics of a "nano-electronic nose". The bar graph summarizes the percentage change in response of the array to acetone (dark gray) and hexane (light gray) vapors. The inset shows the normalized response of the sensor library to acetone and hexane vapors. Each of the four axes represents the four unique surface functionalities. Reproduced with permission from Ref. [64].

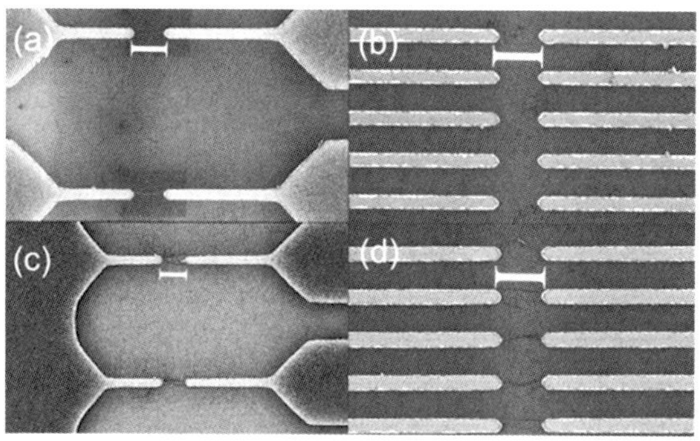

Figure 5.22 SEM images after nanotube deposition on chips with a designed electrode configuration. Effect of oxide thickness and counter electrode area of the chip on the directed assembly of nanotube devices, in terms of number of bridging nanotubes per electrode pair. Single nanotubes were assembled for 800-nm thick oxide layer and 10-μm^2 counter electrode area (a). No nanotubes were assembled for 800-nm oxide and 1-μm^2 electrode area (b). Only a few nanotubes were assembled for 50-nm oxide and 10-μm^2 electrode area (c). One or two nanotubes were assembled for 50-nm oxide and 1-μm^2 electrode area (d). Scale bars equal 1 μm in all images. Reprinted with permission from Ref. [55].

dimensions of the CNT itself. However, this seems to be the case for other approaches also [66]. So far the electrical device characteristics of over 100 CNTs have thus been measured individually. Only up to 10% of the electrodes were bridged by multiple CNTs [55].

Taking further into account the established routes for functionalizing CNTs [67], this approach seems viable for applying CNTs in a comparable way to that described above for Si wire arrays, for example, as sensor devices for the multiple detection of analytes. The field of immobilization of biomolecules is a sector of intense research activity [68]. Due to the electronic properties of individual CNTs (e.g., as semiconducting or metallic tubes), their sensing towards biomolecules could be very selective. Therefore, combining nanotubes with biosystems may provide access to nanosized biosensors. The first step for biomolecular interaction is the attachment of the biomolecule to the tube. This could be achieved either via a covalent bonding interaction or via a wrapping mechanism which uses non-covalent interactions, for example, on the basis of van der Waals interactions. The former needs functionalization of the CNTs followed by covalent bonding of the target biomolecule and is already well established. The latter rely on physisorption mechanisms and work typically for proteins.

For the development of covalent CNT functionalization techniques, a tool box of methods which have shown success in fullerene functionalization have been advantageously employed so far [67]. For example, for the case of functionalization of CNTs with bromomalonates containing thiol groups, attachment to gold surfaces is possible, allowing the synthesis of electrode arrays for sensor applications [69].

Chemical functionalization of CNTs has also paved the way to DNA–CNT adducts and subsequently to studies of their properties as biosensors. However, it seems that DNA attachment occurs mainly towards the end of the nanotubes [70]. This site-specific interaction points towards a sequence-specific poly(nucleic acid)–DNA base pairing rather than an unspecific interaction. This might be helpful for differentiation between two DNA sequences [68]. Pyrene functionalization of DNA itself can lead to a very specific DNA–CNT adduct. Its stability is mediated via hydrophobic interactions of the graphite-type sidewalls of the nanotubes and the pyrene anchors of the DNA [71].

DNA–CNT adducts are highly water soluble and therefore DNA functionalization of nanotubes avoids the use of surfactants, which are often used to solubilize CNTs alone. As with native DNA, the adducts with CNTs are still charged species. Specific DNA sequence detection with electrochemical CNT–DNA biosensors has been reported [72]. So far still a theoretical promise, the specific size-selective major groove binding mechanism of B-DNA towards single-walled CNTs could lead to a composite structure with unique sensor properties for ultrafast DNA sequencing or electronic switching based on the combination of the individual electronic properties of the single components CNT and DNA joined together in this composite structure (Figure 5.23). [73].

In this respect, it is intriguing to see the results that a strong binding capability of the major groove of DNA towards 0D nanoparticles has already been proven for gold nanoclusters, both theoretically and experimentally [74].

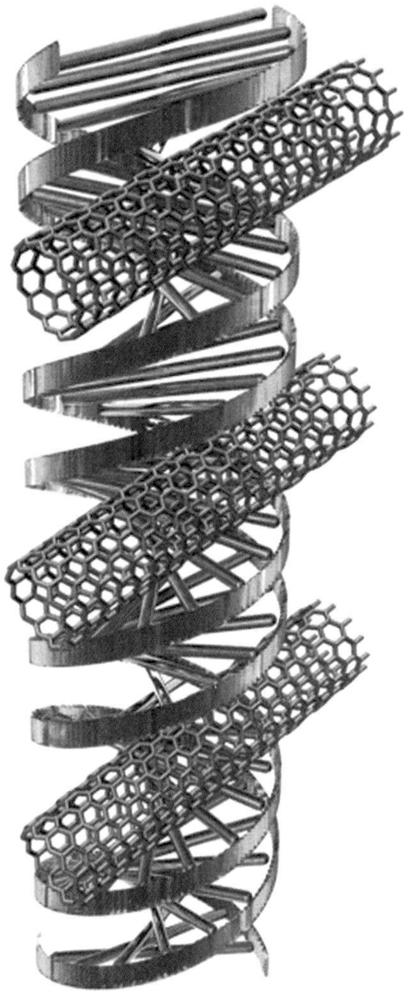

Figure 5.23 Theoretically proposed DNA–CNT composite structure. The single-wall CNTs are intercalated within the major groove of DNA. Adapted from Ref. [73].

5.4.2
Piezoelectrics Based on Nanowire Arrays

Converting mechanical energy into electric energy or signals is the area of piezoelectronics and relies on specific structures and morphologies of materials. Recently, nanomaterials have come into the focus of this application-driven area. ZnO is a material which probably exhibits the most diverse morphological configurations of any nanomaterial so far studied in detail. In addition to particles and wires there are nanobelts, nanosprings, nanorings, nanobows and nanohelices known and

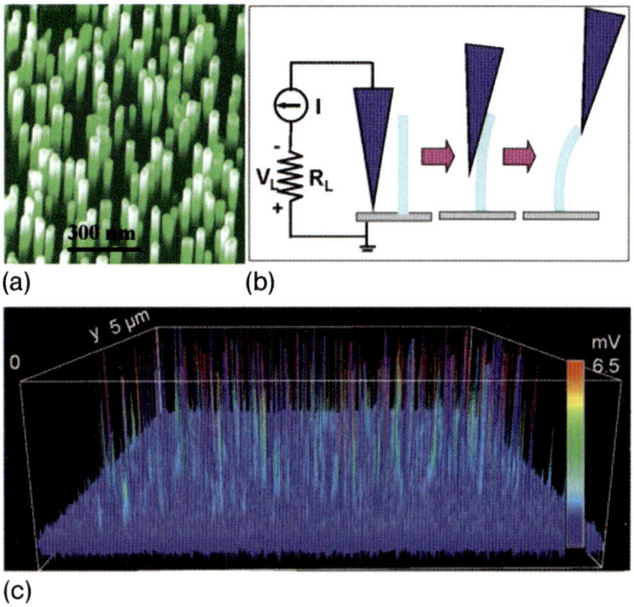

Figure 5.24 Scanning electron microscopy (SEM) images of aligned ZnO nanowires grown on an α-Al_2O_3 substrate (a). Experimental setup for generating electricity through the deformation of a semiconducting and piezoelectric nanowire using a conductive AFM tip. The root of the nanowire is grounded and an external load of $R_{load} = 500$ MW is applied, which is much larger than the inner resistance R_{inner} of the nanowire (b). The AFM tip is scanned across the nanowire array in contact mode. Output voltage image obtained when the AFM tip scans across the nanowire array (c). Reprinted with permission from Ref. [78].

characterized for ZnO [75]. Apart from use as a catalyst and sensor material which can be considered as more traditional areas for 1D nanoscale ZnO [76], the semiconducting and piezoelectric properties of ZnO nanowire arrays have recently been probed as piezoelectrics with unique properties. Such a ZnO-based nanogenerator converts mechanical energy into electric power and vice versa using massively aligned ZnO nanowires (Figure 5.24) [77, 78]. An electric field is created by deformation of a ZnO nanowire within the array by the piezoelectric effect. Across the top of the nanowire the potential distribution varies from negative at the strained side of the surface (V_s^-) to positive at the stretched surface (V_s^+). This potential difference is measured in millivolts and is due to the piezoelectric effect. On an atomic scale, the displacement of Zn^{2+} ions in the wurtzite lattice with respect to the O^{2-} counterions is responsible for that. The charges cannot move or combine without releasing strain via mechanical movement. As long as the potential difference is maintained (via constant deformation), the system is stable. When external free charges (e.g., via a metal tip) are induced, the wire is discharged. As an effect, the current which then flows is the result of the ΔV (V_s^+/V_s^-)-driven flow of electrons from the semiconducting ZnO wire to the metal tip. This flow of electrons neutralizes the ionic charges in the volume of the ZnO wire and reduces the potentials V_s^+ and V_s^- [78, 77].

The potential technological impact of such a nano-piezoelectronic effect might be in converting various type of energies (mechanical, vibration, hydraulic) into electrical energy. Thus extremely large deformations are possible, which are interesting for flexible electronics as a power source and realizing a much larger power/volume density output. The effect has already been shown to work for field effect transistor devices [79], piezoelectric gated diodes [80] and piezoelectric resonators [81], all on a ZnO material basis.

5.4.3 With Nanowires and Nanotubes to Macroelectronics

Using top-down techniques for scaling conventional microelectronic devices has by now reached the size regime of about 50 nm with a gate length of about 30 nm for a FET. This demonstrates impressively how far the conventional top-down technique using lithography, deposition techniques and etching methods have developed. Nevertheless, scaling down characteristic feature sizes and the channel gate length even further seems limited using these techniques.

On the other hand, despite tremendous success in recent years, incorporating nanosized materials into typical FET device architectures is still in its infancy. Exciting new developments have been reported over the last few years which might have the potential to bridge the nano–micro–macro gap for certain device architectures. Due to their aspect ratio, 1D materials can be produced in micrometer-long structures, but displaying nanometer size diameters. Hence they are *a priori* ideally suited to be incorporated in current microdevice technology.

Due to the possibility that a variety of semiconductor nanowires, including the workhorse element Si, show great promise to be processable from solution, a couple of areas of high technological input might appear on the horizon for such materials (e.g., flat panel displays or flexible solar cells). Especially when these nanomaterials are available as single-crystalline matter in 1D morphology, they show electronic performances even exceeding those of high-quality macro (bulk)-sized single crystals or thin films. Combining their high electronic performance with the possibility of arranging 1D inorganic nanowires over larger dimensions offers the ability finally to bridge the gap from nano- over micro- to macroelectronics and thus puts these materials one step closer to the realm of application.

5.4.3.1 Inorganic Nanowire and Nanotube Transistors

In general, inorganic semiconductors have the intrinsic advantage over organic-based semiconductors that the electronic mobility μ (electron–hole) usually drastically exceeds those of organic materials. The mobility μ is the proportionality constant between an applied electric field and the corresponding average charge carrier drift velocity. It is therefore a direct measure of the switching speed of an electronic device. The carrier mobility thus depends critically on the crystallinity of the material. In inorganic semiconductor wires of high crystallinity, a high order of atomic or molecular building blocks over long distances is maintained due to strong ionic or covalent bonds, whereas in organic semiconductors the mobility μ seems fundamentally

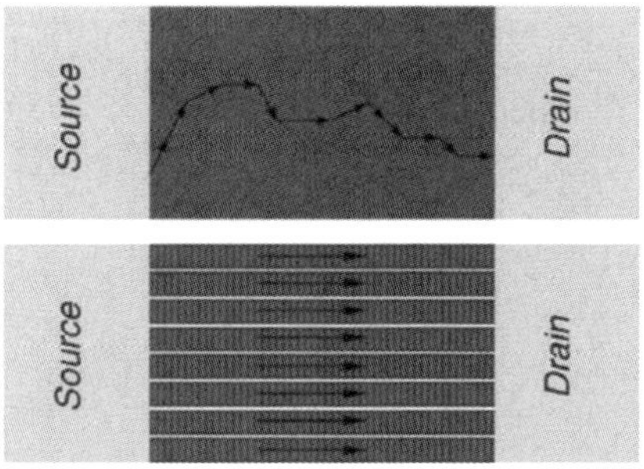

Figure 5.25 In polycrystalline silicon thin field effect transistors (TFTs), electrical carriers have to travel across multiple grain boundaries, resulting in low carrier mobility. Nanowire TFTs have conducting channels consisting of multiple single-crystal nanowires in parallel. Therein charges travel from source to drain within single crystals, ensuring high carrier mobility. Reprinted with permission from Ref. [82].

limited due to weak van der Waals bonds between electronically active molecular building blocks. This situation is a general one for an inorganic crystalline semiconductor compared with crystalline organic ones.

It is important for the exploitation of inorganic semiconductor nanowires to integrate them in massive arrangements in semiconductor devices, for example, thin-film transistors (TFTs), in which they might offer an increased electronic performance and future device integration. The former is due to the fact that drastically higher charge carrier mobilities μ can be obtained when using multiple single-crystal nanowires in FET devices instead of polycrystalline thin films (Figure 5.25) [82].

In a poly-Si TFT channel material, the electrical carriers have to travel across multiple grain boundaries (curved pathway), whereas in a massive parallel-arranged single-crystal nanowire array, charge carriers travel from source to drain within a single-crystalline structure, which ensures a high carrier mobility μ. The same effect has been observed, for example, for compound semiconductors such as ZnO. First sintering of isolated ZnO 0D nanoparticles has to be employed to obtain coarse grain polycrystalline thin films (Figure 5.26) [83]. When comparing the electronic performance of polycrystalline ZnO thin films with that of single-crystalline ZnO nanowires deposited between the source and drain of a FET device, the performance of ZnO nanowires is intriguing (Table 5.1).

Calculations have shown that for a channel length (geometric source to drain distance) of 20 μm in a FET device, the number of grain boundary contacts is diminished by a factor of 6 compared with a polycrystalline ZnO semiconductor electrode against one composed of a 1D wire morphology. In addition, the reduced hoping frequency for the charge carriers across the grain boundaries in the 1D ZnO structures compared with the polycrystalline ZnO thin-film morphology adds towards their increased mobility (Figure 5.27) [85]. One may imagine how this can

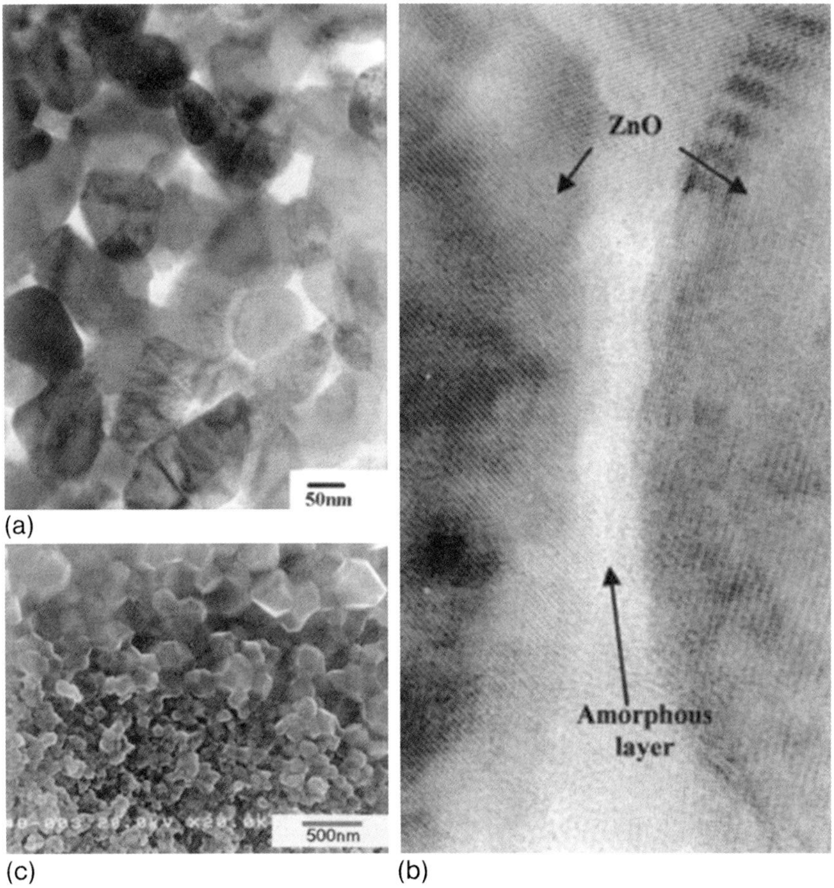

Figure 5.26 Transmission electron microscope (TEM) image of a nanocrystalline ZnO film. Processing temperature 550 °C (a); TEM of the grain boundary of two ZnO nano particles (b), Processing temperature 600 °C, sintered ZnO nanostructures can be observed (c). Reprinted with permission from Ref. [84].

Table 5.1 Charge carrier mobilities μ and on/off ratios of FET architectures with ZnO as active material (2D thin films vs. 1D rods).

Material	Morphology (diameter) (nm)	Mobility μ ($cm^2 V^{-1} s^{-1}$)	On/off ratio	Ref.
Si	Rods 20 or 40	119	10^8	92
Si	Particles 300	50–100; 6.5 (printed)	100	84a
Si	Rods 2000	180	1000	89
ZnO	Particles	0.23	1.6×10^5	84b
ZnO	Particles 6	0.00023	5×10^3	85
	Rods 10	0.023	1×10^5	85

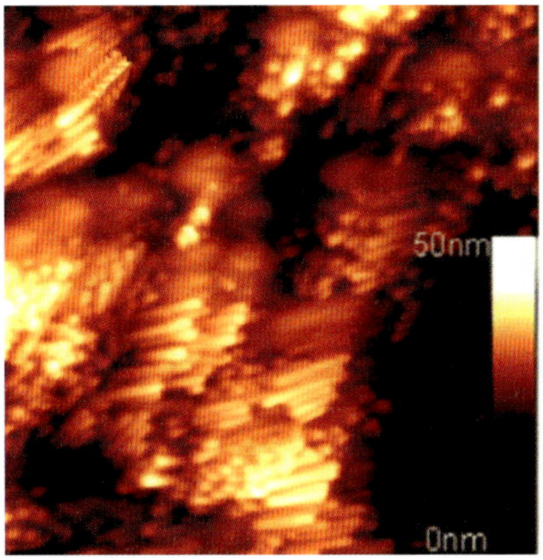

Figure 5.27 Atomic force microscope (AFM) picture of ZnO nanorods, length 65 nm, diameter 10 nm, deposited between source and drain of a FET device structure. Reprinted with permission from Ref. [85].

be further improved if it were possible to arrange such wires in a massive parallel fashion.

Dramatically higher carrier mobilities for well-aligned single-walled CNTs compared with randomly oriented networks of CNTs [86] have been measured [87]. The CNT alignment was achieved along the $[21\bar{0}]$ plane of a right-handed α-quartz substrate. The process is attractive since Y-cut α-quartz is a commercial substrate. The tube orientation is so far not fully understood, but is directed along step edges and/or micro/nanofacets on the surface of the quartz (Figure 5.28). Device mobilities of up to $125\,cm^2\,V\,s^{-1}$ have been measured, which correspond to individual tube

Figure 5.28 AFM image of terraced quartz surface structures after thermal annealing. The steps are 0.7–1 nm in height with spacing 30–35 nm. The dots seen are ferritin-derived Fe catalyst particles. Scale bar, 5 nm height. Reprinted with permission from Ref. [87].

mobilities within the array comparable to those already measured in pristine single tubes. These results clearly indicate that alignment of 1D nanostructures is crucial for the most effective device integration of CNTs [87].

The channel conductivity for nanowire and nanotube transistor devices has been treated theoretically and a universal analytical description has been developed. It has been found that the transconductance of the channel differs from classical device theory because of the specific nanowire charge distribution. Mainly different electrostatics for the 1D channel structure are responsible for the different device characteristics [88].

Due to the microscale length dimensions in which many semiconductor nanowire materials are obtainable, large-area substrates such as standard FET device structures can already be applied in device fabrication using 1D nanomaterials. An intriguing example is the arrangement of crystalline, several tens of micrometer long Si nanowires, which have been arranged between the source and drain of a FET device from a solution-derived process (Figure 5.29) [84]. Therein growth and integration of the particular nanowire material are separated from the device fabrication. Wire synthesis is done via a CVD approach, which of course is not compatible with the device fabrication process. However, the wires can be solution processed and in a second follow-up step deposited in a highly aligned fashion. The two-step route is applicable to sensitive FET substrates such as plastics and points towards a general route for future integration of nanowires into micro–macro integrated device architectures. Always a single-crystalline Si wire connects the source and drain; the distance between such individual wires is 500–1000 nm. This assures that

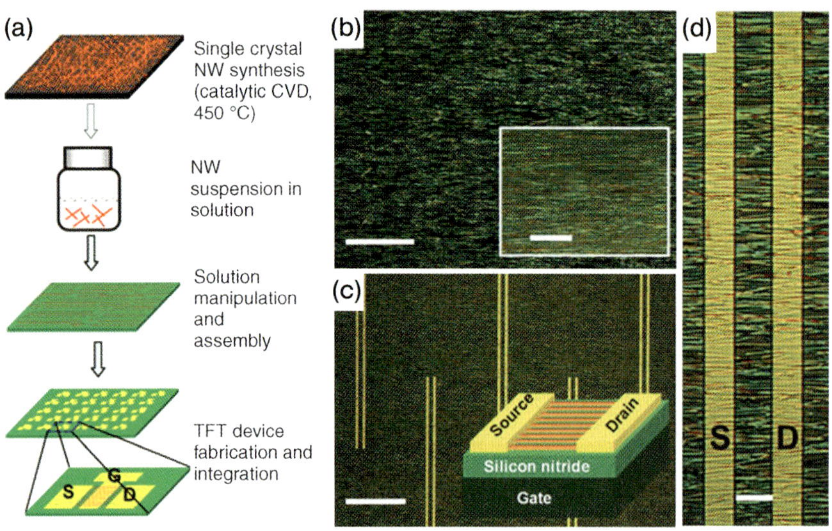

Figure 5.29 Scheme of parallel aligned crystalline Si nanowires between source and drain structure of a FET (b–d). Distance between individual wires is 500–1000 nm and can be adjusted via the solution deposition process (a). Reprinted with permission from Ref. [89c].

the wires have no contact. Extraordinary high charge carrier mobilities have been realized [82, 89].

The idea of separating the two individual processes, (i) material synthesis and (ii) materials processing, into device architectures has been used to arrange semiconductor nanowires on flexible polymer-based substrates. Seminal to the development of this field of printed inorganic devices was probably the report of the first printed all-inorganic thin-film field effect transistor based on 0D CdSe nanoparticles [90].

Synthesis of semiconductor rods or wires was performed, for example, via a CVD process on a silicon wafer. First the wafer-based source material was structured via a lithographic etching procedure. This technique works for a variety of 1D materials such as single-walled CNTs, GaN, GaAs and Si wires [84a]. An independent dry transfer process of the previously synthesized nanowires using an elastomeric stamp-based printing technique involves transfer of the 1D structures to a flexible substrate, for example, polyimide. It was thus possible to fabricate transistor devices with the highest electronic performance and also very high mechanical flexibility (Figure 5.30).

A general applicable condensed-phase approach for the alignment of nanowires uses the Langmuir–Blodgett technique [91]. It allows a massive parallel arrangement on planar substrates. The technique can be used in a layer-by-layer process allowing crossed nanowire structures with defined pitch to be formed (Figures 5.31 and 5.32). Moreover, this approach may allow the use of nanowires of different composition within the layering process. This gives rise to organized nanowire heterostructures.

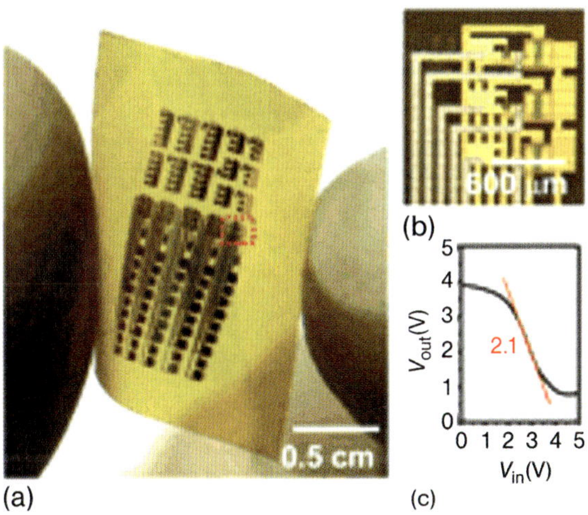

Figure 5.30 (a) Image of a printed array of 3D silicon n-channel metal oxide semiconductor inverters on a polyimide substrate. The inverters consist of MOSFETs (channel lengths of 4 µm, load-to-driver width ratio of 6.7 and a driver width of 200 µm) on two different levels. (b) View of the region indicated by the red box in (a). (c) Transfer characteristics of a typical inverter. Reprinted with permission from Ref. [89b].

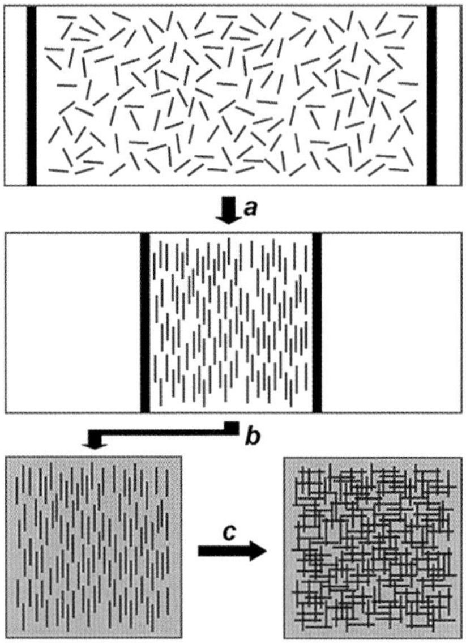

Figure 5.31 Nanowires (blue lines) in a monolayer of surfactant at the air–water interface are (a) compressed on a Langmuir–Blodgett trough to a specified pitch. (b) The aligned nanowires are transferred to the surface of a substrate to make a uniform parallel array. (c) Crossed NW structures are formed by uniform transfer of a second layer of aligned parallel nanowires (red lines) perpendicular to the first layer (blue lines). Reprinted with permission from Ref. [91].

Due to the hierarchical scaling of the structures from the nanometer up to the micrometer regime, reliable electrical contacting is possible.

Molecular precursors for the synthesis of polycrystalline silicon wires represent a promising alternative route to the widely employed synthetic CVD technique to inorganic semiconductor materials. Known since the early days of organosilicon chemistry, hydrogenated silicon compounds are known as chain (Si_nH_{2n+2}) or ring molecules (Si_nH_{2n}). For $n \geq 3$, these molecules are liquids and decompose around

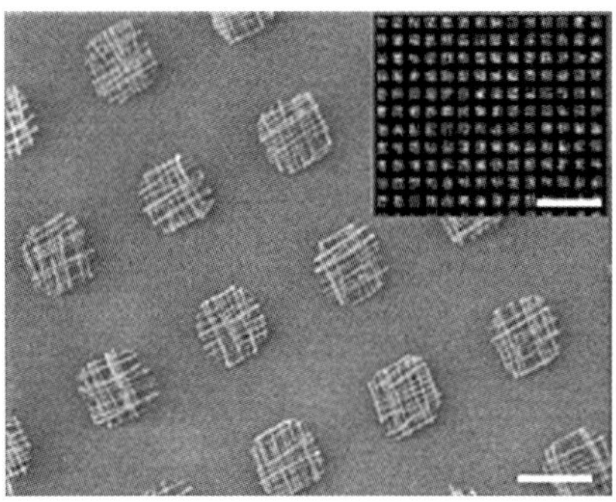

Figure 5.32 SEM image of patterned crossed nanowire arrays; scale bar 10 μm. Inset: Large area dark-field optical micrograph of the patterned crossed nanowire arrays; scale bar 100 μm. Reprinted with permission from Ref. [91].

Figure 5.33 The polycrystalline Si thin film was formed by spin-coating and baking of the liquid single source Si_5H_{10} precursor followed by laser crystallization. The TEM inset picture highlights the atomic image of the silicon crystal. Grain size in the film is about 300 nm, which is comparable to that of conventional CVD-formed poly-Si film. Reprinted with permission from Ref. [92].

300 °C to give Si. Using cyclopentasilane Si_5H_{10}, thin films of this precursor have been solution processed and converted to polycrystalline Si (Figure 5.33) [92].

This route is compatible with inkjet printing and has been used to set up a complete TFT device architecture via printing. Charge carrier mobilities of 6.5 $cm^2\,V^{-1}\,s^{-1}$ have been reported [92]. Other FET device architectures have also been studied using nanowires as active materials. FETs with surrounding gates have been synthesized and their electrical performance studied for Si [93]. Similar structures have been fabricated for InAs as active semiconductor. InAs performs with a high electron mobility and tends to form ideal ohmic contacts to many metals (Figure 5.34) [94]. Such structures are expected to show enhanced transconductance along the wire [95].

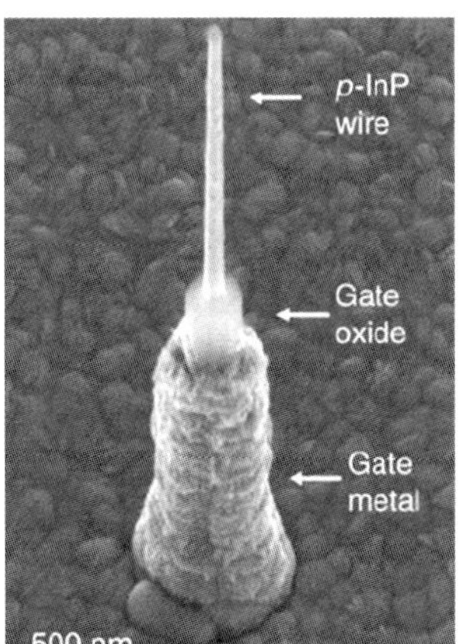

Figure 5.34 SEM micrograph of a vertical device with surrounding gate structure consisting of a p-type InP wire covered with gate oxide and gate metal. Reprinted with permission from Ref. [94].

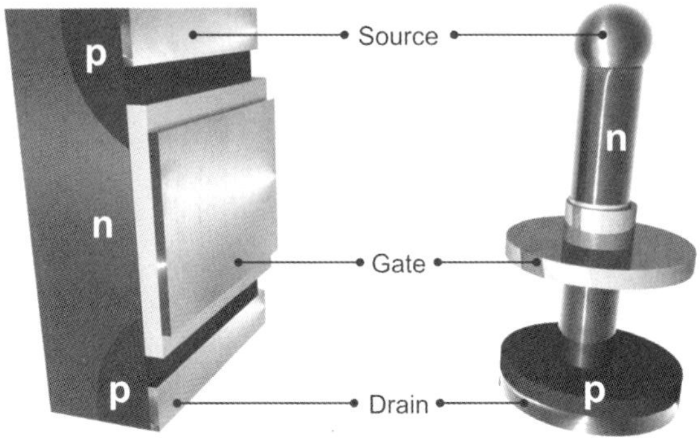

Figure 5.35 Schematics of a conventional p-channel MOSFET and a silicon nanowire surround gate FET. Adapted from Ref. [96].

A complete flow process for the fabrication of a silicon nanowire array with vertical surround gate FETs has been devised (Figure 5.35) [96]. Such an architecture was proposed earlier for CNTs but inorganic nanowires have the advantage that they can be grown as mechanically stiff vertical objects, whereas this is more complicated with single-walled CNTs [97].

Surround gate-type FETs have also been reported for Ge nanowire arrays. First, the nanowire array with the surround gate shell structure is synthesized via a multistep CVD process, followed by patterning of the nanosized core shell Ge/Al_2O_3/Al from solution on to an Si substrate. Finally, the structure is electrically contacted. This leads to a macroscopic device with a multiple number of surround gate nanowires arranged in a quasi-parallel fashion (Figure 5.36) [98].

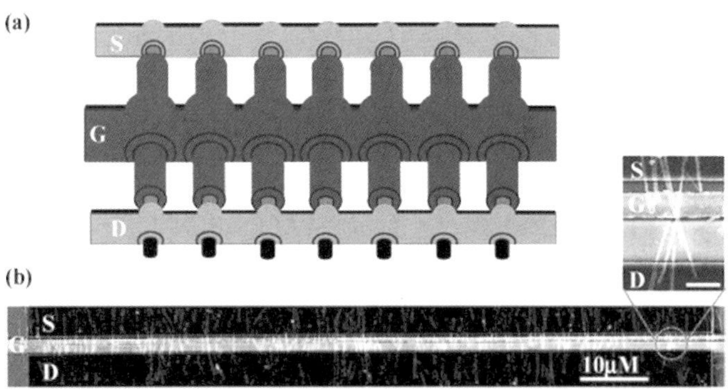

Figure 5.36 Transistor comprised of multiple surround-gate nanowires in parallel. (a) An idealized schematic presentation of a device. (b) SEM image of a device with ~35 surround gate nanowires in parallel. Crossing wires (each with its own gate shell) are seen in the zoomed-in image (scale bar = 1 μm). Reprinted with permission from Ref. [98].

5.4.3.2 Branched Nanowire Structures

Growth of 1D nanowires into certain defined directions leads to branched nanowire structures. In these, a higher degree of complexity is reached than in isolated wires. For example, the number of connection points and interconnecting possibilities to other nanostructures are enhanced compared with classical 1D structures. Their potential as interconnects for nanoelectronics, for example, circuiting, is obvious and certainly drives research into that area to some extent. One prerequisite for forming branched nanowire structures is the wide ability of semiconductors to undergo polymorphism, its characteristic feature to exist in more than one stable crystal structure.

Due to their nanosized dimensions, efficient strain release due to lattice mismatch of element combinations is much more efficient in nanowire architectures than in bulk film heterostructures of semiconductors. Therefore, lattice mismatches of $>3\%$ are not detrimental for a sharp interface growth process of wires [99]. In recent years, the scope of materials within the area of branched nanowire structures has broadened significantly. Condensed-phase (solution) and gas-phase synthetic techniques, the latter based on both physical vapor and chemical vapor deposition, have been used successfully to generate branched structures for many elemental combinations [100]. Compositions such as CdSe, PdSe, CdS, MnS and CdTe have been made available through solution-based techniques [101]. A successful technique to initiate branching of individual nanorods is to add nanoparticles as new growth sites. This has been demonstrated for GaP, InP, Ga As, AlN and GaN [102]. Successful studies even towards the hierarchical growth of interconnected nanobranched structures have been undertaken and reviewed recently [103]. Again, ZnO is a material which has shown an enormous breadth of structures with branched characteristics [104]. The growth of ZnO tetrapods is well understood and may serve as a model system to illustrate some general points of the growth of branched nanostructures. Although several models have been discussed and differ in some detail, it is accepted that growth of the cylindrical arms of a tetrapod proceeds via nucleation of a core structure. As determined by detailed high-resolution transmission electron microscopy studies, a zinc blende structure represents the core structure from which wurtzite arms grow. The former thus serve as seed templates for wurtzite arm growth. Growth of the $[00\bar{1}]$ wurtzite facets is outwards from the four [111] crystal faces of the zinc blende seeds. The same growth mode has been discussed for CdSe and CdTe tetrapod architectures [105]. Although the synthesis and growth of ZnO tetrapods is well studied, the electronic properties of this architecture have only recently been addressed.

Making electrical contacts to the edges of the tetrapod nanocrystal is, of course, difficult. A diode configuration has been realized with rectification characteristics (W forms an ohmic contact, Pt forms a Schottky contact to the tetrapod) (Figure 5.37) [106].

Branching out of nanowires has been achieved by using Au particles as catalyst seeds, resulting in the formation of Group III and V compound semiconductor rods (Figure 5.38).

Au catalyst particles are prepared via a gas-phase aerosol process and are size selected using a differential mobility analyzer [107]. The wire growth process is

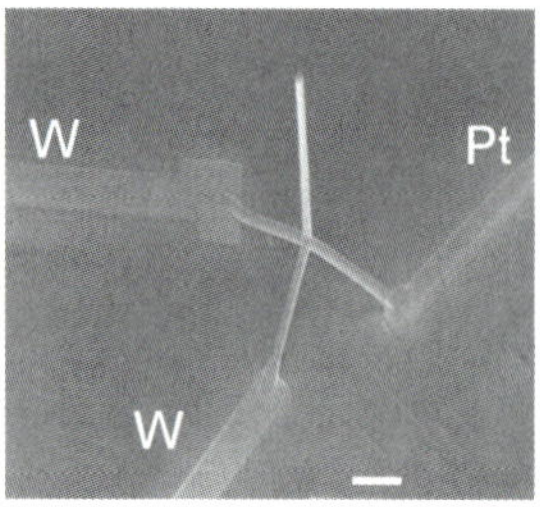

Figure 5.37 SEM image of a ZnO tetrapod with two W contacts and one Pt contact made to each arm of the structure. Scale bar = 1 μm. Reprinted with permission from Ref. [106].

repeated on the initially formed nanowire trunks to yield finally the branch structure in a follow-up process step. In this second process step, individual synthesis conditions, for example, the molecular wire precursor, can be varied, giving rise to branched nanowire heterostructures. The individual nanowire diameter of the branches is determined by the Au catalyst particle size; however, additional deposition of material on the side facets of the wires leads to thickening of the wires, and a tapered structure is therefore often observed [108].

The hierarchical growth of already branched nanowire structures to higher organized ensembles can also be realized in a solution-based growth process [105]. The precursor (e.g., an elemental chalcogenide) and a capping agent [e.g., an alkyl

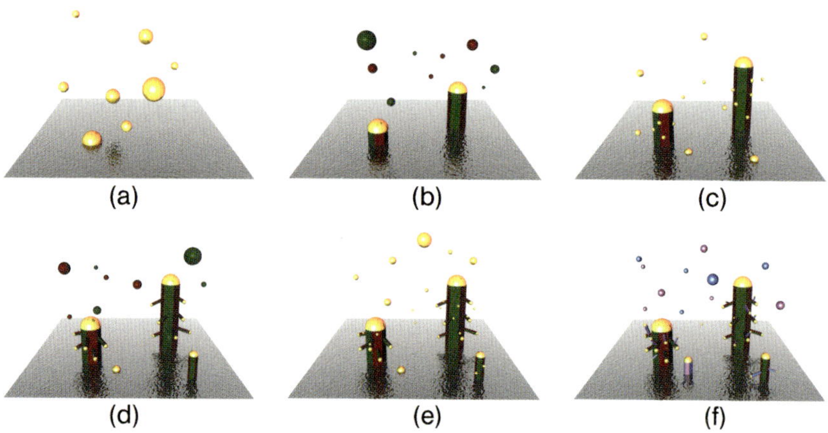

Figure 5.38 Branched nanowire and hetero nanowire formation process. Au particles are deposited on a substrate (a); precursor molecules (arrows) are introduced at elevated temperatures and combine beneath seed particles to form compound nanowire trunks (b); a second set of Au particles is deposited onto these trunks (c); branched nanowires grow in the same way as for (b) in (d); a third set of Au particles is introduced (e), followed by a new set of precursor molecules, which results in the growth of branched hetero nanowires (f). Adapted from Ref. [108].

(aryl)phosphine] first initiates the nanorod growth, followed by branching due to the facetted growth mechanism into a nanotetrapod structure. An end selective extension and further branching of such nanotetrapods are then possible by adding a new precursor to the same mixture. However, removal of the nano-objects from the initial synthesis mixture of the grown nanostructures, redispersion and regrowth, for example, with a different precursor, are also possible and allow further hierarchical growth at the ends of the original tetrapod structures. This growth gives rise to the formation of branched heterostructures. Thus, interfaces with different compositions can be designed in these structures, allowing control over charge carriers (electrons or holes) due to tailoring of the interface.

A detailed catalyst-driven (via Au aerosol particles) growth study of branched heterostructures by combining Group III–V materials, GaAs, GaP, InP and AlAs, has led to the conclusion that the growth mechanisms of such heterostructures depend on the relationship of the interface energies between the growing materials and the catalyst particle. It turned out that the growth of straight heterostructures in a particular direction seems favored over growth in another direction.

Finally, comparing gas-phase techniques and solution-based routes to branched nanowire structures, both methods show great potential towards the synthesis of higher organized morphologies based on 1D wires. It is remarkable that in both synthetic approaches the growth can be initiated on already formed branched structures just by adding fresh catalyst particles. However, the solution-based methods surely are favored due to the relative ease of formation of such structures and their further assembly potential towards higher organized multibranched wire structures. One further step towards this end has been demonstrated by tipping the ends of individual CdSe rods or tetrapods on both sides with Au nanoparticles, in solution (Figure 5.39) [109].

Although already a truly composite structure on its own (e.g., the excitonic spectra of the CdSe nanorod structure and the plasmonic Au spectra are not a superposition of the individual features of the individual semiconductor material and the metal nanoparticle), functionalization of the nanoparticle ends with alkanedithiols has led to assembled lines of individual composite nanowires (Figure 5.39) [109].

Figure 5.39 Schematics of dumbbell-like structures of a CdSe nanorod with Au gold tips at both ends. Single CdSe dumbbell doubly tipped with Au nanoparticles (a); typical unordered arrangement of Au tipped CdSe nanorods, where the Au tip size can be varied by increasing the $AuCl_3$ concentration during tip growth (b); self-assembled chain of nano CdSe dumbbells (real size 29 × 4 nm) formed by adding hexanedithiol bifunctional linker molecules which connect Au tipped ends of CdSe dumbbells. Adapted from Ref. [110].

5.5 Outlook

Studies towards functionality in inorganic 1D materials are an interesting and interdisciplinary field, bringing together fundamental science and opening up opportunities towards possible applications of these materials in various areas. Directed, organized growth of such 1D objects is surely a key in most of the envisioned areas of application of nanoscience.

In recent years, it has become clear that in addition to a bottom-up synthetic approach to 1D nanomaterials, a top-down approach which allows structuring, assembly and integration of 1D nanomaterials on the next higher length scale is necessary to organize and use 1D materials as building blocks in functional devices. The question no longer seems to be which of the two techniques, bottom-up or top-down, is the more powerful approach to new functional devices, but rather how we can make use in the most efficient way of both technologies to bridge the dimension gap: nano–micro–macro. In areas where current microtechnologies and materials are best compatible with bottom-up techniques for the synthesis and handling of 1D nanomaterials, they already work hand in hand and the unique properties of, for example, 1D materials have to be exploited successfully. This has already been demonstrated in current electronic and optoelectronic devices such as field effect transistors, light-emitting diodes, gas sensors and nanoresonators [110]. A key to this development of the field is certainly strong interdisciplinary research efforts between chemists, physicists and engineers.

References

1 (a) Rao, C.N.R. and Govindaraj, A. (2005) *Nanotubes and Nanowires*, Royal Society of Chemistry, Cambridge; (b) Yang, P. and Poeppelmeier, K.R. (2006) Inorganic Chemistry Forum: Special Issue on Nanowires. *Inorganic Chemistry*, **45**, 7509–7510; (c) Tenne, R. (2006) *Nature Nanotechnology*, **1**, 103–111; (d) Murphy, C.J., Gole, A.M., Hunyadi, S.E. and Orendorff, C.J. (2006) *Inorganic Chemistry*, **45**, 7544–7554; (e) Kline, T.R., Tan, M., Wang, J., Sen, A., Chan, M.W.H. and Mallouk, T.E. (2006) *Inorganic Chemistry*, **45**, 7555–7565; (f) Xiang, X., Yang, P., Sun, Y., Wu, Y., Mayers, B., Gates, E., Yin, Y., Kim, F. and Yan, H. (2003) *Advanced Materials*, **15**, 353–389; (g) Goldberger, J., Fan, R. and Yang, P. (2006) *Accounts of Chemical Research*, **39**, 239–248; (h) Pokropivnyi, V.V. (2001) *Powder Metallurgy and Metal Ceramics*, **40**, 485–496; (i) Pokropivnyi, V.V. (2001) *Powder Metallurgy and Metal Ceramics*, **40**, 582–594; (j) Remškar, M. and Mrzel, A. (2003) *Science Direct*, **71**, 177–183; (k) Tenne, R. (2006) *Journal of Materials Research*, **21**, 2726–2743; (l) Rao, C.N.R. and Nath, M. (2003) *Dalton Transactions*, 1–24; (m) Remškar, M. (2004) *Advanced Materials*, **16**, 1497–1504; (n) Monthioux, M., Flahaut, E. and Cleuziou, J.-P. (2006) *Journal of Materials Research*, **21**, 2774–2793; (o) Yan, Y., Chan-Park, M.B. and Zhang, Q. (2007) *Small*, **3**, 24–42; (p) Tenne, R., *Angewandte Chemie*, **2003**, **115**, 5280–5289; (2003) *Angewandte Chemie-International Edition*, **42**, 5124–5132; (q) Patzke, G.R., Krumeich, F. and Nesper, R.

(2002) *Angewandte Chemie*, **114**, 2554–2571; (2002) *Angewandte Chemie-International Edition*, **42**, 2446–2461.

2 (a) Beck, J.S., Vartulli, J.C., Kennedy, G.J., Kresge, C.T., Roth, W.J. and Schramm, S.E. (1994) *Chemistry of Materials*, **6**, 1816–1821; (b) Hamley, I.W. (2000) *Introduction to Soft Matter*, Wiley, Chichester; (c) Texter, J. (ed.) (2001) *Reactions and Synthesis in Surfactant Systems*, Vol. 100, Marcel Dekker, New York; (d) Förster, S. and Antonietti, M. (1998) *Advanced Materials*, **10**, 195–217; (e) John, V.T., Simmons, B., McPherson, G.L. and Bose, A. (2002) *Current Opinion in Colloid and Interface Science*, **7**, 288–295; (f) Dabbs, D.M. and Aksay, I.A. (2000) *Annual Review of Physical Chemistry*, **51**, 601–622; (f) Harrison, W.T.A. (2002) *Current Opinion in Solid State & Materials Science*, **6**, 407–413; (g) Liang, Z. and Susha, A.S. (2004) *Chemistry – A European Journal*, **10**, 4910–4914.

3 Ma, D.D.D., Lee, C.S., Au, F.C.K., Tong, S.Y. and Lee, S.T. (2003) *Science*, **299**, 1874–1877.

4 Pokropivnyi, V.V. (2002) *Powder Metallurgy and Metal Ceramics*, **41**, 123–135.

5 (a) Hicks, L.D. and Dresselhaus, M.S. (1993) *Physical Review B-Condensed Matter*, **47**, 16631–16634; (b) Hicks, L.D. and Dresselhaus, M.S. (1993) *Physical Review B-Condensed Matter*, **47**, 12727–12731.

6 Heinzel, T. (2007) *Mesoscopic Electronics in Solid State Nanostructures*, 2nd edn, Wiley-VCH, Weinheim.

7 (a) Xiong, Y., Mayers, B.T. and Xia, Y. (2005) *Chemical Communications)*, 5013–5022; (b) Pokropivnyi, V.V. (2001) *Powder Metallurgy and Metal Ceramics*, **40**, 485–496; (c) Ivanaovskii, A.L. (2002) *Russian Chemical Review*, **71**, 175–194; (d) Tenne, R. (2006) *Journal of Materials Research*, **21**, 2726–2743; (e) Remskar, M. and Mrzel, A. (2003) *Vacuum*, **71**, 177–183; (e) Roveri, N., Falini, G., Foresti, E., Fracasso, G., Lesci, I.G. and Sabatino, P. (2006) *Journal of Materials Research*, **21**, 2711–2725.

8 Lohrengel, M.M. (1993) *Materials Science and Engineering*, **R11** (6), 243–294.

9 (a) Keller, F., Hunter, M.S. and Robinson, D.L. (1953) *Journal of the Electrochemical Society*, **100**, 411–419; (b) Wood, G.C. and O'Sullivan, J.P. (1970) *Proceedings of the Royal Society of London. Series A*, **317**, 511–543; (c) Ono, S. and Masuko, N. (1992) *Corrosion Science*, **33**, 503–505; (d) Diggle, J.W., Downie, T.C. and Goulding, C.W. (1969) *Chemical Reviews*, **69**, 365–405; (e) Kniep, R., Lamparter, P. and Steeb, S. (1989) *Advanced Materials*, **1**, 229–231; (f) Hoar, T.P. and Yahalom, J. (1963) *Journal of the Electrochemical Society*, **110**, 614–621; (g) McDonald, D.D. (1993) *Journal of the Electrochemical Society*, **140**, L27–L30; (h) Thompson, G.E., Furneuax, R.C., Wood, G.C., Richardson, J.A. and Goode, J.S. (1978) *Nature*, **271**, 433; (i) Jessensky, O., Müller, F. and Gösele, U. (1998) *Applied Physics Letters*, **72**, 1173–1175; (j) Ono, S., Ichinose, H. and Masuko, N. (1992) *Corrosion Science*, **33**, 841–850; (k) Uchi, H., Kanno, T. and Alwitt, R.S. (2001) *Journal of the Electrochemical Society*, **148**, B17–B23; (l) Masuda, H., Yamada, H., Satoh, M., Asoh, H., Nakao, M. and Tamamura, T. (1997) *Applied Physics Letters*, **71**, 2770–2772.

10 (a) Schneider, J.J. and Engstler, J. (2008) Mesoscopic Ceramic Structures in 1D, 2D and 3D. (eds Riedel, R. and Chen, J.) *Ceramics Science and Technology*, Vol. 1, Structures, Wiley-VCH, Weinheim.

11 Petkov, N., Platschek, B., Morris, M.A., Holmes, J.D. and Bein, T. (2007) *Chemistry of Materials*, **19**, 1376–1381.

12 Rice, R.L., Arnold, D.C., Shaw, M.T., Iacopina, D., Quinn, A.J., Amenitsch, H., Holmes, J.D. and Morris, M.A. (2007) *Advanced Functional Materials*, **17**, 133–141.

13 (a) Possin, G.E. (1970) *Review of Scientific Instruments*, **41**, 772–774; (b) Price, P.B. and Walter, R.M. (1962) *Journal of Applied*

Physics, **33**, 3407–3412; (c) Kawai, S. and Ueda, R. (1975) *Journal of the Electrochemical Society*, **22**, 32–36.

14 Karim, S., Toimil-Molares, M.E., Maurer, F., Miehe, G., Ensinger, W., Liu, J., Cornelius, T.W. and Neumann, R. (2006) *Applied Physics A*, **84**, 403–407.

15 Ensinger, W. and Vater, P. (2005) *Materials Science and Engineering C*, **25**, 609–613.

16 Ferre, R., Ounadjela, K., George, J.M., Pireuax, L. and Dubois, S. (1997) *Physical Review B-Condensed Matter*, **56**, 14066–14075.

17 (a) Parallel rod (bunch) arrangement, Love, J.C., Urbach, A.R., Prentiss, M.G. and Whitesides, G.M. (2003) *Journal of the American Chemical Society*, **125**, 12696; (b) one after each other arrangement, Volkov, V.V., Schofield, M.A. and Zhu, Y. (2003) *Modern Physics Letters B*, **17**, 791–801.

18 Wetz, F., Soulantica, K., Falqui, A., Respaud, M., Snoeck, E. and Chaudret, B. (2007) *Angewandte Chemie*, **119**, 7209–7211. (2007) *Angewandte Chemie-International Edition*, **46**, 7079–7081.

19 Trahey, L., Becker, C.R. and Stacy, A. (2007) *Nano Letters*, **7**, 2535–2539.

20 Decher, G. (1997) *Science*, **277**, 1232–1237.

21 (a) Lu, Q., Feng, G., Kormaneni, S. and Mallouk, T.E. (2004) *Journal of the American Chemical Society*, **126**, 8650–8651; (b) Wu, Y., Cheng, G., Katsov, K., Sides, S.W., Wang, J., Tiang, J., Frederickson, G.H., Moskovits, M. and Stucky, G.D. (2004) *Nature Materials*, **3**, 816; (c) Wu, Y., Livneh, T., Zhang, Y.X., Cheng, G., Wang, J., Tang, J., Moskovits, M. and Stucky, G. (2004) *Nano Letters*, **4**, 2337–2342.

22 (a) Reneker, D.H. and Chun, I. (1996) *Nanotechnology*, **7**, 216–223; (b) Greiner, A. and Wendorff, H. (2007) *Angewandte Chemie*, **119**, 5770–5805; (2007) *Angewandte Chemie-International Edition*, **46**, 5670–5703.

23 (a) Taylor, G. (1969) *Proceedings of the Royal Society of London. Series A*, **313**, 453–475; (b) Doshi, J. and Reneker, D.H. (1995) *Journal of Electrostatics*, **35**, 151–160; (c) Reneker, D.H., Yarin, A.L., Fong, H. and Koanbhongse, S. (2000) *Journal of Applied Physics*, **87**, 4531–4547; (d) Larsen, G., Velarde-Ortiz, R., Minchow, K., Barrero, A. and Loscertales, I.G. (2003) *Journal of the American Chemical Society*, **125**, 1154–1155.

24 Dai, H., Gong, J., Kim, H. and Lee, D. (2002) *Nanotechnology*, **13**, 674–677.

25 Li, D. and Xia, Y. (2003) *Nano Letters*, **3**, 555–560.

26 (a) Li, D., Wang, Y. and Xia, Y. (2003) *Nano Letters*, **3**, 1167–1171; (b) Li, D. and Xia, Y. (2004) *Advanced Materials*, **16**, 1151–1170.

27 McCann, J.T., Li, D. and Xia, Y. (2005) *Journal of Materials Chemistry*, **15**, 735–738.

28 Hunter, R.J. (1981) *Zeta Potential in Colloid Science: Principles and Applications*, Academic Press, London.

29 Everett, D.H. (1988) *Basic Principles of Colloid Science*, Royal Society of Chemistry London.

30 Schubert, U. and Hüsing, N. (2000) *Synthesis of Inorganic Materials*, Wiley-VCH, Weinheim.

31 Cao, G. (2004) *The Journal of Physical Chemistry B*, **108**, 19921–19931.

32 Antonietti, M. and Ozin, G.A. (2004) *Chemistry – A European Journal*, **10**, 28–41.

33 Wagner, R.S. and Ellis, W.C. (1964) *Applied Physics Letters*, **4**, 89–90.

34 Wu, Y., Cui, Y., Huynh, L., Barrelet, C.J., Bell, D.C. and Lieber, C.M. (2004) *Nano Letters*, **4**, 433–436.

35 Holmes, J.D., Johnston, K.P., Doty, R.C. and Korgel, B.A. (2000) *Science*, **287**, 1471–1473.

36 Kodambaka, S., Hannon, J.B., Tromp, R.M. and Ross, F.M. (2006) *Nano Letters*, **6**, 1292–1296.

37 (a) Kovalev, D. and Fujii, M. (2005) *Advanced Materials*, **17**, 2531–2544; (b) Song, J., Wang, X., Riedo, E. and Wang, Z.L. (2005) *The Journal of Physical Chemistry B*, **109**, 9869–9872; (c) Wang, X., Summers, C.J. and Wang, Z.L. (2004) *Nano Letters*, **4**, 423–426.

38 (a) Schmitt, A.L., Bierman, M.J., Schmeisser, D., Himpsel, F.J. and Jiu, S. (2006) *Nano Letters*, **6**, 1617–1621; (b) Szczech, J.P., Schmitt, A.L., Bierman, M.J. and Jin, S. (2007) *Chemistry of Materials*, **19**, 3238–3243.

39 Aella, P., Ingole, S., Petuskey, W.T. and Picraux, S.T. (2007) *Advanced Materials*, **19**, 2603–2607.

40 Murphy, C.J., Gole, A.M., Hunyadi, S.E. and Orendorff, C.J. (2006) *Inorganic Chemistry*, **45**, 7544–7554.

41 Wang, F., Dong, A., Sun, J., Tang, R., Yu, H. and Buhro, W.E. (2006) *Inorganic Chemistry*, **45**, 7511–7521.

42 Hanrath, T. and Korgel, B.A. (2002) *Journal of the American Chemical Society*, **124**, 1424–1429.

43 Hanrath, T. and Korgel, B.A. (2003) *Advanced Materials*, **15**, 437–440.

44 Smith, P.A., Nordquist, C.D., Jackson, T.N., Mayer, T.S., Martin, B.R., Mbindyo, J. and Mallouk, T.E. (2000) *Applied Physics Letters*, **77**, 1399–1401.

45 Whang, D., Jin, S., Wu, Y. and Lieber, C.M. (2003) *Nano Letters*, **3**, 1255–1259.

46 Huang, Y., Duan, X., Wei, Q. and Lieber, C.M. (2001) *Science*, **291**, 630–633.

47 Snow, E.S., Novak, J.P., Campbell, P.M. and Park, D. (2003) *Applied Physics Letters*, **82**, 2145–2147.

48 Patolsky, F. and Lieber, C.M. (2005) *Materials Today*, **8**, 20–28.

49 Patolsky, F., Timko, B.P., Zheng, G. and Lieber, C.M. (2007) *MRS Bulletin*, **32**, 142–149.

50 Bergveld, P. (1972) *IEEE Transactions on Bio-Medical Engineering*, **BME-19**, 342–351.

51 Blackburn, G.F. (1987) in *Biosensors: Fundamentals and Applications* (ed. A.P.F. Turner), Oxford University Press, Oxford, p. 481.

52 Hafeman, D.G., Parce, J.W. and McConnell, H.M. (1988) *Science*, **240**, 1182–1185.

53 (a) Smith, P.A., Nordquist, C.D., Jackson, T.N., Mayer, T.S., Martin, B.R., Mbindyo, J. and Mallouk, T.E. (2000) *Applied Physics Letters*, **77**, 1399–1401; (b) Barsotti, R.J., Vahey, M.D., Wartena, R., Chiang, Y.M., Voldman, J. and Stellacci, F. (2007) *Small*, **3**, 488–499; (c) Shang, L., Clare, T.L., Eriksson, M.A., Marcus, M.S., Metz, K.M. and Hamers, R.J. (2005) *Nanotechnology*, **16**, 2846–2851; (d) Dong, L., Bush, J., Chiarayos, V., Solanki, R., Jiao, J., Ono, Y., Conley, J.F. Jr. and Ulrich, B.D. (2005) *Nano Letters*, **5**, 2112–2115.

54 Marcus, M.S., Shang, L., Li, B., Streifer, J.A., Beck, J.D., Perkins, E., Eriksson, M.A. and Hamers, R.J. (2007) *Small*, **3**, 1610–1617.

55 Vijayaraghavan, A., Blatt, S., Weissenberger, D., Oron-Carl, M., Hennrich, F., Gerthsen, D., Hahn, H. and Krupke, R. (2007) *Nano Letters*, **7**, 1556–1560.

56 Small, W.R. and in het Panhuis, M. (2007) *Small*, **3**, 1500–1503.

57 Li, Q., Koo, S.-M., Richter, C.A., Edelstein, M.D., Bonevich, J.E., Kopanski, J.J., Suehle, J.S. and Vogel, E.M. (2007) *IEEE Transactions on Nanotechnology*, **6**, 256–262.

58 Ingole, S., Aella, P., Haerne, S.J. and Picraux, S.T. (2007) *Applied Physics Letters*, **91**, 033106–033106.

59 Cui, Y., Wei, Q., Park, H. and Lieber, C.M. (2001) *Science*, **293**, 1289–1292.

60 Hahm, J. and Lieber, C.M. (2004) *Nano Letters*, **4**, 51–54.

61 (a) Gao, Z., Agarwal, A., Trigg, A.D., Singh, N., Fang, C., Tung, C.-H., Fan, Y., Buddharaju, K.D. and Kong, J. (2007) *Analytical Chemistry*, **79**, 3291–3297; (b) Patolsky, F. (2005) *Proceedings of the National Academy of Sciences of the United States of America*, **102**, 3208–3212; (c) Stadler, K. (2003) *Nature Reviews Microbiology*, **1**, 209–218; (d) Atlas, R.M. (2003) *Nature Reviews Microbiology*, **1**, 70–74; (e) Nijler, E. (2002) *Nature Biotechnology*, **20**, 21–25.

62 Patolsky, F., Zheng, G., Hayden, O., Lakadamyali, M., Zhuang, X. and Lieber, C.M. (2004) *Proceedings of the National Academy of Sciences of the United States of America*, **101**, 14017–14022.

63 (a) Whang, D., Jin, S. and Lieber, C.M. (2004) *Japanese Journal of Applied Physics*, **43**, 4465–4470; (b) Jin, S., Whang, D., McAlpine, M.C., Friedman, R.S., Wu, Y. and Lieber, C.M. (2004) *Nano Letters*, **4**, 915–919.

64 (a) McAlpine, M.C., Ahmad, H., Wang, D. and Heath, J.R. (2007) *Nature Materials*, **6**, 379–384; (b) Boukai, A., Diana, F., Gerardot, B., Badolato, A., Petroff, P.M. and Heath, J.R. (2003) *Science*, **300**, 112–115.

65 (a) Krupke, R., Hennrich, F., Weber, H.B., Beckmann, D., Hamper, O., Malik, S., Kappes, M.M. and Löhneysen, H.v. (2003) *Applied Physics A*, **76**, 397–400; (b) Chung, J., Lee, K.-H. and Ruoff, R.S. (2004) *Langmuir*, **20**, 3011–3017.

66 see e.g. Ingole, S., Aella, P., Haerne, S.J. and Picraux, S.T. (2007) *Applied Physics Letters*, **91**, 033106–033106.

67 (a) Hirsch, A. (1994) *The Chemistry of Fullerenes*, Georg Thieme, Stuttgart; (b) Krüger, A. (2007) *Neue Kohlenstoffmaterialien – eine Einführung*, Teubner, Wiesbaden.

68 Daniel, S., Rao, T.P., Rao, K.S., Rani, S.U., Naidu, G.K.R., Lee, H.Y. and Kawai, T. (2007) *Sensors and Actuators B-Chemical*, **122**, 672–682.

69 Coleman, K.S., Bailey, S.R., Fogden, S. and Green, M.L.H. (2003) *Journal of the American Chemical Society*, **125**, 8722–8723.

70 Williams, K.A., Peter, T.M., Veenhuizen Torre, B.G., Eritja, R. and Dekker, C. (2002) *Nature*, **420**, 761–762.

71 Taft, B.J., Lazareck, A.D., Withey, G.D., Yin, A., Xu, J.M. and Kelley, S.O. (2004) *Journal of the American Chemical Society*, **126**, 12750–12751.

72 Cai, H., Cao, X., Jiang, Y., He, P. and Fang, Y. (2003) *Analytical and Bioanalytical Chemistry*, **375**, 287–293.

73 Lu, G., Maragakis, P. and Kaxiras, E. (2005) *Nano Letters*, **5**, 897–900.

74 Liu, Y., Meyer-Zaika, W., Franzka, S., Schmid, G., Leis, M. and Kuhn, H. (2003) *Angewandte Chemie*, **115**, 2959–2963; (2003) *Angewandte Chemie-International Edition*, **42**, 2853–2857.

75 (a) Huang, M.H., Wu, Y., Feick, H., Tran, N., Weber, E. and Yang, P. (2001) *Advanced Materials*, **13**, 113–116; (b) Pan, Z.W., Dai, Z.R. and Wang, Z.L. (2001) *Science*, **291**, 1947–1949; (c) Kong, X.Y. and Wang, Z.L. (2003) *Nano Letters*, **3**, 1625–1631; (d) Kong, X.Y., Ding, Y., Yang, R. and Wang, Z.L. (2004) *Science*, **303**, 1348–1351; (e) Hughes, W.L. and Wang, Z.L. (2004) *Journal of the American Chemical Society*, **126**, 6703–6709; (f) Gao, P.X., Ding, Y., Mai, W., Hughes, W.L., Lao, C. and Wang, Z.L. (2005) *Science*, **309**, 1700–1704.

76 (a) Glaspell, G.G., Hassan, H.M.A., Elzatahry, A., Fuoco, L., Radwan, N.R.E. and El-Shall, M.S. (2006) *The Journal of Physical Chemistry B*, **110**, 21387–21393; (b) Yang, H.M. and Liao, P.H. (2007) *Applied Catalysis A-General*, **317**, 226–233; (c) Dalal, S.H., Baptista, D.L., Teo, K.B.K., Lacerda, R.G., Jefferson, D.A. and Milne, W.I. (2006) *Nanotechnology*, **17**, 4811–4818; (d) Comini, E., Faglia, G., Ferroni, M. and Sberveglieri, G. (2007) *Applied Physics A*, **88**, 45–48; (e) Jeong, M.C., Oh, B.Y., Nam, O.H., Kim, T. and Myoung, J.M. (2006) *Nanotechnology*, **17**, 526–530.

77 Wang, Z.L. (2007) *Materials Today*, **10**, 20–28.

78 Wang, Z.L. and Song, J. (2006) *Science*, **312**, 242–246.

79 Gao, P.X., Song, J., Liu, J. and Wang, Z.L. (2007) *Advanced Materials*, **19**, 67–72.

80 He, H., Hsiu, C.L., Liu, J., Chen, L.J.V. and Wang, Z.C. (2007) *Advanced Materials*, **19**, 781–784.

81 Buchine, B.A., Hughes, W.L., Degertekin, F.L. and Wang, Z.L. (2006) *Nano Letters*, **6**, 1155–1159.

82 Duan, X. (2007) *MRS Bulletin*, **32**, 134–141.

83 Gao, L., Li, Q., Luan, W., Kawaoka, H., Sekino, T. and Niihara, K. (2002) *Journal of the American Ceramic Society*, **85**, 1016–1018.

84 (a) Duan, X., Niu, Ch., Sahi, V., Chen, J., Parce, J.W., Empedocles, St. and Goldman, J.L. (2003) *Nature*, **425**, 274–278; (b) Ong, B.S., Li, Ch., Li, Y., Wu, Y. and Loutfy, R. (2007) *Journal of the American Chemical Society*, **129**, 2750–2751.

85 Sun, B. and Sirringhaus, H. (2005) *Nano Letters*, **5**, 2408–2413.

86 (a) Durkop, T., Getty, S.A., Cobas, E. and Fuhrer, M.S. (2004) *Nano Letters*, **4**, 35–39; (b) Snow, E.S., P: Novak, J., M: Campbell, P. and Park, D. (2003) *Applied Physics Letters*, **82**, 2145–2147; (c) Xiao, K., Liu, Y., Hu, P., Yu, G., Wang, X. and Zhu, D. (2003) *Applied Physics Letters*, **82**, 2145–2147; (d) Bradley, K., Gabriel, J.C.P. and Grüner, G. (2003) *Nano Letters*, **3**, 1353–1355; (e) Seidel, R., Graham, A.P., Unger, E., Duesberg, G.S., Liebau, M., Steinhoegl, W., Kreupl, F. and Hoehnlein, W. (2004) *Nano Letters*, **4**, 831–834; (f) Zhou, Y., Gaur, A., Hur, S.-H., Kocabas, C., Meitl, M.A., Shim, M. and Rogers, J.A. (2004) *Nano Letters*, **4**, 2031–2035.

87 Kocabas, C., Hur, S.H., Gaur, A., Meitl, M.A., Shim, M. and Rogers, J.A. (2005) *Small*, **1**, 1110–1116.

88 Rotkin, S.V., Ruda, H.E. and Shik, A. (2003) *Applied Physics Letters*, **83**, 1623–1626.

89 (a) Ridley, B.A., Nivi, B. and Jacobson, J.M. (1999) *Science*, **286**, 746–748; (b) Ahn, J.-H., Kim, H.-S., Lee, K.J., Jeon, S., Kang, S.J., Sun, Y., Nuzzo, R.G. and Rogers, J.A. (2006) *Science*, **314**, 1754–1757; (c) Menard, E., Lee, K.J., Khang, D.-Y., Nuzzo, R.G. and Rogers, J.A. (2004) *Applied Physics Letters*, **84**, 5398–5400.

90 Ridley, B.A., Nivi, B. and Jacobson, J.M. (1999) *Science*, **286**, 746–749.

91 Whang, D., Jin, S., Wu, Y. and Lieber, C.M. (2003) *Nano Letters*, **3**, 1255–1259.

92 Shimoda, T., Matsuki, Y., Furusawa, M., Aoki, T., Yudasaka, I., Tanaka, H., Iwasawa, H., Wang, D., Miyasaka, M. and Taakeuchi, Y. (2006) *Nature*, **440**, 783–786.

93 Becker, J.S., Suh, S. and Gordon, R.G. (2003) *Chemistry of Materials*, **15**, 2969–2976.

94 (a) Doh, Y.-J., van Dam, J.A., Roest, A.L., Bakkers, E.P.A.M., Kouwenhoven, L.P. and De Franceschi, S. (2005) *Science*, **309**, 272–275; (b) van Dam, J.A., Nazarov, Y.V., Bakkers, E.P.A.M., De Franceschi, S. and Kouwenhouven, L.P. (2006) *Nature*, **442**, 667–670.

95 Wong, H.S.P. (2002) *IBM Journal of Research and Development*, **46**, 133–167.

96 Schmidt, V., Riel, H., Senz, S., Karg, S., Riess, W. and Gösele, U. (2006) *Small*, **2**, 85–88.

97 For a survey on growth of nanotubes for electronics: Robertson, J. (2007) *Materials Today*, **10**, 36–43.

98 Zhang, L., Tu, T. and Dai, H. (2006) *Nano Letters*, **6**, 2785–2789.

99 (a) Zakharov, N.D., Werner, P., Gerth, G., Schubert, L., Sokolov, L. and Gösele, U. (2006) *Journal of Crystal Growth*, **290**, 6–10; (b) Borgström, M.T., Verheijen, M.A., Immink, G., de Smet, T. and Bakkers, E.P.A.M. (2006) *Nanotechnology*, **17**, 4010–4013.

100 Mieszawska, A.J., Jalilian, R., Sumanasekera, G.U. and Zamborini, F.P. (2007) *Small*, **3**, 722–756.

101 (a) Manna, L., Scher, E.C. and Alivisatos, A.P. (2000) *Journal of the American Chemical Society*, **122**, 12700–12706; (b) Grebinski, J.W., Hull, K.L., Zhang, J., Kosel, T.H. and Kuno, M. (2004) *Chemistry of Materials*, **16**, 5260–5272; (c) L: Hull, K., Grebinski, J.W., Kosel, T.H. and Kuno, M. (2005) *Chemistry of Materials*, **17**, 4416–4425; (d) Yun, Y.W., Lee, S.M., Kang, N.J. and Cheon, J. (2001) *Journal of the American Chemical Society*, **123**, 5150–5151; (e) Jun, Y.W., Jun, Y.Y. and Cheon, J. (2002) *Journal of the American Chemical Society*, **124**, 615–619; (f) Manna, L., Milliron, D.J., Mesiel, A., Scher, E.C. and Alivisatos, A.P. (2003) *Nature Materials*, **2**, 382–385.

102 (a) Dick, K.A., Deppert, K., Larsson, M.W., Martenssson, T., Seifert, W., Wallenberg, L.R. and Samuelson, L. (2004) *Nature Materials*, **3**, 380–384; (b) Dick, K.A., Deppert, K., Karlsson, L.S., Wallenberg,

L.R., Samuelson, L. and Seifert, W. (2005) *Advanced Functional Materials*, **15**, 1603–1610; (c) Dick, K.A., Getertovszky, Zs., Mikkelsen, A., Karlsson, L.S., Lundgren, E., Malm, J.-O., Andersen, J.N., Samuelson, L., Seifert, W., Waacaser, B.A. and Deppert, K. (2006) *Nanotechnology*, **17**, 1344–1350; (d) Su, J., Cui, G., Gherasimova, M., Tsukamoto, H., Han, J., Ciuparu, D., Lim, S., Pfeferrle, L., He, Y., Nurmikko, A.V., Broadbridge, C. and Lehman, A. (2005) *Applied Physics Letters*, **86**, 013105.

103 Dick, K.A., Deppert, K., Karlsson, L.S., Larsson, M.W., Seifert, W., Wallenberg, L.R. and Samuelson, L. (2007) *MRS Bulletin*, **32**, 127–133.

104 (a) Bae, S.Y., Seo, H.W., Choi, H.C., Park, J. and Park, J. (2004) *The Journal of Physical Chemistry B*, **108**, 12318–12326; (b) Zhang, T., Dong, W., Keeter-Brewer, M., Konar, S., Njabon, R.N. and Tian, Z.R. (2006) *Journal of the American Chemical Society*, **128**, 10960–10968.

105 (a) Milliron, D.J., Hughes, S.M., Cui, Y., Manna, L., Li, J.B., Wang, L.W. and Alivisatos, A.P. (2004) *Nature*, **430**, 190–194; (b) for a discussion in more breadth, see Ozin, G.A. and Arsenault, A.C. (2005) *Nanochemistry, a Chemical Approach to Nanomaterials*, Royal Society of Chemistry, Cambridge, Chapter 6, pp. 297–303.

106 Newton, M.C., Firth, S. and Warburton, P.A. (2006) *Applied Physics Letters*, **89**, 072104.

107 (a) Magnusson, M.H., Deppert, K., Malm, J.-O., Bovin, J.-O. and Samuelson, L. (1999) *Nanostructured Materials*, **12**, 45–48; (b) Magnusson, M.H., Deppert, K., Malm, J.-O., Bovin, J.-O. and Samuelson, L. (1999) *Journal of Nanoparticle Research*, **1**, 243–251; (c) Karlsson, M.N.A., Deppert, K., Karlsson, L.S., Magnusson, M.H., Malm, J.-O. and Srinivasan, N.S. (2005) *Journal of Nanoparticle Research*, **7**, 43–49.

108 Dick, K.A., Deppert, K., Karlsson, L.S., Larsson, M.W., Seifert, W., Wallenberg, L.R. and Samuelson, L. (2007) *MRS Bulletin*, **32**, 127–133.

109 Mokari, T., Rothenberg, E., Popov, I., Costi, R. and Banin, U. (2004) *Science*, **304**, 1787–1790.

110 Lieber, C.M. and Wang, Z.L., guest ed., Special issue on Functional Nanowires. (2007) *MRS Bulletin*, **32**, 99–142.

6
Biomolecule–Nanoparticle Hybrid Systems

Maya Zayats and Itamar Willner

6.1
Introduction

Metal and semiconductor nanoparticles (NPs) or quantum dots (QDs) exhibit unique electronic, optical and catalytic properties. The comparable dimensions of NPs or QDs and biomolecules such as enzymes, antigens/antibodies and DNA suggest that by the integration of the biomolecules with NPs (or QDs) combined hybrid systems with the unique recognition and catalytic properties of biomolecules and with the electronic, optical and catalytic features of NPs might yield new materials with predesigned properties and functions. For example, metallic NPs, such as Au or Ag NPs, exhibit size-controlled plasmon excitons. These plasmon absorbance bands are sensitive to the dielectric properties of the stabilizing capping layers of the NPs [1, 2] or to the degree of aggregation of the NPs that leads to interparticle-coupled plasmon excitons [3, 4]. Thus, spectral changes occurring in metal NP assemblies as a result of biomolecule-induced recognition events that occur on NP surfaces and alter the surface dielectric properties of the NP or stimulate aggregation might be used for optical biosensing. Similarly, semiconductor QDs reveal size-controlled absorption and fluorescence features [5–7]. The high fluorescence quantum yields of QDs and the stability of QDs against photobleaching can be used to develop new fluorescent labels for optical biosensors [8, 9]. Alternatively, metallic NPs exhibit catalytic functions reflected by their ability to catalyze the reduction and growth of the NP seeds by the same metal or a different metal to form core–shell NPs. These properties may be applied to form NP-functionalized proteins or nucleic acids that provide hybrid systems that act as electroactive labels for amplified biosensing [10, 11] or as templates for growing nanostructures [12]. Furthermore, the coupling of biomolecules to metallic or semiconductor NPs might allow the use of the electron-conducting properties of metal NPs or the photoelectrochemical functions of semiconductor NPs to develop new electrical or photoelectrochemical biosensors. Indeed, tremendous scientific advances have been achieved in the last few years by conjugating biomolecules and NPs to functional hybrid systems. Numerous new

Nanotechnology. Volume 1: Principles and Fundamentals. Edited by Günter Schmid

ISBN: 978-3-527-31732-5

biosensors and bioanalytical paradigms were developed, and substantial progress in the use of these hybrid systems as building units of nanodevices was achieved. Several comprehensive review articles addressed different aspects and future perspectives of biomolecule–NP hybrid systems [13–18]. This chapter is aimed at summarizing the different venues where the unique optical and electronic properties of biomolecule–NP hybrid systems have been applied and to discuss future opportunities in the area. Naturally, this review is not aimed at providing full bibliographic coverage of the different topics, but it will highlight the different scientific directions that use biomolecule–NP hybrid systems, and address some specific examples.

6.2 Metal Nanoparticles for Electrical Contacting of Redox Proteins

The electrical contacting of redox proteins with electrodes is a key issue in bioelectronics. Numerous redox enzymes exchange electrons with other biological components such as other redox proteins, cofactors or molecular substrates. The exchange of electrons between the redox centers of proteins and electrodes could activate the bioelectrocatalytic functions of these proteins and thus provide a route to design different amperometric biosensors. Most of the redox proteins lack, however, direct electron transfer communication with electrodes, and hence, the bioelectrocatalytic activation of the redox enzymes is prohibited. The lack of electrical contact between the redox centers and the electrode surfaces is attributed to the spatial separation of the redox sites from the electrode by means of the protein shell [19]. Different methods to electrically communicate redox enzymes with electrodes were developed, including the application of diffusional electron mediators [20], tethering redox relays to the proteins [21–23] and the immobilization of redox proteins in electroactive polymers [24–26]. A recently developed procedure for the electrical contacting of redox proteins with electrodes involved the extraction of the native redox cofactor from the protein and the reconstitution of the resulting apo-enzyme on a surface modified with a monolayer consisting of a relay tethered to the respective cofactor units [27–30]. The reconstitution process aligned the protein on the electrode surface in an optimal orientation, while the relay units electrically contacted the cofactor sites with the conductive support, by shortening the electron transfer distances [31]. All of these methods permitted the bioelectrocatalytic activation of the respective enzymes and the development of amperometric biosensors and biofuel cells, [32, 33].

The availability of metal nanoparticles exhibiting conductivity allowed the generation of nanoparticle–enzyme hybrid systems for controlled electron transfer [34, 35]. Recently, highly efficient electrical contacting of the redox enzyme glucose oxidase (GOx) through a single Au nanoparticle (Au NP) was demonstrated [36]. The GOx–Au NP conjugate was constructed by the reconstitution of an apo-flavoenzyme, apo-glucose oxidase (apo-GOx) on a 1.4-nm Au_{55} nanoparticle functionalized with N^6-(2-aminoethyl) flavin adenine dinucleotide (FAD cofactor, amino derivative, **1**). The resulting enzyme–NP conjugate was assembled on a thiolated monolayer by using different dithiols (**2–4**) as linkers [Figure 6.1(A), route a]. Alternatively, the FAD-functionalized Au nanoparticle was assembled on a thiolated monolayer associated with an electrode, and apo-GOx was subsequently reconstituted on the functional nanoparticles [Figure 6.1(A), route b]. The enzyme electrodes prepared by these two routes revealed similar protein surface coverage of about 1×10^{-12} mol cm^{-2}. The Au NP was found to act as a nanoelectrode that acted as relay units transporting the electrons from the FAD cofactor embedded in the protein to the electrode with no additional mediators, thus activating the bioelectrocatalytic functions of the enzyme. Figure 6.1(B) shows the cyclic voltammograms generated by the enzyme-modified electrode, in the presence of different concentrations of glucose. The electrocatalytic currents increase as the concentrations of glucose are elevated, and the appropriate calibration curve was extracted [Figure 6.1(B), inset]. The resulting nanoparticle-reconstituted enzyme electrodes revealed unprecedented efficient electrical communication with the electrode (electron transfer turnover rate about 5000 s^{-1}). This effective electrical contacting, far higher than the turnover rate of the enzyme with its native electron acceptor, oxygen (about 700 s^{-1}), made the enzyme electrode insensitive to oxygen or to ascorbic acid or uric acid, which are common interferents in glucose biosensing. The rate-limiting step in the electron transfer communication between the enzyme redox center and the electrode was found to be the charge transport across the dithiol molecular linker that bridges the particle to the electrode. The conjugated benzenedithiol (**4**) was found to be the most efficient electron-transporting unit among the linkers (**2–4**).

A similar concept was applied to electrically contact pyrroloquinoline quinone (PQQ)-dependent glucose dehydrogenase [37]. Apo-glucose dehydrogenase (apo-GDH) was reconstituted on PQQ-cofactor units (**5**) covalently linked to the amino-functionalized Au NPs [Figure 6.2(A)]. The electrocatalytic anodic currents developed by the enzyme-modified electrode, in the presence of variable concentrations of glucose, are depicted in Figure 6.2(B). The resulting electrocatalytic currents imply that the system is electrically contacted and that the Au NPs mediate the electron transfer from the PQQ-cofactor center embedded in the protein to the electrode. Using the saturation current value generated by the system [Figure 6.2(B), inset] and knowing the surface coverage of the reconstituted enzyme, 1.4×10^{-10} mol cm^{-2}, the electron transfer turnover rate between the biocatalyst and the electrode was estimated to be 1180 s^{-1}, a value that implies effective electrical communication between the enzyme and the electrode, which leads to the efficient bioelectrocatalytic oxidation of glucose.

6.3 Metal Nanoparticles as Electrochemical and Catalytic Labels

The possibility to functionalize metallic nanoparticles with different biomolecules allows the use of the biomolecule–-NP conjugates as labels for the amplified detection of biorecognition events. The NPs may be modified by their direct functionalization with the biomolecules, for example, the binding of thiolated nucleic acids to Au or Ag

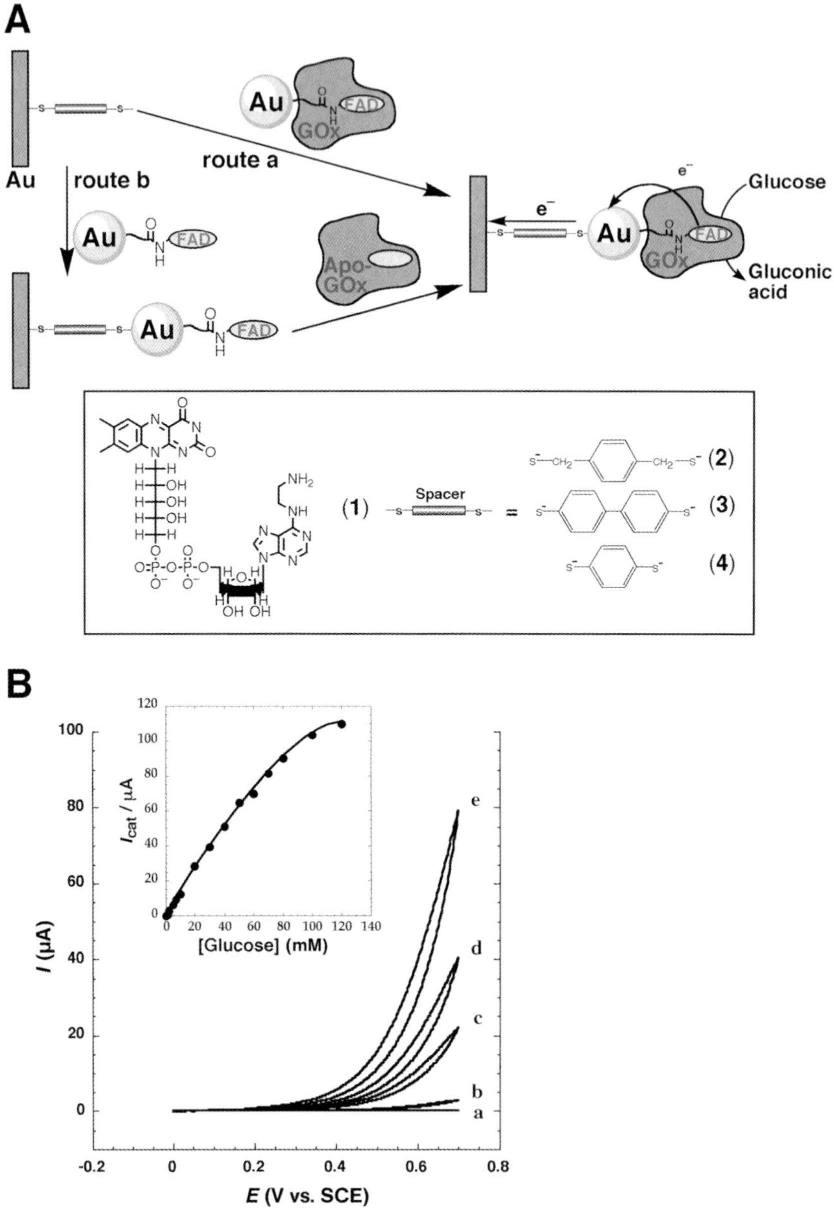

Figure 6.1 (A) The assembly of an electrically-contacted glucose oxidase (GOx) monolayer-functionalized electrode by the reconstitution of the apo-enzyme on an Au NP modified with the flavin adenine dinucleotide (FAD) cofactor (**1**). (B) Cyclic voltammograms corresponding to the bioelectrocatalyzed oxidation of different glucose concentrations by the GOx-reconstituted electrode: (a) 0, (b) 1, (c) 10, (d) 20 and (e) 50 mM. Scan rate 5 mV s^{-1}. Inset: calibration plot derived from the cyclic voltammograms at $E = 0.6$ V in the presence of different concentrations of glucose. (Reproduced from [36]. Reprinted with permission from AAAS).

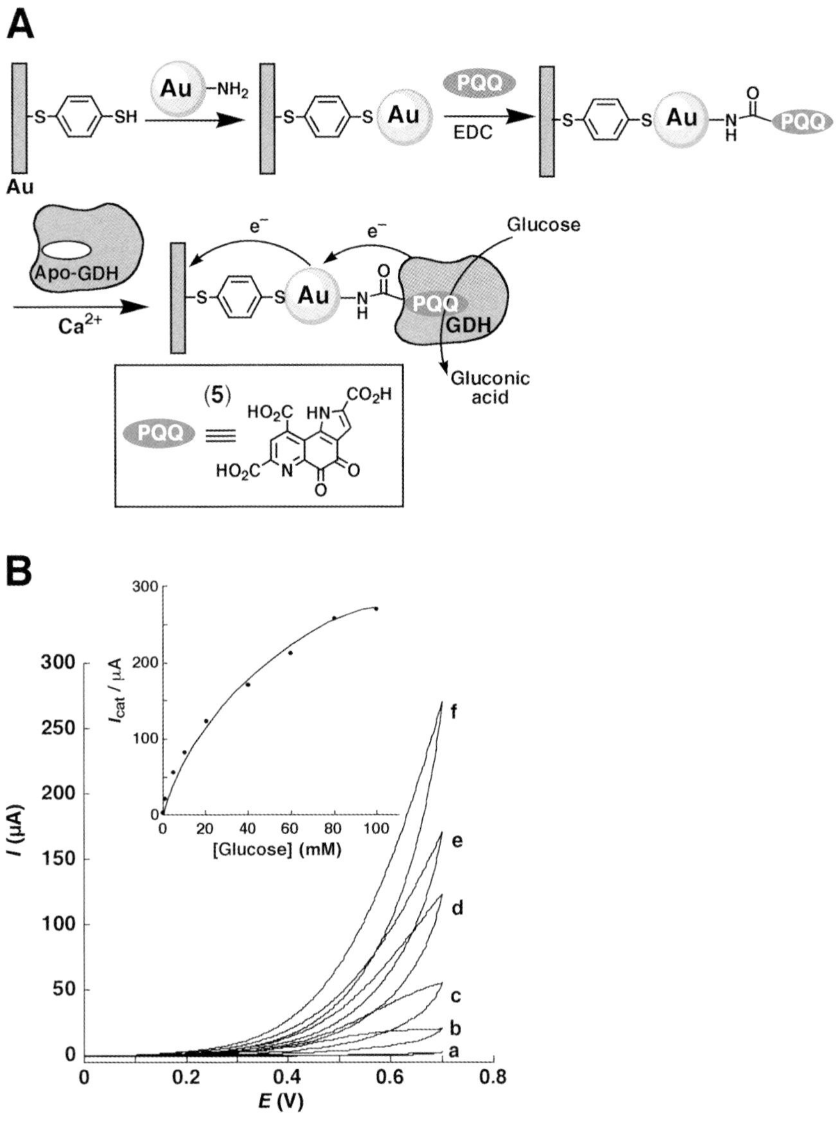

Figure 6.2 (A) The assembly of an electrically contacted glucose dehydrogenase (GDH) enzyme electrode by the reconstitution of the apo-enzyme on a PQQ (**5**)-functionalized Au NP associated with the electrode. (B) Cyclic voltammograms corresponding to the bioelectrocatalyzed oxidation of glucose by the GDH reconstituted on the PQQ-functionalized Au NPs associated with an Au electrode in the presence of different concentrations of glucose: (a) 0, (b) 1, (c) 5, (d) 20, (e) 40 and (f) 100 mM. Potential scan rate 5 mV s^{-1}. Inset: calibration plot derived from the cyclic voltammograms at $E = 0.7$ V. (Reprinted with permission from [37]. Copyright 2005 American Chemical Society).

NPs [3, 38, 39], the covalent tethering of the biomolecules to chemically functionalized NPs [40, 41] or the supramolecular binding of biomolecules to functionalized NPs, for example, the association of avidin-tagged nucleic acids with biotinylated NPs [42]. The biomolecule-modified NPs may be employed as electrochemical tracers for the amplified detection of biorecognition events [43, 44]. The chemical dissolution of the NP labels associated with the biorecognition events followed by electrochemical preconcentration of the released ions on the electrode and the subsequent stripping off of the collected metal provide a general means for the use of the particles as amplifying units for the biorecognition events [45]. Alternatively, the direct electrochemical stripping off of the NPs bound to the biorecognition complex was employed to transduce the biorecognition events [46–48]. These methods for stripping of the electrochemically pre-concentrated metals or the direct electrochemical dissolution of the metals led to 3–4 orders of magnitude improved detection limits, compared with normal pulse voltammetric techniques used to monitor DNA hybridization.

The method for the detection of DNA by the capturing the gold [49, 50] or silver [51] nanoparticles on the hybridized target, followed by the anodic stripping off of the metal tracer is depicted in Figure 6.3. Picomolar and sub-nanomolar levels of the DNA target have thus been detected. For example, the electrochemical method was employed for the Au NP-based quantitative detection of the 406-base human cytomegalovirus DNA sequence (HCMV DNA) [50]. The HCMV DNA was immobilized on a microwell surface and hybridized with the complementary oligonucleotide-modified Au nanoparticles as labels. The resulting surface-immobilized Au nanoparticle double-stranded assembly was treated with HBr–Br_2, resulting in the oxidative dissolution of the gold particles. The solubilized Au^{3+} ions were then electrochemically reduced and accumulated on the electrode and subsequently analyzed by anodic stripping voltammetry. The same approach was applied for analyzing an antigen [52] using Au nanoparticle labels and stripping voltammetry measurement. Further sensitivity enhancement can be obtained by catalytic enlargement of the gold tracer in connection with nanoparticle-promoted precipitation of gold [49] or silver [53–55]. Combining such enlargement of the metal particle tags, with the effective "built-in" amplification of electrochemical stripping analysis, paved the way to sub-picomolar detection limits. The silver-enhanced Au NP stripping method was used for the detection of DNA sequences related to the BRCA1 breast cancer gene [53]. The method showed substantial signal amplification as a result of

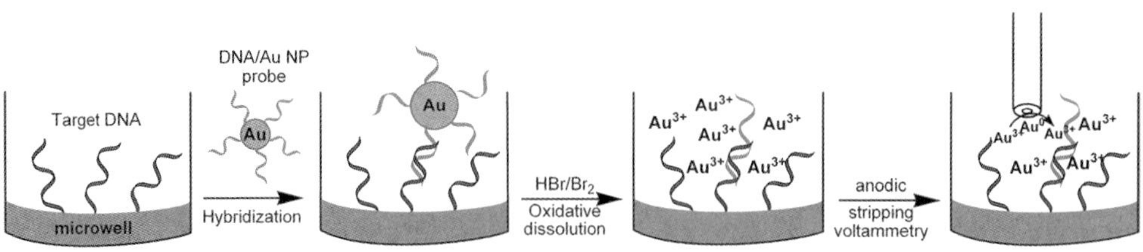

Figure 6.3 Electrochemical detection of a DNA by its labeling with a complementary nucleic acid-functionalized Au NP and the subsequent dissolution of the NPs and the electrochemical stripping of the Au^{3+} ions.

the catalytic deposition of silver on the gold tags; the silver signal was 125 times greater than that of the gold tag. A detection limit of 32 pM was achieved.

A further modification of the metal NP-induced amplified detection of DNA involved combination of magnetic particles and biomolecule-functionalized metallic NPs as two inorganic units that operate successfully in biosensing events. The formation of a biorecognition complex on the magnetic particles followed by labeling of the complex with the metallic NPs allowed the magnetic separation of the metal-labeled recognition complexes and their subsequent electrochemical detection by stripping voltammetry [49, 53]. Whereas the magnetic separation of the labeled recognition complexes enhanced the specificity of the analytical procedures, the stripping off of the metallic labels contributed to specific analysis [56, 57]. Furthermore, the use of the metallic NP labels accumulated on the magnetic NPs as catalytic seeds for the electroless enlargement of the particles by metals provided a further amplification path for the electrochemical stripping of the labels [57].

Figure 6.4 depicts the amplified detection of DNA by the application of nucleic acid-functionalized magnetic beads and Au NPs as catalytic seeds for the deposition of silver [57]. A biotin-labeled nucleic acid (**6**) was immobilized on the avidin-functionalized magnetic particles and hybridized with the complementary biotinylated nucleic acid (**7**). The hybridized assembly was then reacted with the Au-nanoparticle-avidin conjugate (**8**). Treatment of the magnetic particles–DNA–Au nanoparticle conjugate with silver ions (Ag^+) in the presence of hydroquinone results in the electroless catalytic deposition of silver on the Au nanoparticles, acting as catalytic labels. The latter process provided the amplification path since the catalytic accumulation of silver on the Au nanoparticle originates from a single DNA recognition event. The magnetic separation of the particles by an external magnet concentrated the hybridized assembly from the analyzed sample. The current originated by the voltammetric stripping off of the accumulated silver then provided the electronic

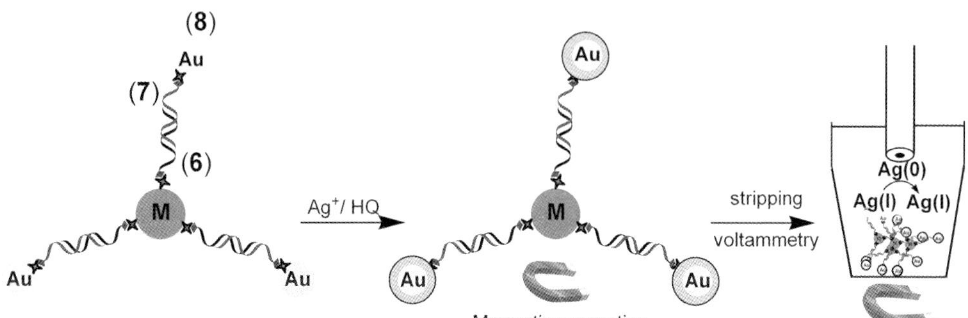

Figure 6.4 Electrochemical analysis of a DNA (**7**) by its hybridization with the complementary nucleic acid (**6**)-functionalized magnetic particle and the hybridization with the complementary nucleic acid (**8**)-modified Au NPs, followed by the deposition of Ag^0 on the particles. The analysis is performed by the magnetic separation of the particles aggregates, followed by the electrochemical stripping of the metallic NPs.

signal that transduced the analysis of the target DNA. Also, Au nanoparticle-based detection of DNA hybridization based on the magnetically induced direct electrochemical detection of the 1.4-nm Au_{67} NP tag linked to the target DNA was reported [58]. The Au_{67} NP tag was directly detected after the hybridization process, without the need for acid dissolution.

An additional method to enhance the sensitivity of electrochemical DNA detection involved the use of polymeric microparticles (carriers) loaded with numerous Au nanoparticle tags [59]. The Au NP-loaded microparticles were prepared by binding of biotinylated Au NPs to streptavidin-modified polystyrene spheres. The hybridization of target DNA immobilized on magnetic beads with the nucleic acid functionalized with Au nanoparticle-carrier polystyrene spheres, followed by the catalytic enlargement of gold labels, magnetic separation and then detection of the hybridization event by stripping voltammetry (Figure 6.5) allowed the determination of DNA targets at a sensitivity corresponding to 300 amol. A further method for the amplified detection of biorecognition complexes included the use of Au NPs as carriers of electroactive tags [60]. That is, the redox-active units capping the Au NPs linked to the biorecognition complexes allowed the amperometric transduction of the biosensing process. A detection limit of 10 amol for analyzing DNA with the functionalized NPs was reported.

The metal NP labels might be linked along double-stranded DNA, rather than be tethered by hybridization to the analyzed DNA, for the amplified electrical analysis of DNA. The generation of the metal nanoclusters along the DNA and their use for the amplified electrical analysis of DNA are depicted in Figure 6.6 and they follow the concepts used for the fabrication of metallic nanowires (see Section 6.10) [61]. The method involves the immobilization of a short DNA primer on the electrode that

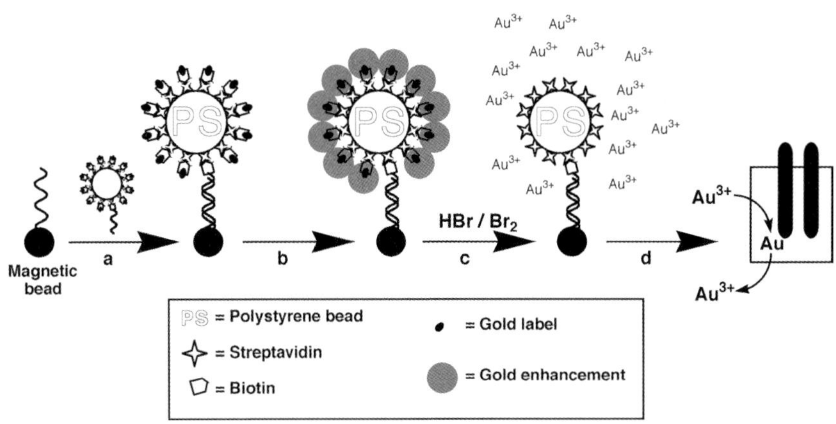

Figure 6.5 Amplified electrochemical detection of DNA by using nucleic acid–Au NP-functionalized microparticles as labels, and electroless catalytic deposition of gold on the NPs as a means of amplification: (a) hybridization of the nucleic acid–Au NP-functionalized microparticles with the target DNA, which is associated with a magnetic bead; (b) enhanced catalytic deposition of gold on the NPs; (c) dissolution of the gold clusters; (d) detection of the Au^{3+} ions by stripping voltammetry. (Reproduced with permission from [59]. Copyright 2004 Wiley-VCH).

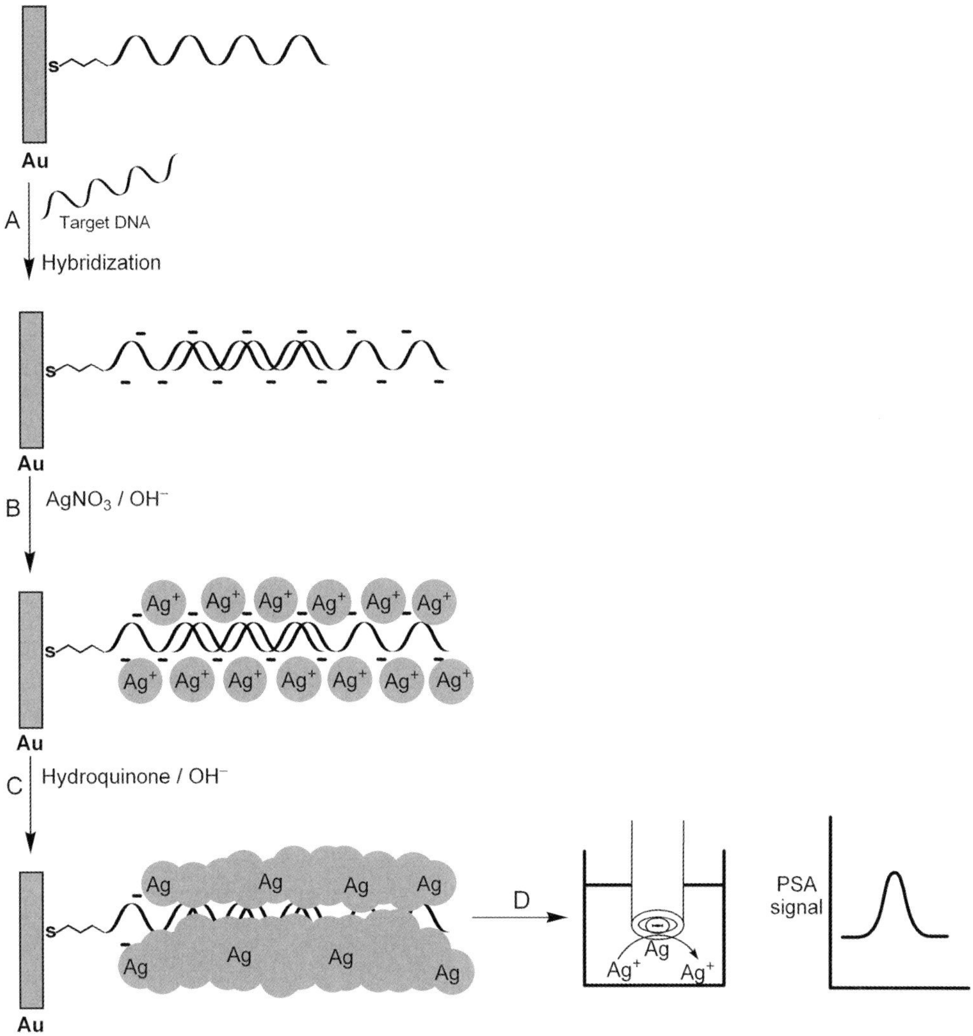

Figure 6.6 Outline of the steps involved in the amplified electrochemical detection of DNA by the deposition of catalytic silver clusters on the DNA strand: (A) hybridization of the complementary target DNA with the short DNA primer, which is covalently linked to the electrode surface through a cystamine monolayer; (B) loading of the Ag^+ ions on to the immobilized DNA; (C) reduction of Ag^+ ions by hydroquinone to form silver aggregates on the DNA backbone; (D) dissolution of the silver aggregates in acidic solution and transfer of the solution to the detection cell for stripping potentiometric measurement (PSA = potentiometric stripping analysis).

hybridizes with the target DNA (step A). The negative charges associated with the phosphate units of the long target DNA collect Ag^+ ions from the solution to form phosphate–Ag^+ complexes (step B). The bound Ag^+ ions are then reduced by hydroquinone, resulting in the formation of metallic silver aggregates along the

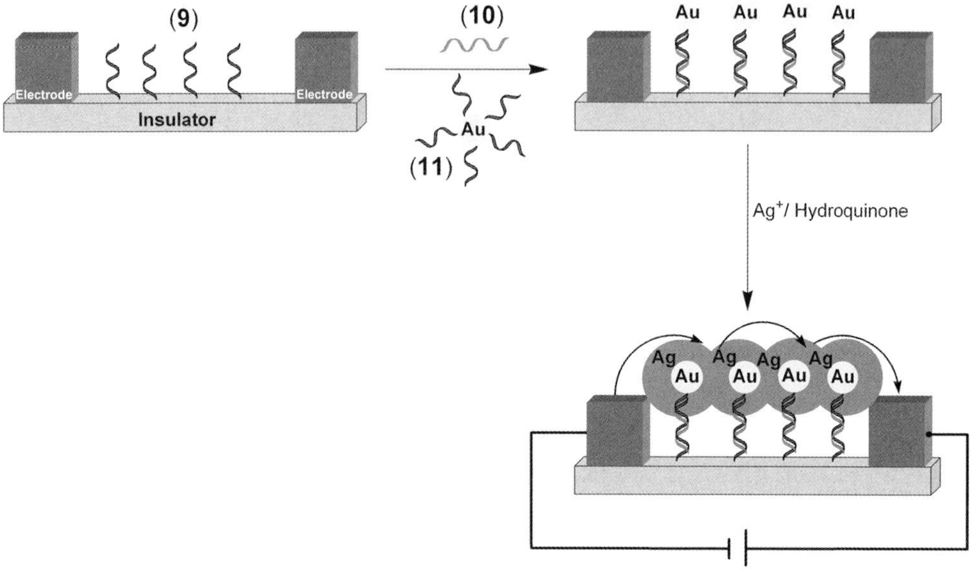

Figure 6.7 The use of a DNA–Au NP conjugate and subsequent silver deposition to connect two microelectrodes, as a means of sensing a DNA analyte.

DNA (step C). The subsequent dissolution and electrochemical stripping of the dissolved silver clusters (step D) then provide the route to detect the hybridized DNA.

The catalytic features of metal nanoparticles permit the subsequent electroless deposition of metals on the nanoparticle clusters associated along the DNA and the formation of enlarged, electrically interconnected, nanostructered wires. The formation of conductive domains as a result of biorecognition events then provides an alternative path for the electrical transduction of biorecognition events. This was exemplified by the design of a DNA detection scheme by using microelectrodes fabricated on a silicon chip [62] (Figure 6.7). The method relied on the generation of a DNA–Au NP sandwich assay within the gap separating two microelectrodes. A probe nucleic acid (**9**) was immobilized in the gap separating the microelectrodes. The target DNA (**10**) was then hybridized with the probe interface and, subsequently, the nucleic acid (**11**)-functionalized Au nanoparticles were hybridized with the free 3′-end of the target DNA, followed by silver enhancement of the Au NP labels. Catalytic deposition of silver on the gold nanoparticles resulted in electrically interconnected particles, exhibiting low resistance between the electrodes. The low resistances between the microelectrodes were controlled by the concentration of the target DNA, and the detection limit for the analysis was estimated to be about 5×10^{-13} M. A difference of 10^6 in the gap resistance was observed upon analyzing by this method the target DNA and its mutant. A related conductivity immunoassay of proteins was developed, based on Au NPs and silver enhancement [63].

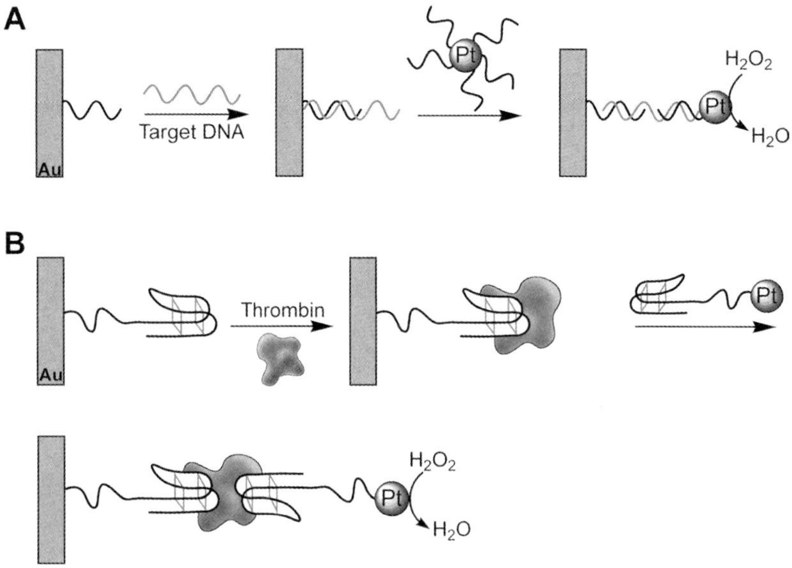

Figure 6.8 (A) Amplified electrochemical analysis of a DNA by nucleic acid-functionalized Pt NPs acting as electrocatalysts for the reduction of H_2O_2. (B) Amplified electrochemical analysis of thrombin by an aptamers monolayer-functionalized electrode and an aptamer-functionalized Pt NP labels as electrocatalysts for the reduction of H_2O_2.

A further approach for the amplified detection of DNA or proteins by employing metal nanoparticle (Pt NP) labels as catalysts for the reduction of H_2O_2 was described [64]. Nucleic acid-functionalized Pt NPs act as catalytic labels for the amplified electrochemical detection of DNA hybridization and aptamer/protein recognition events. Hybridization of the nucleic acid-modified Pt NPs to the nucleic acid–analyte DNA complex associated with an electrode permits the amperometric, amplified, detection of the DNA through the Pt NP electrocatalyzed reduction of H_2O_2 with a sensitivity limit of 1×10^{-11} M [Figure 6.8(A)]. Similarly, the association of the aptamer-functionalized Pt NPs to a thrombin aptamer–thrombin complex associated with the electrode allowed the amplified, electrocatalytic detection of thrombin with a sensitivity limit corresponding to 1×10^{-9} M [Figure 6.8(B)].

6.4
Metal Nanoparticles as Microgravimetric Labels

Nanoparticles provide a "weight label" that may be utilized for the development of microgravimetric sensing methods [quartz crystal microbalance (QCM)] that are ideal for the detection of biorecognition events. Also, the catalytic properties of metallic NPs may be employed to deposit metals on the NP-functionalized biorecognition complexes, thus allowing enhanced mass changes on the transducers (piezoelectric crystals) and amplified biosensing. For a quartz piezoelectric crystal (AT-cut), the crystal resonance frequency changes by Δf when a mass change Δm occurs on the crystal according to the Sauerbrey equation [65]:

$$\Delta f = -2f_0^2[\Delta m/A(\mu_q\rho_q)^{1/2}] \tag{6.1}$$

where f_0 is the fundamental frequency of the quartz crystal, Δm is the mass change, A is the piezoelectrically active area, ρ_q is the density of quartz (2.648 g cm^{-3}) and μ_q is the shear modulus (2.947×10^{11} dyn cm^{-2} for AT-cut quartz). Thus, any mass changes of the piezoelectric crystals are accompanied by a change in the resonance frequency of the crystal.

The microgravimetric QCM method was applied for the amplified detection of DNA using nucleic acid-functionalized Au NPs as "weight labels" [66–68]. A target DNA molecule (**13**) was hybridized to an Au–quartz crystal that was modified with a probe oligonucleotide (**12**) and the **14**-functionalized Au NPs were hybridized to the 3′-end of the duplex DNA associated with the crystal (Figure 6.9). The subsequent secondary dendritic amplification was achieved by the interaction of the resulting interface with the target DNA (**13**) that was pretreated with the (**12**)-functionalized Au NP [67, 69]. Concentrations of DNA (**13**) as low as 1×10^{-10} M could be detected by the amplification of the target DNA by the nucleic acid-functionalized Au NP labels.

Also, the detection of DNA using nucleic acid-functionalized Au NPs and catalytic metal deposition on the NPs labels was reported [70, 71]. The Au nanoparticles act as catalytic "seeds" and catalyze the reduction of $AuCl_4^-$ and the deposition of gold on the Au NPs. Thus, the catalytic enlargement of the nanoparticles increased the mass associated with the piezoelectric crystal and provided an active amplification route for the amplified microgravimetric detection of the DNA. For example, Figure 6.10(A) depicts the amplified detection of the 7249-base M13mp18 DNA by using the catalytic

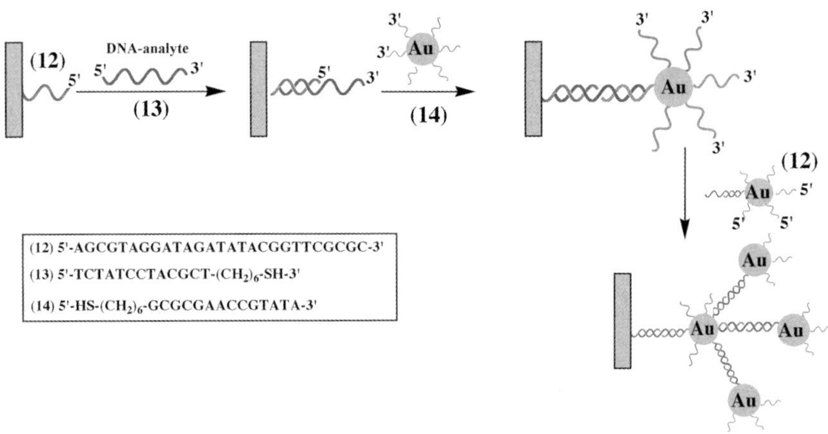

Figure 6.9 Dendritic amplified DNA sensing by the use of oligonucleotide-functionalized Au NPs, which are assembled on a quartz crystal microbalance (QCM) Au–quartz crystal (Reproduced from [67] by permission of The Royal Society of Chemistry).

deposition of gold on an Au NP conjugate [71]. The DNA primer (**15**) was assembled on an Au–quartz crystal. After hybridization with M13mp18 DNA (**16**), the double-stranded assembly was replicated in the presence of a mixture of nucleotides (deoxynucleotide triphosphates, dNTP mixture) that included dATP, dGTP, dUTP, biotinylated dCTP (B-dCTP) and polymerase (Klenow fragment). The resulting biotin-labeled replica was then treated with the streptavidin–Au NP conjugate (Sav–Au NP) (**17**), and the resulting Au-labeled replica was subjected to the Au NP-catalyzed deposition of gold by the NH_2OH-stimulated reduction of $AuCl_4^-$. The replication process represents the primary amplification step as it increases the mass associated with the crystal and simultaneously generates a high number of biotin labels for the association of the Sav-Au NP. The binding of the conjugate represents the secondary amplification step for the analysis of M13mp18 DNA. The third step, which involves the catalyzed precipitation of the metal, led to the greatest amplification in the sensing process as a result of the increase in the mass of the Au NPs. This method enabled the M13mp18 DNA to be sensed with a detection limit of $\sim 1 \times 10^{-15}$ M.

This amplification method was also applied for the analysis of a single-base mismatch in DNA as depicted in Figure 6.10(B) [70, 71]. This was exemplified with the analysis of the DNA mutant **18a**, which differs from the normal gene (**18**) by the substitution of a G base with an A base. The analysis of the mutant was performed by the immobilization of the probe DNA (**19**), which is complementary to the normal gene (**18**) and also to the mutant (**18a**) (up to one base prior to the mutation site), on the Au–quartz crystal. Hybridization of the normal gene or the mutant with this probe interface, followed by the reaction of the hybridized surfaces with biotinylated dCTP (B-dCTP) in the presence of polymerase (Klenow fragment), led to the incorporation of the biotin-labeled base only into the assembly that included the mutant **18a**. The subsequent association of the Sav–Au NP conjugate **17** followed by the catalyzed deposition of gold on the Au NPs amplified the analysis of the single-base mismatch in **18a**. Figure 6.10(C), curve a, shows the microgravimetric detection of the mutant **18a** revealing a frequency change of $\Delta f = -700$ Hz upon analyzing (**18a**), 3×10^{-9} M. The normal gene (**18**) does not alter the frequency of the crystal [Figure 6.10(C), curve b]. The mutant could be detected with a detection limit of 3×10^{-16} M.

Microgravimetric detection method was further extended to analyze proteins by aptamer-functionalized surfaces, using metal NPs as "weight labels" and as catalytic sites for enlargement of nanoparticles and the amplified sensing of the proteins. The use of metal nanoparticle labels for the amplified analysis of thrombin was reported [72]. The fact that thrombin consists of a dimer with two binding sites for aptamers [73, 74] allowed the application of aptamer-functionalized Au NPs for developing a microgravimetric QCM aptasensors for thrombin (Figure 6.11). The thiolated aptamer **20** was linked to an Au–quartz crystal. The interaction of the **20**-modified Au–quartz crystal with thrombin and the subsequent association of the aptamer **20**-functionalized Au NPs provide a primary amplification of the analysis of thrombin by applying the Au NPs as a "weight label". The secondary catalytic enlargement of the Au nanoparticles by the particle-catalyzed reduction of $AuCl_4^-$ by 1,4-dihydronicotinamide adenine dinucleotide (NADH) [75] provided a further

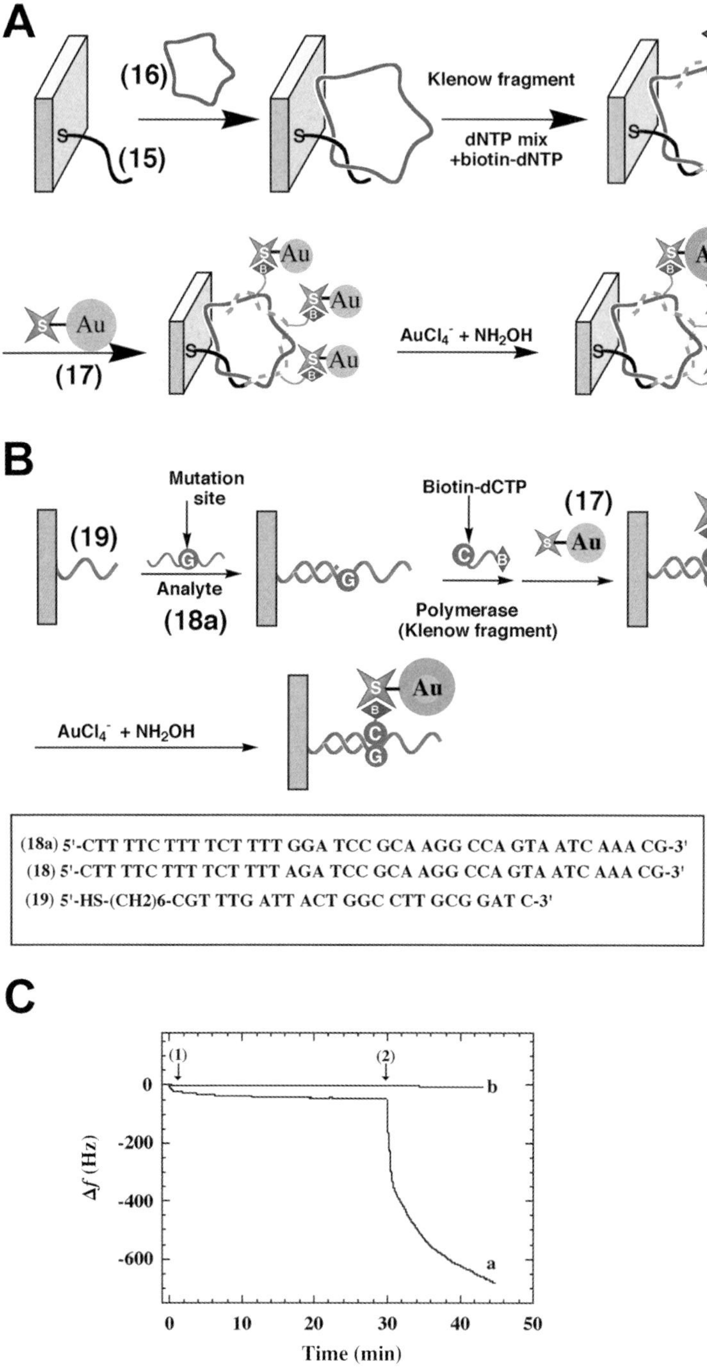
A
(16)
(15)
Klenow fragment
dNTP mix
+biotin-dNTP
(17)
AuCl4- + NH2OH
B
(19)
Mutation site
Analyte
(18a)
Biotin-dCTP
(17)
Polymerase
(Klenow fragment)
AuCl4- + NH2OH
(18a) 5'-CTT TTC TTT TCT TTT GGA TCC GCA AGG CCA GTA ATC AAA CG-3'
(18) 5'-CTT TTC TTT TCT TTT AGA TCC GCA AGG CCA GTA ATC AAA CG-3'
(19) 5'-HS-(CH2)6-CGT TTG ATT ACT GGC CTT GCG GAT C-3'
C
(1)
(2)
Δf (Hz)
Time (min)
a
b

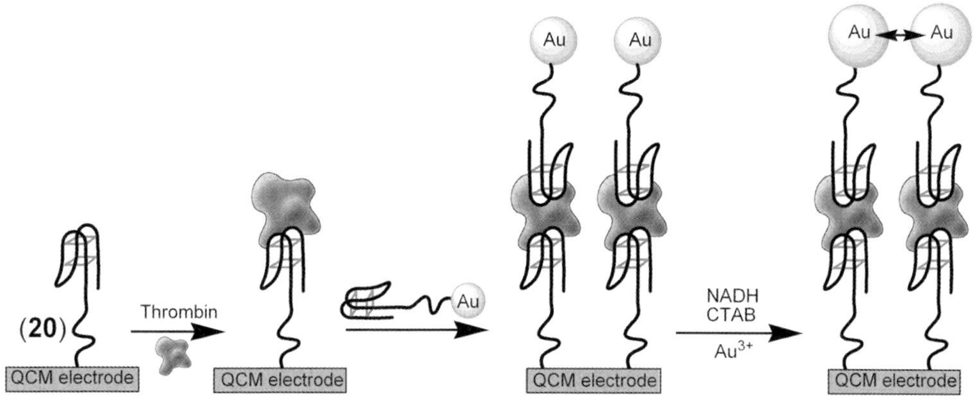

(20) 5'-HS$(CH_2)_6$TTTTTTTTTTTTTTTGGTTGGTGTGGTTGG-3'

Figure 6.11 Microgravimetric analysis of thrombin by an aptamer (**20**)-functionalized Au–quartz crystal and the aptamer-functionalized Au NP as label and the subsequent enlargement of the NPs by the NADH-induced reduction of Au^{3+} ions. The enlarged NPs act as "weight labels" for the amplified detection of thrombin.

amplification for the sensing of thrombin. Upon analyzing thrombin at a concentration of 2×10^{-9} M, the primary amplification step resulted in a frequency change of -30 Hz, whereas the secondary amplification step altered the crystal frequency by -900 Hz.

6.5 Semiconductor Nanoparticles as Electrochemical Labels for Biorecognition Events

Other inorganic nanoparticle composites such as semiconductor NPs (or quantum dots) have been employed as labels (that substitute metallic NPs) for the amplified electrochemical detection of proteins or DNA [16, 43, 44]. The use of different semiconductor NPs allowed the parallel analysis of different targets, using the NPs as codes for different analytes. For example, CdS semiconductor nanoparticles modified with nucleic acids were employed as labels for the detection of hybridization

Figure 6.10 (A) Amplified detection of the 7249-base M13mp18 DNA (**16**) using the catalytic deposition of gold on an Au nanoparticle conjugate. (B) Analysis of a single-base mismatch in DNA (**18a**) using the catalytic deposition of gold on an Au nanoparticle conjugate. (C) Microgravimetric detection of a single-base mutant (**18a**) enhanced by the catalytic deposition of gold on an Au nanoparticle conjugate. The frequency responses were observed with a mutant DNA (**18a**) (a) and with a normal DNA (**18**) (b). Arrow (1) shows the attachment of the Sav–Au NP conjugate (**17**) and arrow (2) shows catalytic deposition of gold on the Au NPs. (Reproduced from [71] by permission of The Royal Society of Chemistry).

events of DNA [76]. Dissolution of the CdS (in the presence of 1 M HNO_3) followed by the electrochemical reduction of the Cd^{2+} to Cd^0 accumulated the metal on the electrode. The subsequent stripping off of the generated Cd^0 (to Cd^{2+}) provided the electrical signal for the DNA analysis. This method was further developed by using magnetic particles functionalized with probe nucleic acids as sensor units that hybridize with the analyte DNA and the nucleic acid-functionalized CdS NPs labels that hybridize with the single-strand domain of the analyte DNA and trace the primary formation of the probe–analyte double-stranded complex. The magnetic separation of the magnetic particle–CdS NP aggregates crosslinked by the analyte DNA, followed by dissolution of the CdS and electrochemical collection and stripping off of the Cd metal, provide the amplified electrochemical readout of the analyte DNA. In fact, this system combined the advantages of magnetic separation of the tracers CdS NPs associated with the DNA recognition events with the amplification features of the electrochemical stripping method. Highly sensitive detection of DNA is accomplished by this method (detection limit 100 fmol, reproducibility = RSD 6%) [76].

By using different semiconductor NPs as labels, the simultaneous and parallel analysis of different antibodies or different DNAs was accomplished. A model system for multiplexed analysis of different nucleic acids with semiconductor NPs was developed [77]. Three different kinds of magnetic particles were modified by three different nucleic acids (**21a–c**) and subsequently hybridized with the complementary target nucleic acids (**22a–c**). The particles were then hybridized with three different kinds of semiconductor nanoparticles, ZnS, CdS, PbS, that were functionalized with nucleic acids (**23a–c**) complementary to the target nucleic acids associated with the magnetic particles [Figure 6.12(A)]. The magnetic particles allowed the easy separation and purification of the analyte samples, whereas the semiconductor particles provided nonoverlapping electrochemical readout signals that transduced the specific kind of hybridized DNA. Stripping voltammetry of the respective semiconductor nanoparticles yielded well-defined and resolved stripping waves, thus allowing simultaneous electrochemical analysis of several DNA analytes. The same strategy was also applied for the multiplexed immunoassay of proteins [78], with simultaneous analysis of four antigens. The arsenal of inorganic labels for the parallel multiplexed analysis of biomolecules and the level of amplification were further extended by using other metal sulfide composite nanostructures. For example, InS nanorods provided an additional resolvable voltammetric wave, while the nanorods configuration of the label increased the amplification efficiency due to the higher content of stripped-off metal from the nanorod configuration as compared with a spherical NP structure [79].

This method to encode biomolecular identity by semiconductor NPs was extended for the parallel analysis of different proteins by their specific aptamers [80]. An Au electrode was functionalized with aptamers specific for thrombin and lysozyme [Figure 6.12(B)]. Thrombin and lysozyme were labeled with CdS and PbS NPs, respectively, and the NP-functionalized proteins acted as tracer labels for the analysis of the proteins. The NP-functionalized proteins were linked to the respective aptamers and subsequently interacted with the nonfunctionalized thrombin or lysozyme. The competitive displacement of the respective labeled proteins associated with the surface by the analytes, followed by dissolution of the metal sulfides associated with the

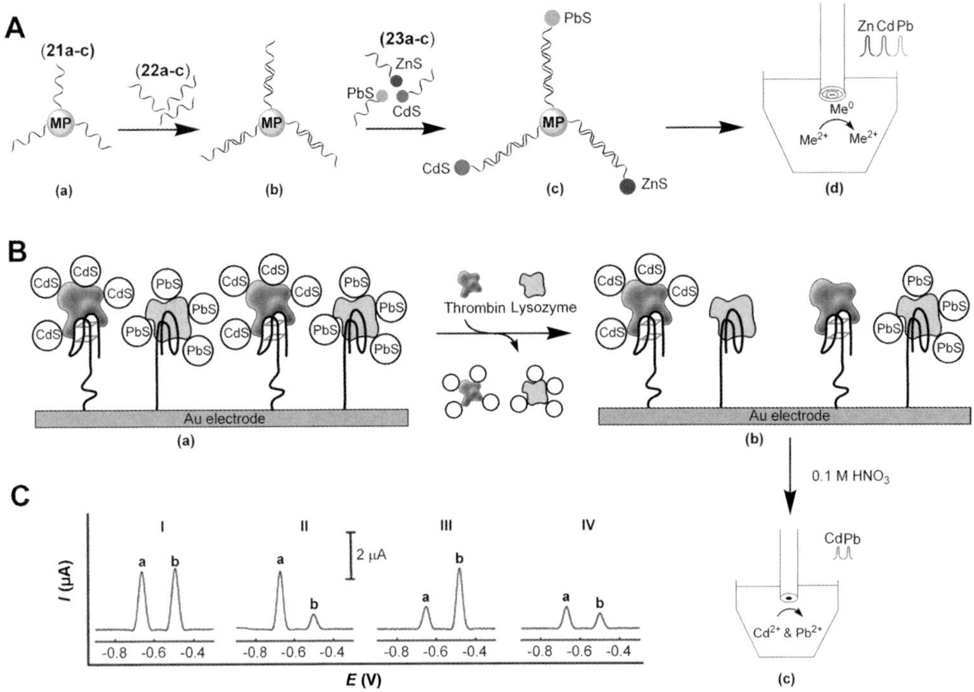

Figure 6.12 (A) Parallel electrochemical analysis of different DNAs using magnetic particles functionalized with probes for the different DNA targets and specific nucleic acid-functionalized metal sulfides as tracers. (B) Simultaneous electrochemical analysis of the two proteins, thrombin and lysozyme, using a competitive assay, where thrombin modified with CdS QDs and lysozyme modified with PbS QDs are used as tracers. (C) Square-wave stripping voltammograms corresponding to the simultaneous detection of lysozyme (a) and thrombin (b): (I) no (a), no (b); (II) 1 $\mu g L^{-1}$ (a), no (b); (III) no (a), 0.5 $\mu g L^{-1}$ (b); (IV) 1 $\mu g L^{-1}$ (a), 0.5 $\mu g L^{-1}$ (b). (Reprinted with permission from [80]. Copyright 2006 American Chemical Society).

surface and the detection of the released ions by electrochemical stripping, then allowed the quantitative detection of the two proteins [Figure 6.12(C)]. This method was further extended for coding unknown single polymorphisms (SNPs) using different encoding QDs [81]. This protocol relied on the ZnS, CdS, PbS and CuS NPs modified with four different mononucleotides and the application of the NPs to construct different combinations for specific SNPs, that yielded a distinct electronic fingerprints for the mutation sites.

6.6 Metal Nanoparticles as Optical Labels for Biorecognition Events

The unique size-controlled optical properties of the metallic NPs reflected by intense localized plasmon excitons [1, 2, 82] turn the NPs into powerful optical tags for biorecognition processes. Furthermore, the electronic interactions of the localized plasmon with other plasmonic waves allow one not only to develop new optical amplification paths for biosensing, but also the use these electronic coupling phenomena to follow dynamic processes associated with biorecognition events. For example, surface plasmon resonance (SPR) is a common technique for following biorecognition events at metallic surfaces [83–85]. The changes in the dielectric properties of the metallic surfaces and the changes in the thickness of the dielectric films associated with the metallic surfaces alter the resonance features of the surface plasmon wave and provide the basis for SPR biosensors. The electronic coupling between an Au NP conjugated to the biorecognition complex and the plasmon wave may lead to amplification of the detection processes [86].

A colorimetric detection method for nucleic acids is based on the distance-dependent optical properties of DNA-functionalized Au NPs. The aggregation of Au NPs leads to a red shift in the surface plasmon resonance of the Au NPs as a result of an interparticle coupled plasmon exciton. Thus, the hybridization-induced aggregation of DNA-functionalized Au NPs changes the color of the solution from red to blue [3]. Changes in the optical properties of the Au NPs upon their aggregation provide a method for the sensitive detection of DNA and provide a way to design optical DNA biosensors [87–90]. Specifically, two batches of 13-nm diameter Au NPs were separately functionalized with two thiolated nucleic acids that acted as labels for the detection of the analyte DNA [Figure 6.13(A)]. Each of the NP labels is modified with nucleic acid complementary to the two ends of the analyte DNA. Since each of the nucleic acid-functionalized Au NPs includes many modifying oligonucleotides, the addition of the target DNA to a solution of the two DNA-functionalized Au NPs resulted in the crosslinking and aggregation of the nanoparticles through hybridization. Aggregation changed the color of the solution from red to purple as a result of interparticle coupled plasmon absorbance [Figure 6.13(B)]. The aggregation process was found to be temperature-dependent, and the aggregated Au NPs can reversibly dissociate upon elevation of the temperature through the melting of the double strands and reassociate with a decrease in the temperature through the rehybridization process, which results in the reversible changes of the spectrum [91]. These melting transitions, aggregation and deaggregation, occur in a narrow temperature range [Figure 6.13(C)] and allow the design of selective assays for DNA targets and high discrimination of the mismatched targets. The color changes in the narrow temperature range lies in the background of the simplest test to follow the aggregation of Au NPs, the "Northwestern" spot test [87]. This is an extremely sensitive method to discriminate between aggregated and nonaggregated gold NPs in aqueous solutions, and it relies on a detectable color change from red to blue upon aggregation. The test consists of spotting of a droplet of an aqueous solution of particles on a reversed-phase thin-layer chromatographic plate. A blue spot indicates aggregation in the presence of the target DNA, whereas a red spot indicates the presence of freely dispersed particles [Figure 6.13(C)]. The sharp melting transitions of DNA-functionalized gold nanoparticles were applied to discriminate the target DNA from DNA with single-base-pair mismatches simply by following the changes in the nanoparticle absorption as a function of temperature [87, 88]. The melting properties of DNA-linked nanoparticle aggregates are affected by a number of factors, which include DNA

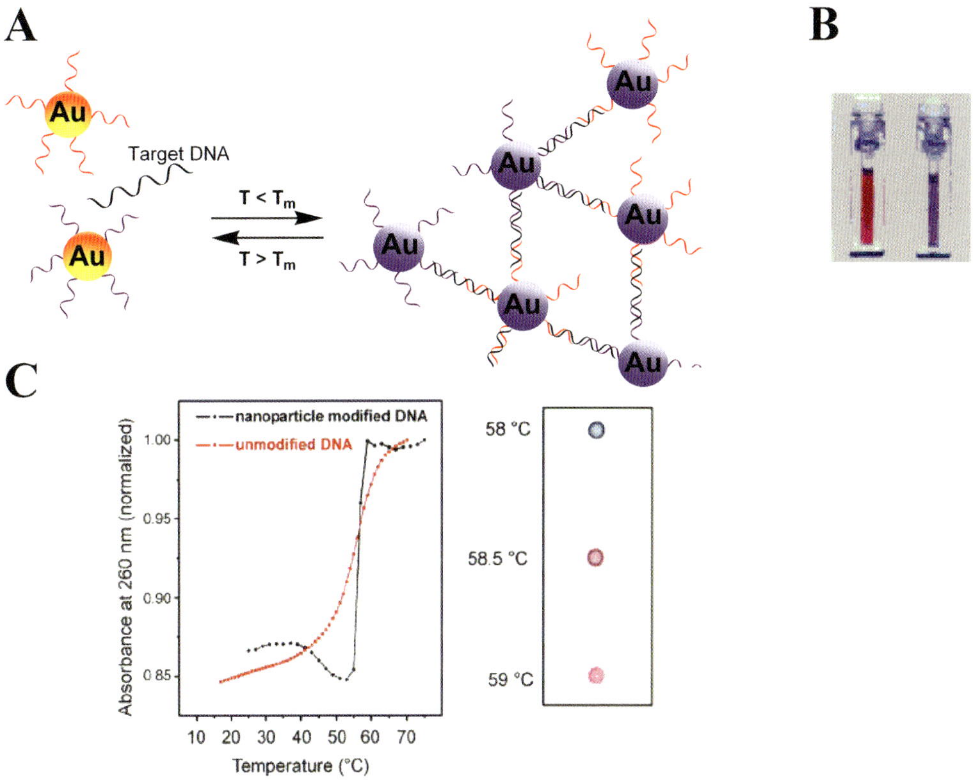

Figure 6.13 Optical detection of a target DNA through the hybridization of two kinds of nucleic acid-functionalized Au NPs complementary to the ends of the target DNA. The hybridization leads to the aggregation of the NPs (A) and to a red-to-purple color transition (B). The deaggregation of the nanoparticles is stimulated by the melting of the crosslinked DNA duplexes (C). Part (B) (Reprinted with permission from [15]. Copyright 2005 American Chemical Society). Part (C) (Reprinted from [87] with permission from AAAS).

surface density of the modifying nucleic acids, nanoparticle size, interparticle distances and salt concentration [91].

Recently, the use of DNA-based machines for the amplified detection of DNA was developed [92, 93]. The aggregation of Au NPs was used as a readout signal that follows the operation of the machine and the detection of the respective DNA [94] [Figure 6.14 (A)]. The machine consists of a nucleic acid "track" (**24**) that includes three domains, I, II and III. Domain I acts as the recognition site, and hybridization with the target DNA (**24a**) triggers, in the presence of polymerase and the dNTP mixture, the replication of the DNA track. Formation of the duplex, and specifically the formation of the duplex region II, generates the scission site for nicking enzyme Nb BbvC I. The enzyme-induced scission of the duplex activates the autonomous operation of the machine, where the replication and strand displacement of the complementary nucleic acid of region III proceed continuously. The displaced nucleic acid **25** may be considered as the "waste product". In the presence of two kinds of nucleic acids **26**- and **27**-

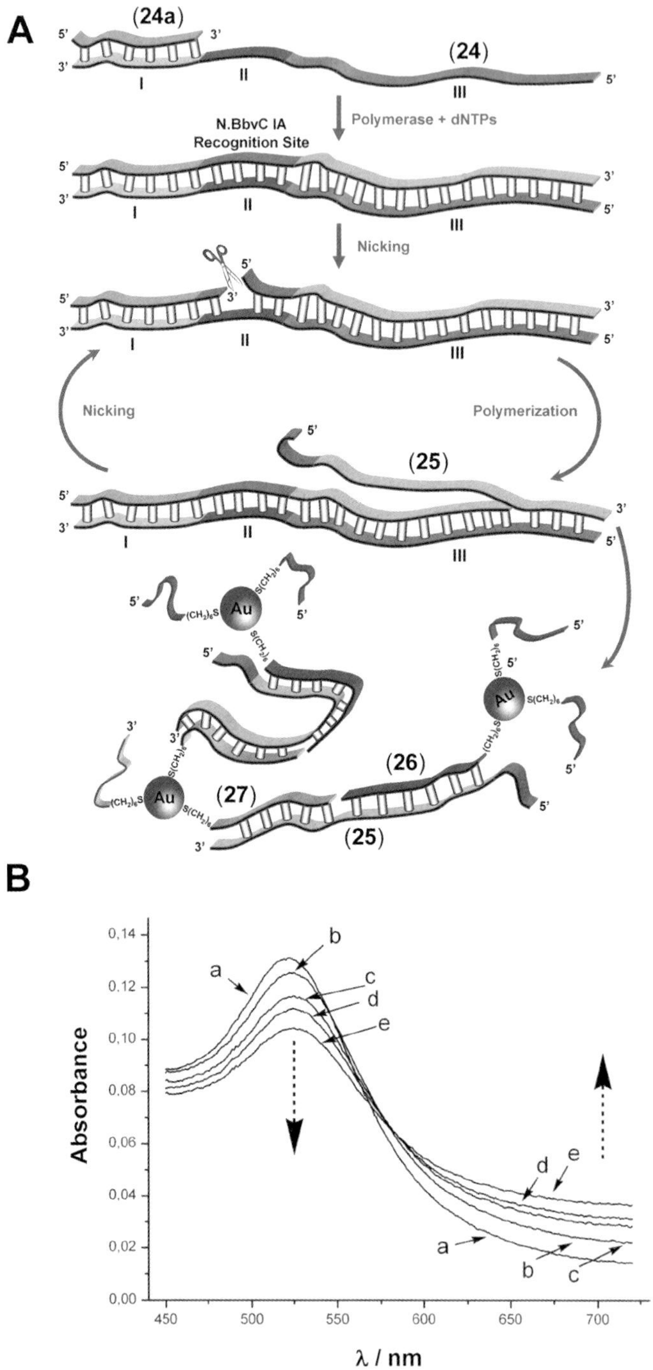
A
(24a)
(24)
Polymerase + dNTPs
N.BbvC IA
Recognition Site
Nicking
Nicking
Polymerization
(25)
(26)
(27)
(25)
Au
B
Absorbance
λ / nm

functionalized Au NPs, which are complementary to the two ends of the "waste product", aggregation of Au NPs proceeds. The color changes as readout method of aggregation of the Au NPs are depicted in Figure 6.14(B). The method allowed the optical detection of the target DNA with a sensitivity that corresponded to 1×10^{-12} M.

A different metal NP-induced analysis of DNA through the aggregation of the NPs employed the salt effect on the double layer potential of the NPs [95, 96]. Au NPs (15 nm) were functionalized with a probe nucleic acid, and the effect of the addition of the salt (up to 2.5 M) on the stability of the modified particles, as compared with unmodified (bare) NPs was examined. While the nucleic acid-functionalized Au NPs revealed stability in the presence of added salt, the nonfunctionalized Au NPs precipitated at a salt concentration of 0.1 M. The probe-functionalized Au NPs, when hybridized with the complementary target DNA, revealed a rapid red to purple color transition (<3 min) upon addition of 0.5 M NaCl. This color change was attributed to the lowering of the Au NP surface potential upon addition of the salt, which resulted in a decrease in the electrostatic repulsive interactions between the particles and consequently to shorter interparticle distances and a coupled plasmon absorbance.

The effect of salt on the stability of unmodified Au NP upon interaction with a probe nucleic acid before and after hybridization was used for the colorimetric detection of specific sequences in amplified genomic DNA [97, 98]. The method relied on the different effects of single- and double-stranded DNA on unmodified citrate-coated Au NPs. The adsorption of short ss-DNA probes on Au NPs stabilizes the citrate-coated NPs against salt-induced aggregation. The exposure of unmodified gold nanoparticles to a saline mixture containing amplified genomic DNA and short ss-DNA complementary to the regions in genomic DNA resulted in the aggregation of Au NPs and a color change from red to blue. If the short oligomers were not complementary to the regions in the genomic DNA, no color change occurred due to the stabilization of Au NPs by these oligomers. The method permitted the sequence-specific detection of label-free oligonucleotides at the level of 100 fmol and was adapted to detect single-base mismatches.

The hybridization-induced aggregation of metallic NPs was extended to analyze ions and small molecules using aptamers and DNAzymes. Aptamers are nucleic acids with specific recognition properties towards small molecules or proteins. They are prepared by the Systematic Evolution of Ligands by Exponential Enrichment (SELEX) procedure, which involves the selection and amplification of a sequence-specific nucleic acid to a target molecule from a library of 10^{15}–10^{16} nucleic acids [99–101]. Similarly, catalytic nucleic acids (DNAzymes or ribozymes) are prepared by eliciting

Figure 6.14 (A) Analysis of a DNA target (**24a**), by a nucleic acid "track" (**24**), consisting of regions I, II and III. The replication followed by nicking of the primary duplex (**24a**)/(**24**) yields displacement of the "waste product" (**25**). The displaced product (**25**) stimulates the aggregation of (**26**)- and (**27**)-functionalized Au NPs that are complementary to the two ends of (**25**). (B) Spectral changes upon analyzing different concentrations of the target DNA (**24a**) through the aggregation of the Au NPs by the "waste product" (**25**) generated by the DNA machine operated for a fixed time interval of 120 min: (a) 0, (b) 1×10^{-6}, (c) 1×10^{-7}, (d) 1×10^{-8} and (e) 1×10^{-9} M. (Reproduced with permission from [94]. Copyright 2007 Wiley-VCH).

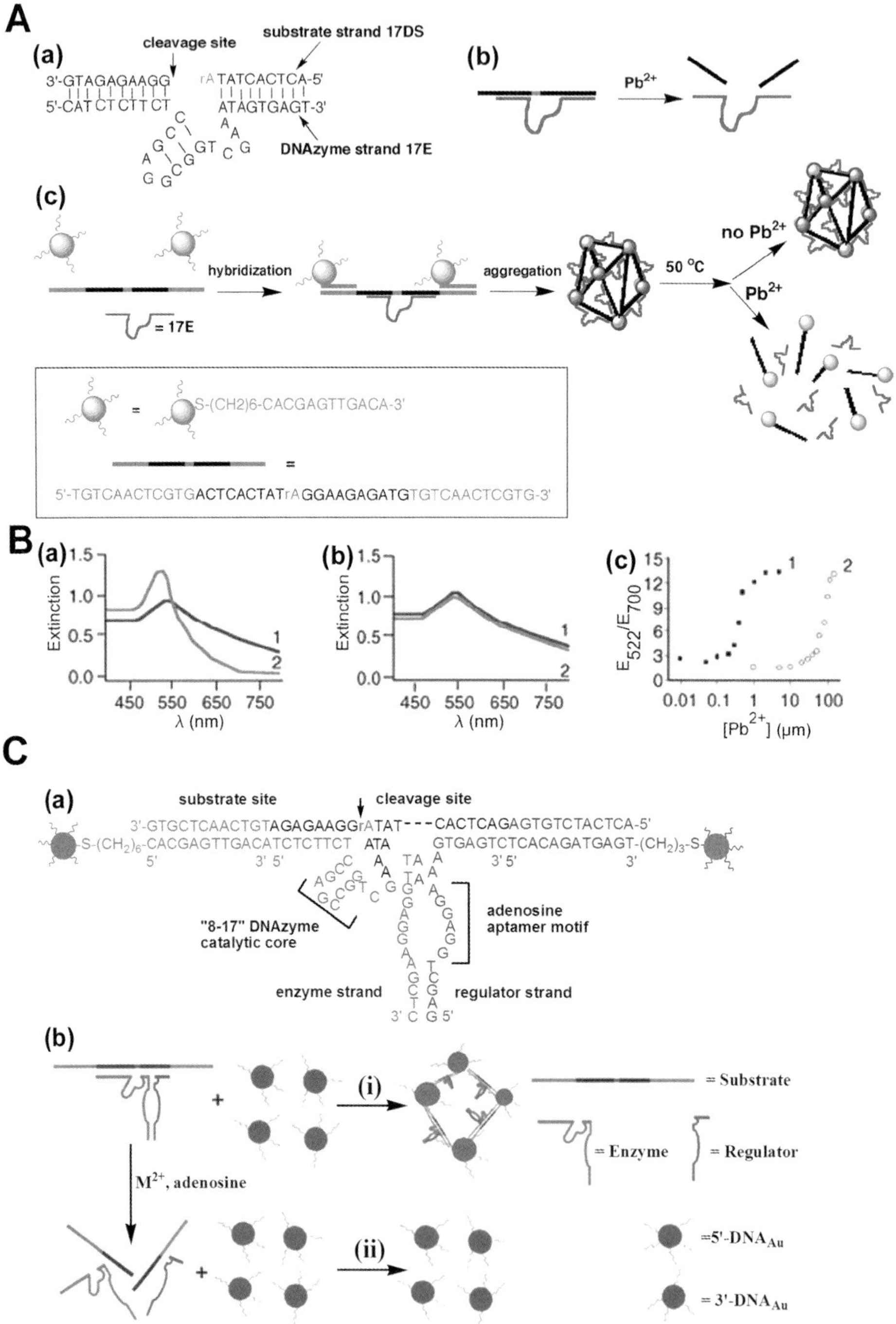

A
(a)
cleavage site
substrate strand 17DS
3'-GTAGAGAAGG rA TATCACTCA-5'
5'-CATCTCTTCT ATAGTGAGT-3'
DNAzyme strand 17E
(b)
Pb2+
(c)
hybridization
aggregation
50 °C
no Pb2+
Pb2+
= 17E
S-(CH2)6-CACGAGTTGACA-3'
5'-TGTCAACTCGTGACTCACTATrAGGAAGAGATGTGTCAACTCGTG-3'
B
(a)
(b)
(c)
Extinction
λ (nm)
E522/E700
[Pb2+] (μm)
C
(a)
substrate site
cleavage site
3'-GTGCTCAACTGTAGAGAAGGrATAT---CACTCAGAGTGTCTACTCA-5'
S-(CH2)6-CACGAGTTGACATCTCTTCT ATA
GTGAGTCTCACAGATGAGT-(CH2)3-S
"8-17" DNAzyme catalytic core
adenosine aptamer motif
enzyme strand
regulator strand
(b)
(i)
(ii)
M2+, adenosine
= Substrate
= Enzyme
= Regulator
=5'-DNAAu
= 3'-DNAAu

nucleic acids by *in vitro* selection towards transition-state analogues of chemical reactions [102] or by the selection of nucleic acids with binding affinities to metal ions [103–105] or active site analogues such as a heme [106, 107]. For example, the "8–17" DNAzyme has demonstrated high activity and specificity towards Pb^{2+} ions and revealed catalytic activity towards the specific scission of the complementary nucleic acid that included the respective cleavage site [108, 109]. Au NPs were used as optical labels for the detection of Pb^{2+} ions based on the activity of the Pb^{2+}-dependent DNAzyme that effects the separation of Au NP aggregates [110] (Figure 6.15). The method used the "8–17" DNAzyme that reveals high activity and specificity toward Pb^{2+} ions. The system consists of a substrate strand **17**DS that includes the cleavage site and is complementary to enzyme strand **17**E, which recognizes Pb^{2+} ions. The enzyme strand revealed catalytic activity in the presence of Pb^{2+} ions and resulted in scission of the substrate strand at the cleavage site. The complementary substrate strand of the DNAzyme was elongated with nucleic acid residues that are complementary to secondary nucleic acids labeled with Au NPs [Figure 6.15(A)]. Mixing of all three components of the DNAzyme – the enzyme strand, the complementary nucleic acid substrate and the nucleic acid-labeled Au NPs – resulted in the hybridization of the respective components, the aggregation of the Au NPs [Figure 6.15(A), part (c)] and the appearance of blue color, which indicated interparticle-coupled plasmon absorbance [Figure 6.15(B), part (a), curve 1]. The addition of Pb^{2+} ions resulted in the DNAzyme-assisted cleavage of the substrate strand, and this led to the separation of the Au NP aggregates and to a red color corresponding to the deaggregated Au NPs [Figure 6.15(B), part (a), curve 2]. This method allowed the colorimetric detection of Pb^{2+} ions in the concentration range 0.1–4 μM [Figure 6.15(B), part (c), curve 1]. The function of DNAzyme as an active component for the detection of Pb^{2+} ions was confirmed by using an inactive DNAzyme (**17**Ec) with poor scission activity of the substrate strand in the presence of Pb^{2+} ions [Figure 6.15(B), part (b), curves 1 and 2] and with low sensitivity [Figure 6.15(B), part (c), curve 2].

Figure 6.15 (A) DNAzyme system for the analysis of metal ions. (a) Secondary structure of the "8–17" DNAzyme system that consists of the enzyme strand **17**E and a substrate strand **17**DS. Except for a ribonucleoside adenosine (rA) at the cleavage site, all other nucleosides are deoxyribonucleosides; (b) cleavage of **17**DS by **17**E in the presence of Pb^{2+} ions; (c) DNAzyme-directed assembly of the oligonucleotide-functionalized Au NPs and the application of the assemblies for Pb^{2+} sensing. (B) UV–visible absorbance spectra of (a) the active **17**E DNAzyme–NP sensor and (b) an inactive **17**Ec DNAzyme–NP sensor; the spectra were recorded in the absence (curve 1) or presence (curve 2) of Pb^{2+} ions (5 μM); (c) calibration plots for the analysis of Pb^{2+} ions, in which the enzyme strand is the active **17**E only (curve 1) and when the **17**E:**17**Ec ratio is 1:20 (curve 2). (C) Colorimetric detection of adenosine monophosphate by a DNAzyme–aptamer conjugate, through the deaggregation of Au NPs aggregates: (a) the construct of the blocked Pb^{2+}-stimulated nucleotide cleaving DNAzyme tethered to the anti-adenosine aptamer that bridges Au NPs; (b) schematic colorimetric detection of adenosine by the DNAzyme–aptamer construct: (i) aggregated structure of blocked DNAzyme-aptamer construct; (ii) adenosine-induced release of the aptamer unit from the blocked construct followed by the activation of the DNAzyme, cleavage of the bridging units and deaggregation of the NPs. (Parts A and B reprinted with permission from [110]. Copyright 2003 American Chemical Society. Part C reprinted with permission from [111]. Copyright 2004 American Chemical Society).

The DNAzyme-controlled aggregation of Au NPs was expanded to a broader range of analytes, and a colorimetric biosensor for adenosine was tailored on the basis of the aptazyme-directed assembly of gold NPs [111]. The aptazyme is based on an "8–17" DNAzyme that is allosterically activated by the adenosine aptamer [Figure 6.15(C), part (a)]. In the absence of adenosine, an inactive aptazyme is formed, and the substrate strand operates as a linker for the assembly of a blue colored aggregate of 13-nm Au NPs [Figure 6.15(C), part (b), route (i)]. In the presence of adenosine, however, the aptazyme is activated and the substrate strand is cleaved [Figure 6.15(C), part (b), route (ii)]. This prevents the aggregation of the Au NPs, which results in a red-colored system characteristic of the individual nonaggregated nanoparticles. Concentrations of up to 1 mM of adenosine could be semiquantitatively analyzed by the extent of blue-to-red color changes or quantitatively measured by the ratio of the absorbance values at 520 and 700 nm. The addition of guanosine, cytidine or uridine (all 5 mM) instead of adenosine did not affect the aggregation of the DNA-bound Au NPs, indicating the specificity of the system towards adenosine.

Aptamer- and oligonucleotide-induced aggregation processes of Au NPs were used to develop colorimetric sensors for low molecular weight substrates such as adenosine and cocaine [112]; for example, a nucleic acid strand (**28**) that includes the specific anti-adenosine aptamer sequence and tether units I and II complementary to two kinds of **29**- and **30**-functionalized Au NPs. The mixture of (**28**) with the functionalized Au NPs resulted in the aggregation of the Au NPs through bridging of the NPs by hybridization of **28** with the functionalized NPs (Figure 6.16). In the presence of added adenosine, the adenosine–aptamer complex folds to a configuration that destabilizes the double-stranded structure, resulting in deaggregation of the particles and a purple-to-red color transition. The method allowed the detection of adenosine in the concentration range 0.3–2 mM. The generality of this method was further demonstrated by the construction of a colorimetric sensor for cocaine based on a specific cocaine aptamer [112].

The aggregation of the Au NPs is not limited to DNA detection and the process was applied to develop optical biosensing assays for sensing enzyme activities. For example, the detection of proteases (thrombin and letal factor) by the enzyme-induced cleavage of peptides was reported [113]. In this protocol, acetylated cysteine residues were added to two termini of the peptide substrate. As the cysteine units bind to Au NPs, the intact peptide bridged the NPs, resulting in their aggregation and the formation of a violet–blue color. The pretreatment of the modified peptide with target protease led to the cleavage of the peptide to the monofunctionalized cysteine residues that did not lead to the aggregation of the NPs. The assay demonstrated high sensitivity and it allowed the sensing of thrombin with a detection limit as low as 5 nM and the detection of letal factor with a detection limit of 25 nM. An additional method for the detection of proteases (thermolysin and nACT–PSA) activity was described [114]. This method is based on the protease-induced deaggregation of Au NP clusters crosslinked by π-stacking of the Fmoc residues linked to the peptide coating of the Au NPs. The cleavage of the peptide by target protease resulted in deaggregation of NPs and a color change from blue to red. Similarly, the activity of alkaline phosphatase (ALP) was monitored through the aggregation of Au NPs [115].

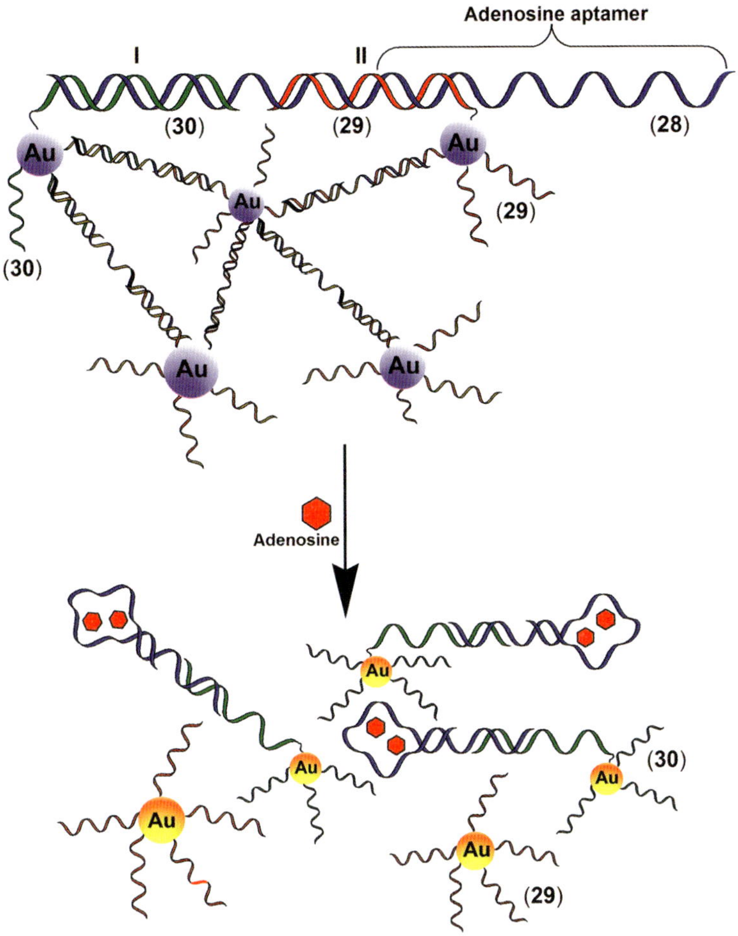

Figure 6.16 Colorimetric analysis of adenosine monophosphate through the deaggregation of Au NPs bridged by the blocked adenosine aptamer tethered to a nucleic acid (**28**), that is hybridized with nucleic acids (**29**)- and (**30**)-functionalized Au NPs. Formation of the aptamer–substrate complex separates the (**29**)-modified Au NPs and stimulates the deaggregation of the NPs.

The phosphorylated peptide H_2N-Cys-Tyr(PO_3^{2-})-Arg-OH was assembled on Au NPs through the cysteine group. Electrostatic repulsion of the negatively charged Au NPs retained the nonaggregated configuration of the NPs. The hydrolytic dephosphorylation of the peptide by ALP removed the phosphate group, thus permitting the bridging and aggregation of the NPs by the free amino group on the peptide chains.

The aggregation of Au NPs as colorimetric test for biorecognition events was extended to numerous other biorecognition complexes. For example, the process was used to detect the cholera toxin [116]. Au NPs were modified with a lactose capping layer, and the cholera toxin (B-subunit) linked to the lactose derivative induced aggregation of the Au NPs. Upon aggregation, the color of solution of Au NPs changed from red to deep purple. The selectivity of the bioassay arises from the fact that thiolated lactose mimics the GM1 ganglioside, the native receptor of the cholera toxin. The detection limit of the assay was 54 nM. Similarly, the colorimetric sensing of platelet-derived growth factors (PDGFs) and their receptors (PDGFRs) based on aggregation of aptamer-functionalized Au NPs was also developed [117].

In addition to the absorbance features of metallic NPs that were used to follow biorecognition events in solution and on surfaces, other optical methods have been employed to detect the association of biomolecule-functionalized Au NPs on biochips. These methods included scanometric detection by light scattering, surface plasmon resonance spectroscopy, resonance-enhanced absorption by NPs and enhanced Raman scattering.

A scanometric DNA detection method was developed, and this was based on a sandwich assay format involving a DNA-functionalized glass slide, the target DNA and Au NP probes [118]. In a typical setup for scanometric detection, the modified glass slide was illuminated in the plane of the slide with white light. The slide served in such a configuration as a planar waveguide that prevents any light from reaching the microscope objective by total internal reflectance. Wherever NP probes were attached to the surface, evanescently coupled light was scattered from the slide, and the NP labels were imaged as bright, colored spots. This approach was used for the detection of target DNA molecules that were specifically bound to a DNA-functionalized surface. The resulting DNA hybrid was labeled with gold nanoparticles that allowed the scanometric detection of the DNA (Figure 6.17). At high target concentrations ($\geq$1 nM), the Au NPs on the surface could be visualized with naked eye. At low target concentrations ($\leq$100 pM), the coverage of the surface-bound Au NPs was too low, and an enhancement process was needed. Enlargement of the Au NPs by the catalytic reduction of silver ions and the deposition of silver metal on the Au NPs

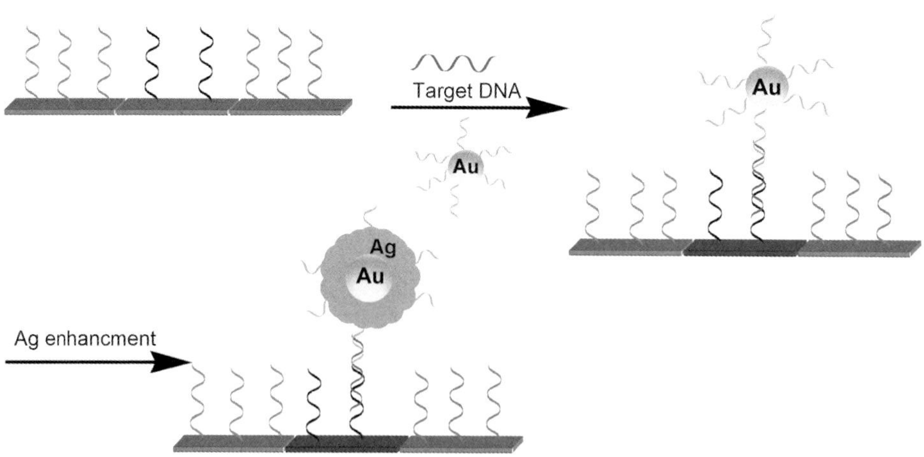

Figure 6.17 Scanometric detection of DNA on a surface using gold–silver core–shell NPs.

resulted in a 100-fold increase in the light-scattered signal and thus increased the sensitivity detection of target DNA (50 fM) [118]. This method was used to detect single-base mismatches in oligonucleotides, which were hybridized to DNA probes and were immobilized at different domains of a glass support. High sensitivities were provided by the deposition of silver, whereas the selectivity was achieved by examination of the melting properties of the spots: the mismatched spot reveals a lower melt temperature owing to its lower association constant. The scattering of light is size-dependent, and hence, by using different sized nanoparticles the simultaneous detection of different DNA sequences is feasible. Accordingly, the light scattered by DNA-functionalized 50- and 100-nm Au NP probes was used to identify two different target DNAs in solution [119]. The scanometric method was successfully applied to detect the MTHFR gene from genomic DNA at concentration as low as 200 fM without PCR amplification of the target, by the application of improved optical imaging instruments [120]. A similar approach was used to identify single nucleotide polymorphisms (SNPs) in unamplified human genomic DNA samples representing all possible genotypes for three genes involved in thrombotic disorders [121].

The scanometric method was applied as an optical detection means in a series of systems that employed nucleic acid-functionalized Au NPs as barcodes for biorecognition events such as antigen–antibody complex formation or DNA hybridization. In one system (Figure 6.18), prostate-specific antigen (PSA) was detected by nucleic acid-functionalized Au NPs that acted as signaling barcodes [122]. The Au NPs were modified with the polyclonal anti-PSA Ab that was further functionalized with thiolated nucleic acid (**31**), which were hybridized with the complementary nucleic acid (**31′**), and its sequence acted as a barcode for the sensing process. In the presence of PSA and magnetic particles functionalized with the monoclonal anti-PSA Ab, an aggregate consisting of the "sandwich" structure of the Au NPs and the magnetic particles was formed. The magnetic separation of the aggregate was followed by the thermal displacement of the barcode DNA. The released barcode nucleic acid was then amplified by the polymerase chain reaction (PCR), and the product was used to bridge the Au NPs with the complementary nucleic acid to a glass surface. The Ag-enhanced Au NPs were then analyzed by the scanometric method. Although this analytical protocol involves many steps, the PCR amplification step leads to an ultrasensitive detection method, and PSA at a level of 30 aM was analyzed.

An analogous process was used to analyze DNA (Figure 6.19) [123]. By this method, the functionalized Au NPs include the duplex DNA barcode (**32/32′**) and the nucleic acid units (**33**) that recognize the target DNA. In the presence of the target DNA (**35**), the magnetic particles modified with the nucleic acids **34** and the **32/33**-functionalized Au NPs, the magnetic particle–Au NP aggregate is formed through crosslinking the particles by **35**. The subsequent magnetic separation of the aggregate and the thermal separation of the DNA barcode were followed by scanometric detection of the released DNA code on surfaces. A PCR-like sensitivity that corresponded to 500 zM was claimed for the analysis of DNA. The DNA barcode-based sensing protocols were used to analyze protein cancer markers [124], the amyloid biomarkers for Alzheimer's disease [125] and the genes of various

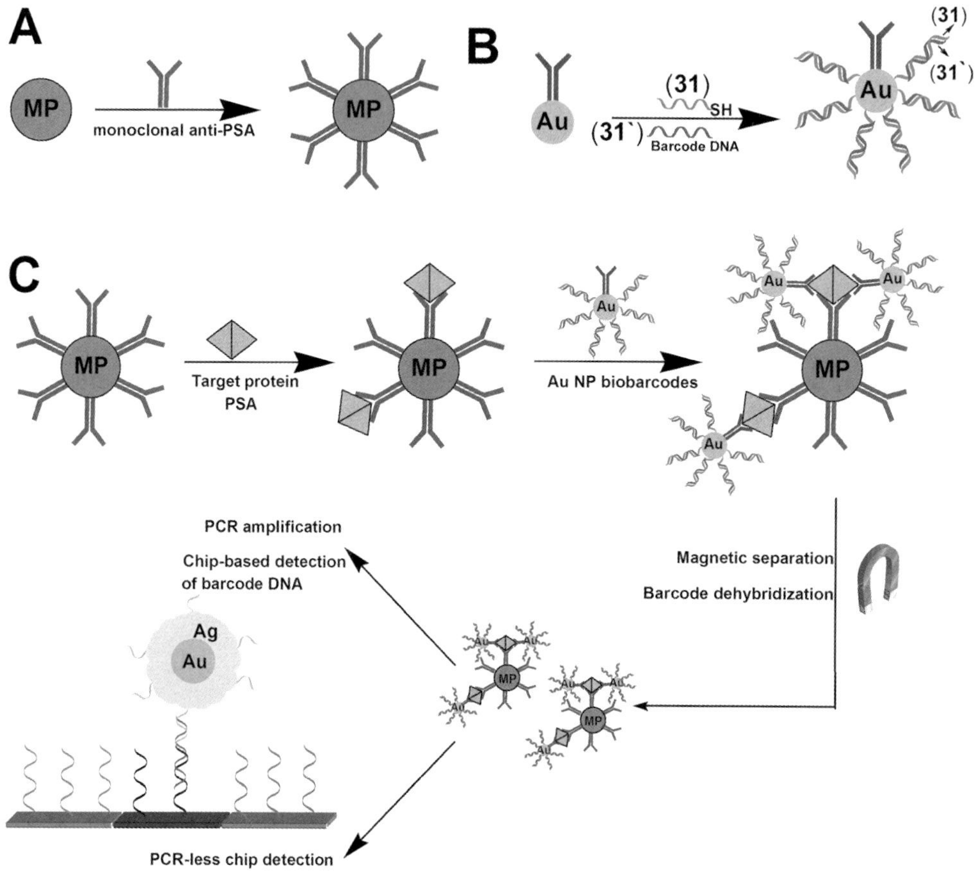

Figure 6.18 Scanometric detection of a target protein (PSA) by an immunoassay that amplifies the analysis by the PCR-induced generation of a DNA barcode: (A) preparation of the antibody-modified magnetic particles; (B) synthesis of antibody and duplex DNA (**31**/**31**′)-functionalized Au NPs; (C) formation of the DNA barcode-labeled Au NP immunocomplex on the magnetic particles, separation of the complex and the PCR amplification of the DNA barcode (**31**′). The amplified production is scanometrically detected on the surface.

pathogens, such as hepatitis B, Ebola virus, variola virus (smallpox, VV) and human immunodeficiency virus (HIV) [126]. Furthermore, different modifications of the method that use fluorescence detection [127] and colorimetric assay [128] have been reported.

The colorimetric scattering assay of gold nanoparticles was applied for the rapid detection of *mecA* gene in unamplified genomic DNA sequences [129]. The method was based on hybridization of DNA-functionalized Au NPs probes with the complementary sequences of target DNA in solution. The resulting solution was then spotted on a glass waveguide, which was illuminated with white light in the plane of the slide. The color of the light scattered by the oligonucleotide-functionalized Au NPs probes is different from that of the Au NPs aggregates generated by the hybridization process

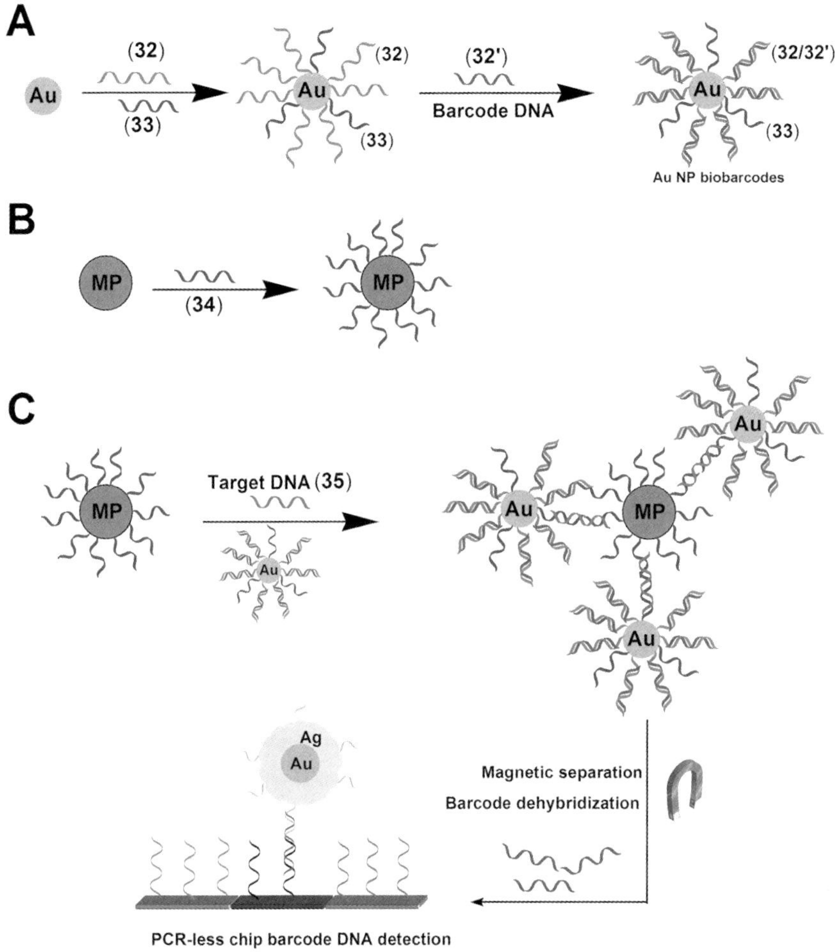

Figure 6.19 Scanometric detection of a DNA by a nucleic acid barcode-functionalized Au NP: (A) synthesis of Au NP labeled with the nucleic acid (**32**), complementary to the DNA barcode (**32′**) and the nucleic acid (**33**) complementary to the target; (B) preparation of the magnetic particles modified with the nucleic acid (**34**) complementary to the other end of the target; (C) aggregation of the functionalized Au NPs and the magnetic particles through hybridization with the target, magnetic separation of the aggregates, thermal separation of the DNA barcode and its scanometric detection.

between the probes and target DNA. In the absence of the target DNA, individual 40–50-nm Au NPs scatter green light, whereas in the presence of the target DNA aggregated nanoparticles scatter yellow to orange light. The method showed enhanced detection sensitivity (about four orders of magnitude) compared with the colorimetric assay, thus allowing the detection of zeptomole quantities of target DNA, 333 fM of synthetic DNA and 33 fM of genomic DNA. An improved light-scattering strategy using gold nanoparticles as labels for the detection of specific target DNA in a homogeneous solution was reported [130]. The method is based on the aggregation of DNA-functionalized Au NPs through the hybridization of the target DNA with probe-labeled Au NPs and enhanced light scattering that originated from the aggregates in solution. The light-scattering assay demonstrated high sensitivity, and the human p53 gene, exon 8 DNA, was detected at a concentration as low as 0.1 pM. Moreover, the assay showed a high degree of specificity and was used for discrimination between perfectly matched targets and targets with single base-pair mismatches.

The hyper-Rayleigh scattering (HRS) technique (nonlinear light scattering) was used to monitor DNA hybridization on unmodified gold nanoparticles in a saline solution, and the target DNA was analyzed at a concentration of 10 nM [131]. The HRS assay for DNA detection was based on the differences in the electrostatic interactions between ssDNA and dsDNA with the particles. The method permitted the analysis of single-base mismatch in DNA by monitoring the HRS intensity from the different DNAs interacting with the gold nanoparticles.

Surface-enhanced Raman scattering (SERS) of nanoparticle-bound substrates allows the amplification of molecular vibrational spectra by up to 10^6-fold [132–134]. Modification of metal NPs with different Raman dyes was used to generate multiply coded NPs [135–138] and for the preparation of thousands of codes to be written and read by means of surface-enhanced Raman resonance (SERR) scattering without the need for spatial resolution of components of the code [135, 137]. The use of SERS for the analysis of biorecognition events was demonstrated with the application of Au NPs that were functionalized with Raman dyes and recognition elements [139, 140]. Formation of the complementary recognition complex on surfaces, followed by the electroless deposition of Ag on the Au NPs, allowed the enhanced readout of the biorecognition events by SERS. The concept was applied for the parallel detection of various analytes on surfaces in an array configuration [Figure 6.20(A)]. For example, six different thiol-functionalized Raman-active dyes, Cy3, Cy3.5, Cy5, TAMRA, Texas Red and Rhodamine 6G, were linked through an oligonucleotide spacer (10 adenosine units) to six different oligonucleotides and coupled to Au NPs (13 nm) to yield six different Raman dye-labeled Au NP probes [139] [Figure 6.20(B)]. These Au NP probes were then employed as labels for hybridization with the complementary targets. The capture DNA strands were spotted on a surface, and the specific binding of Au NPs by hybridization with target DNAs was followed by the silver enhancement of the Au NPs and analysis by SERS [Figure 6.20(C) and (D)]. The detection limit of this method was 20 fM. The assay demonstrated the ability to discriminate SNPs in DNA and RNA targets. A similar concept was also used to identify protein–protein and protein–small molecule interactions by using Au NPs that were functionalized with specific antibodies and encoded with specific Raman dyes [140]. Compared with colorimetric and scanometric detection methods, this method offers enhanced multiplex sensing capabilities afforded by the narrow spectral bands – fingerprints of Raman dyes. Further developments of the SERS technique allowed highly sensitive immunoassay procedures [141–143].

Au NPs have been widely employed for signal amplification of biorecognition events based on nanoparticle-enhanced SPR spectroscopy [82, 86]. The changes in the dielectric properties at thin films of metals, such as gold films, as a result of

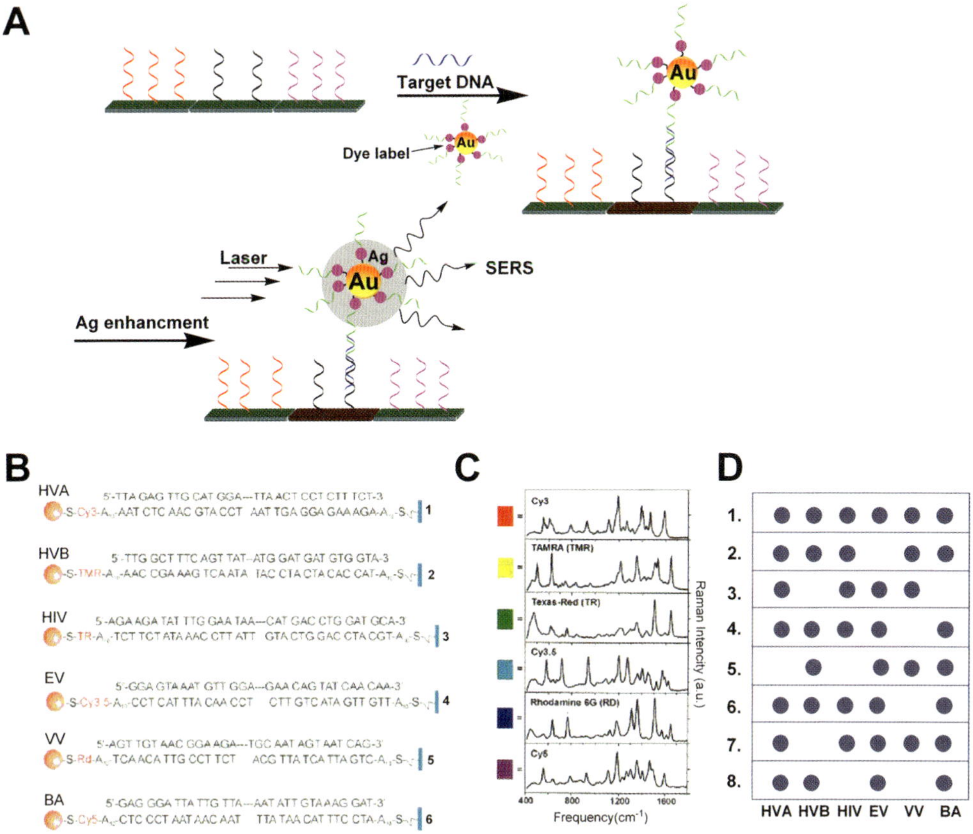

Figure 6.20 (A) Detection of a target DNA by the use of dye-functionalized Au NPs as labels for the SERS imaging. (B) Dye-labeled sequences of nucleic acids complementary to target DNAs of different pathogens. (C) Raman spectra of the dye-labeled Au NPs probes after silver enhancement upon the parallel analysis of the different pathogens on surfaces. (D) Flatbed scanner images of silver-enhanced microarrays upon the sensing of different pathogens. (Parts B, C and D reprinted from [139] with permission from AAAS).

biomolecular recognition processes, are the basis for the SPR technique. Labeling of the biorecognition complexes with Au or Ag NPs besides changing the dielectric properties, results in the electronic coupling between the localized NPs plasmon and the surface plasmon wave of the metal film. This electronic coupling significantly affects the resonance frequency of the surface wave, thus leading to the enhanced amplified optical transduction of the biorecognition events. Accordingly, Au NPs were used as labels in immunosensing [144–146] and DNA sensing [147–149] applications. The binding of Au NPs to the immunosensing interface led to a large shift in the plasmon angle, a broadening of the plasmon resonance region and an increase in the minimum reflectance, and these effects allowed the detection of the antigen with picomolar sensitivities [144]. Similarly, the sensitivity of DNA analysis was enhanced 1000-fold (10 pM) when Au NP-functionalized DNA molecules were used as labels [147]. Also, the sensitive detection of DNA hybridization by applying catalytic growth of Au NPs as a means to enhance the SPR shifts was demonstrated [150].

The charging of Au NPs that are coupled to thin gold films affects the electronic coupling between the localized NP plasmon and the surface plasmon wave, resulting in a shift in the SPR spectra. Thus, biocatalytic electron transfer reactions that involved NP–enzyme conjugates may be monitored by SPR spectroscopy [151]. Au NPs (1.4 nm) were functionalized with N^6-(2-aminoethyl) flavin adenine dinucleotide (FAD cofactor, amine derivative; (**1**), and apo-glucose oxidase (apo-GOx) was reconstituted onto the cofactor sites. The nanoparticle–GOx conjugates were then assembled on a gold thin film (SPR electrode) by using a long-chain dithiol, HS$(CH_2)_9$SH, monolayer as a bridging linker. This yielded the biocatalytically active glucose oxidase (GOx) bound to the Au NPs in an aligned configuration [Figure 6.21 (A)]. The biocatalyzed oxidation of glucose resulted in the formation of the reduced form of the cofactor, $FADH_2$. In the absence of O_2 that acts as the natural electron acceptor for GOx, electron transfer proceeded from the reduced cofactor to the Au

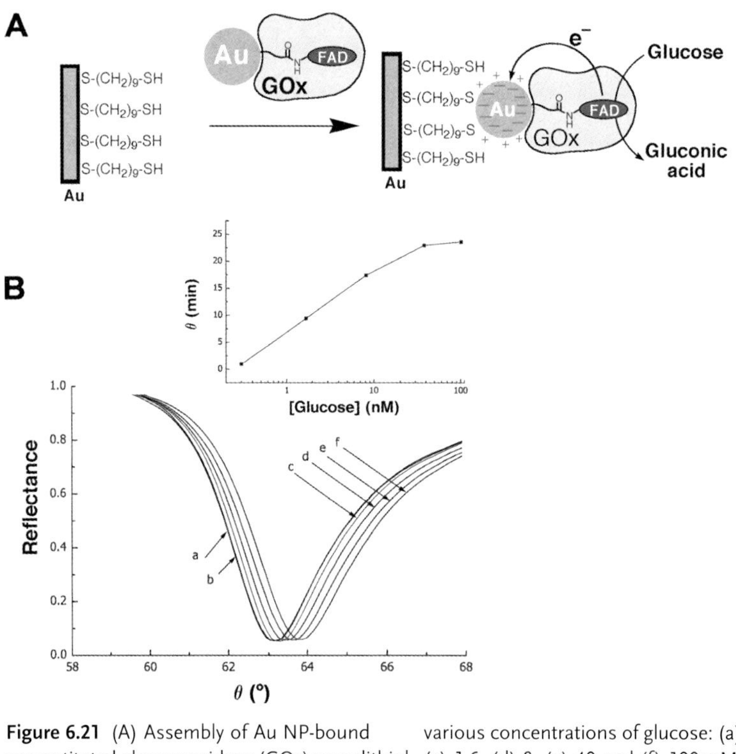

Figure 6.21 (A) Assembly of Au NP-bound reconstituted glucose oxidase (GOx) on a dithiol monolayer that is associated with an SPR-active surface and biocatalytic charging of the Au NPs in the presence of glucose. (B) SPR spectra of the Au NP–GOx hybrid system upon the addition of various concentrations of glucose: (a) 0, (b) 0.3, (c) 1.6, (d) 8, (e) 40 and (f) 100 mM. Inset: calibration plot of the SPR spectra minimum shift as a function of glucose concentration. (Reprinted with permission from [151]. Copyright 2004 American Chemical Society).

NPs, resulting in their charging. The long-chain dithiol monolayer provided a barrier for the electron tunneling from the Au NPs to the bulk Au electrode, and this preserved the electrical charge that was produced on the Au NPs. Potentiometric measurements indicated that the long-chain dithiol linkers provided a resistance layer that generated a potential gradient between the charged Au NPs and the conductive support. This potential gradient originated from the charging of the Au NPs by the biocatalytic process and thus, charge accumulated on the NPs was controlled by the biocatalytic process (the concentration of glucose). Accordingly, the plasmon coupling between the charged Au NPs and the Au support resulted in a shift of the SPR spectra, and these were dependent on the charge generated on the NPs [Figure 6.21(B)]. As the charge value was controlled by the rate of the biocatalytic reaction, the shift in the SPR spectrum was enhanced upon elevation of the glucose concentration [Figure 6.21(B), inset].

Metal nanoparticles exhibit a strong UV–visible absorption band that is not present in bulk material [1, 2, 152–154]. This absorption band, known as the localized surface plasmon resonance (LSPR), occurs when the incident photon frequency matches the collective oscillations of the conduction electrons. It is well established that the maximum extinction wavelength, λ_{max}, of the LSPR is controlled by the composition, size, shape and interparticle spacing of the nanoparticles, and also by the dielectric properties of their local environment (i.e., substrate, solvent and surface-bound molecules) [82]. The sensitivity of λ_{max} of the LSPR to the molecular environment of the NPs allowed the development of a new class of optical biosensors [155, 156] that operate in an analogous manner to their SPR counterparts by transducing small changes in the refractive index near the noble metal surface into a measurable wavelength-shift response. Triangular silver nanoparticles, ~100 nm wide and 25 nm high, fabricated by the nanosphere lithography (NSL) technique [Figure 6.22(A)], showed very unique optical properties [157, 158]. In particular, the λ_{max} of their LSPR spectrum is unexpectedly sensitive to the size, shape and local external dielectric environment of the nanoparticles (10–40 nm shift). The Ag nanotriangles were used to follow the streptavidin–biotin interactions as a model system for LSPR-based biosensors [157]. Triangular silver NPs were functionalized with biotin units and the changes in the LSPR spectra were followed upon binding of streptavidin molecules. The detection by LSPR was further amplified by the secondary coupling of biotinylated Au NPs to the streptavidin-saturated interface. A similar method was applied to design an immunosensor [158]. In addition, LSPR spectroscopy was applied for the detection and diagnosis of biomarkers, for example, the amyloid-β-derived diffusible ligands (ADDLs) that are markers for Alzheimer's disease [159, 160]. The sandwich assay composed of antibody/ADDLs/antibody was developed to amplify the LSPR response, and thus to improve the detection limit [160] [Figure 6.22(B)]. The LSPR nanosensor demonstrated sensitivity and selectivity in the detection of ultralow concentrations of ADDLs in synthetic and human samples [human brain extracts, cerebrospinal fluid (CSF)] [Figure 6.22(C)]. A detection limit corresponding to 100 fM was achieved for synthetic ADDLs. Also, modified LSPR-based biosensors consisting of gold-capped silica nanoparticle-layered substrates for monitoring DNA hybridization and antigen–antibody interactions were fabricated [161, 162].

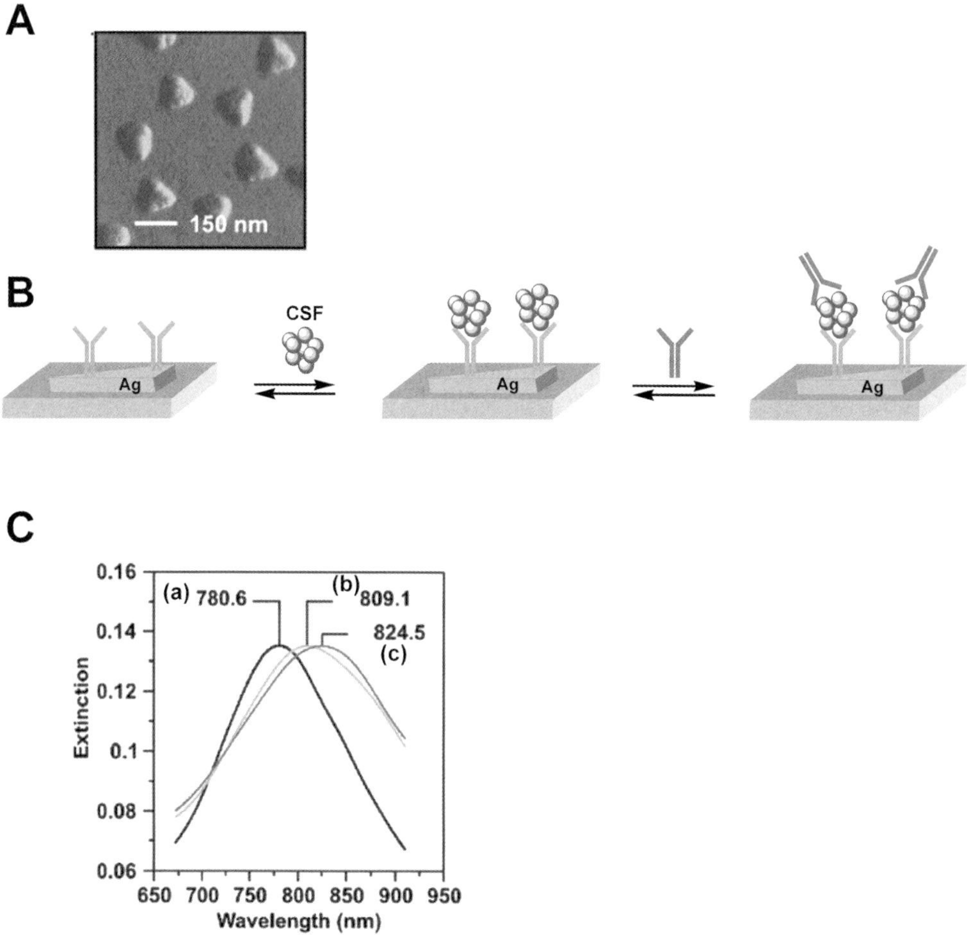

Figure 6.22 (A) AFM image of Ag^0-triangle NPs assembled on a mica surface. (B) Schematic analysis of the amyloid-β-derived diffusible ligands (ADDLs) by an immunoassay on Ag triangle NPs using LSPR. (C) LSPR curves corresponding to: (a) the capture anti-ADDL antibody-modified NPs; (b) the modified surface after treatment with the antigen (ADDL, from CSF); (c) the antibody–antigen complex after interaction with the secondary anti-ADDL antibody. (Reprinted with the permission from [160]. Copyright 2005 American Chemical Society).

6.7 Semiconductor Nanoparticles as Optical Labels

The unique optical properties of semiconductor nanoparticles or quantum dots, QDs, offer a number of advantages for biosensing, including size-controlled luminescence properties, impressive photostability and high quantum yields [5–7, 163, 164]. Efficient methods for the preparation of semiconductor nanoparticles

Figure 6.23 Fluorescent analysis of (A) an antigen by antibody-functionalized QDs and (B) a DNA by a nucleic acid-functionalized QD.

and their functionalization with biomolecules were developed recently [165, 166]. Unlike molecular fluorophores, which typically have very narrow excitation spectra, semiconductor QDs absorb light over a very broad spectral range. This makes it possible to excite optically a broad spectrum of quantum dot "colors" using a single excitation laser wavelength, which may enable one to probe simultaneously several markers in biosensing and assay applications. Indeed, functionalized semiconductor QDs have been used as fluorescence labels for numerous biorecognition events [9, 167–170], including their use in immunoassays for protein detection [Figure 6.23(A)] and nucleic acid detection [Figure 6.23(B)]. For example, CdSe/ZnS QDs were functionalized with avidin and these were used as fluorescent labels for biotinylated antibodies. Fluoroimmunoassays utilizing these antibody-conjugated NPs were successfully used in the detection of protein toxins (staphylococcal enterotoxin B and cholera toxin) [171, 172].

Similarly, CdSe/ZnS QDs conjugated to appropriate antibodies were applied for the multiplexed fluoroimmunoassay of toxins [Figure 6.24(A)]. Sandwich immunoassays for the simultaneous detection of the four toxins (cholera toxin, (CT), ricin, shiga-like toxin 1 (SLT), and staphylococcal enterotoxin B, (SEB)) by using different sized QDs were performed in single wells of a microtiter plate in the presence of mixture of all four QD–antibody conjugates [Figure 6.24(B)] [173], thus leading to the fluorescence that encodes for the toxin. In another example, multiplexed immunoassay formats based on antibody-functionalized QDs were used for simultaneous detection of *Escherichia coli* O157:H7 and *Salmonella typhimurium* bacteria [174] and for the discrimination between diphtheria toxin and tetanus toxin proteins [175].

Fluorescent QDs were used for the detection of single-nucleotide polymorphism in human oncogene p53 and for the multiallele detection of the hepatitis B and hepatitis C virus in microarray configurations [176]. DNA-functionalized CdSe/ZnS QDs of different sizes were used to probe hepatitis B and C genotypes in the presence of a background of human genes. The discrimination of a perfectly matched sequence of p53 gene in the presence of background oligonucleotides including different single-nucleotide polymorphism sequences was detected with true-to-false signal ratios higher than 10 (under stringent buffer conditions) at room temperature within minutes. Also, DNA–QD conjugates were used as fluorescence probes for *in situ*

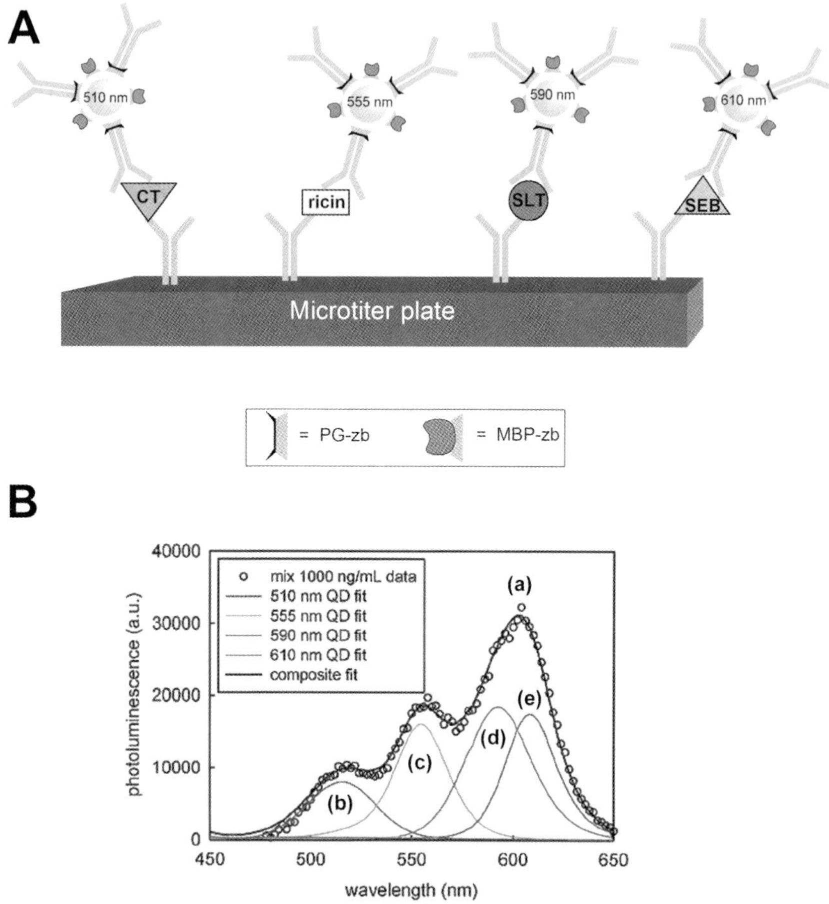

Figure 6.24 (A) Parallel optical analysis of different antigens in a well-array format using the fluorescence of different sized quantum dots. (B) Fluorescence spectrum observed upon analyzing the four analytes, 1 $\mu g\, mL^{-1}$, by the different sized QDs (a), and deconvoluted spectra of individual toxins: (b) CT, (c) ricin, (d) SLT and (e) SEB. (Reprinted with permission from [173]. Copyright 2004 American Chemical Society).

hybridization assays (FISH). For example, the QD-based FISH labeling method was used for the detection of Y chromosome in human sperm cells [177]. A QD-based FISH method to analyze human metaphase chromosomes was reported [178] by using QD-conjugated total genomic DNA as a probe for the detection of EBRB2/HER2/neu gene. Also, the FISH technique was used for the multiplex cellular detection of different mRNA targets [179].

Nonetheless, the superior photophysical features of semiconductor QDs are demonstrated in organic solvents, and their introduction into aqueous media, where biorecognition and biocatalytic reactions proceed, is accompanied by a severe or even complete loss of their fluorescence properties. Different methods to stabilize the fluorescence properties of semiconductor QDs in aqueous media were reported, including their surface passivation with protective layers [180, 181] and the coating of the QDs with protecting glass films [182, 183] or polymers [184]. Although these methods preserve the photophysical properties of the QDs, the protective layers limit important useful applications of the QDs. Quantum dots could be employed as optical labels for dynamic biological processes such as biocatalyzed transformations or structurally induced biomolecular changes, for example, opening of a hairpin nucleic acid by the hybridization of DNA, using fluorescence resonance energy transfer (FRET) or electron transfer quenching as photophysical probe mechanisms [185, 186]. The sensitivity of these photophysical processes to the distance separating the donor–-acceptor or chromophore–quencher pairs prevents, however, the use of fluorescent QDs passivated by relatively thick protecting layers as optical probes for dynamic bioprocesses. Hence, a very delicate nanostructuring of the capping layer is essential to allow the use of QDs as active components in energy/electron transfer reactions.

Different sensing schemes have been developed that use QDs as a FRET donor. For example, CdSe/ZnS QDs conjugated to nucleic acids have been used to follow the biocatalyzed replication of DNA [187]. CdSe/ZnS NPs were functionalized with the DNA primer **36**, which is complementary to a domain of M13mp18 DNA. Hybridization of the M13mp18 DNA with the nucleic acid-functionalized QDs, followed by the replication of the assembly in the presence of polymerase and the nucleotide (dNTP) mixture that included Texas Red-functionalized dUTP (**37**), resulted in the incorporation of the dye labels into the DNA replica [Figure 6.25(A)]. The FRET process from the semiconductor NPs to the incorporated dye units resulted in emission from the dye with concomitant quenching of the fluorescence of the QDs [Figure 6.25(B)], and it allowed the identification of the primary hybridization.

A similar approach was used to follow telomerase activity, a biocatalytic ribonucleoprotein that is a versatile marker for cancer cells. The CdSe/ZnS QDs were modified with a nucleic acid primer that is recognized by telomerase. In the presence of telomerase and the nucleotide mixture dNTPs, that included Texas Red-functionalized dUTP (**37**), the telomerization of the nucleic acid associated with the QDs was initiated, while incorporating the Texas Red-labeled nucleotide into the telomers [Figure 6.25(C)]. The FRET process from the QDs to the dye units then enabled the dynamics of the telomerization process to be followed [187].

Also, the association of maltose with the hybrid composed of maltose-binding protein was examined by the application of a CdSe/ZnS QD linked to the maltose binding protein (MBP) [188]. CdSe/ZnS QDs were functionalized with MBP, and these were interacted with a β-cyclodextrin–QSY-9 dye conjugate [Figure 6.26(A)]. The β-cyclodextrin–QSY-9 dye conjugate resulted in quenching of the luminescence of the QDs by the dye units. Addition of maltose displaced the quencher units, and this regenerated the luminescence of the QDs [Figure 6.26(B)]. This method allowed the development of a competitive QD-based sensor for maltose in solution. Similarly, a competitive QD-based assay for the detection of the explosive trinitrotoluene (TNT) was developed [189]. CdSe/ZnS QDs were functionalized with a single-chain antibody fragment that selectively binds TNT. The analogue substrate trinitrobenzene (TNB) covalently linked to the quencher dye BHQ10 was bound to the QD–antibody conjugate and quenched the QD fluorescence. In the presence of the TNT analyte, the quencher

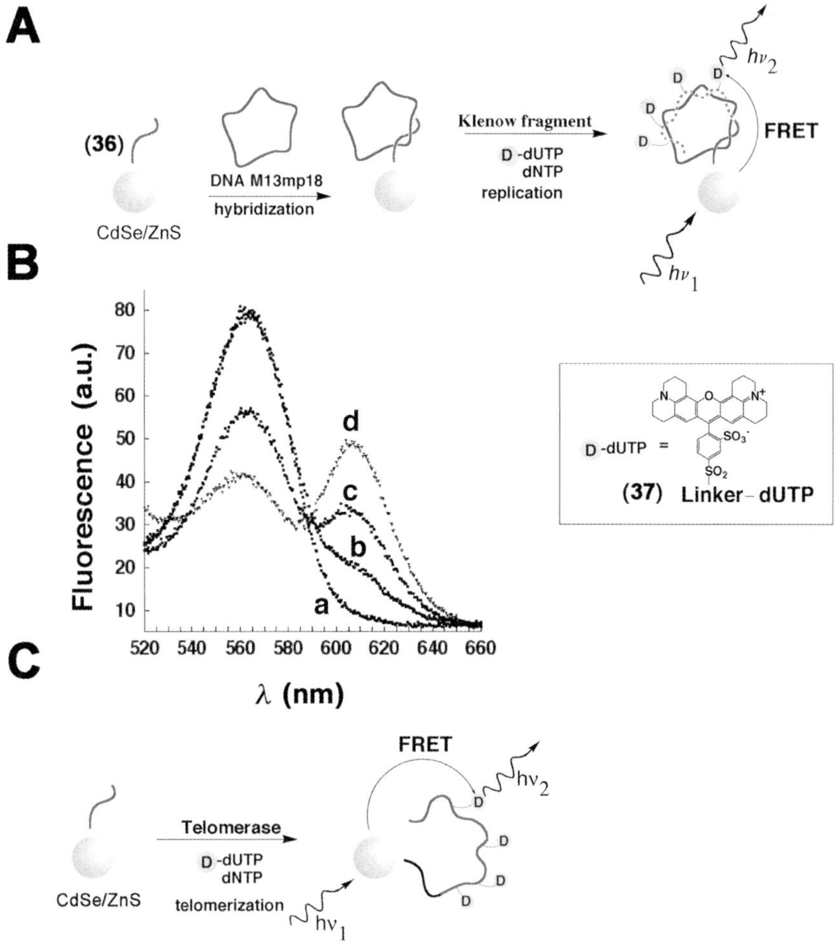

Figure 6.25 (A) Optical detection of M13 phage DNA by nucleic acid-functionalized CdSe/ZnS QDs. The replication of the analyte in the presence of the dNTP mixture that includes the Texas Red-labeled dUTP (**37**) results in the incorporation of the dye into the replica and the stimulates of a FRET process. (B) Time-dependent fluorescence changes upon incorporation of the dye (**37**) into the DNA replica and the analysis of the M13 phage DNA according to (A). (C) Optical analysis of telomerase activity by the incorporation of the Texas Red-dUTP (**37**) into the telomers associated with CdSe/ZnS QDs. (Reprinted with permission from [187]. Copyright 2003 American Chemical Society).

TNB–BHQ10 conjugate was competitively displaced. This eliminated the FRET interactions between the QD and the dye, and the fluorescence of the QDs was restored.

The hydrolytic functions of a series of proteolytic enzymes were followed by the application of QDs modified with peptides as reporter units and the FRET process as a readout mechanism [190, 191]. CdSe QDs were modified with different peptides that included specific cleavage sequences for different proteases, and quencher units were tethered to the peptide termini. The fluorescence of the QDs was quenched in

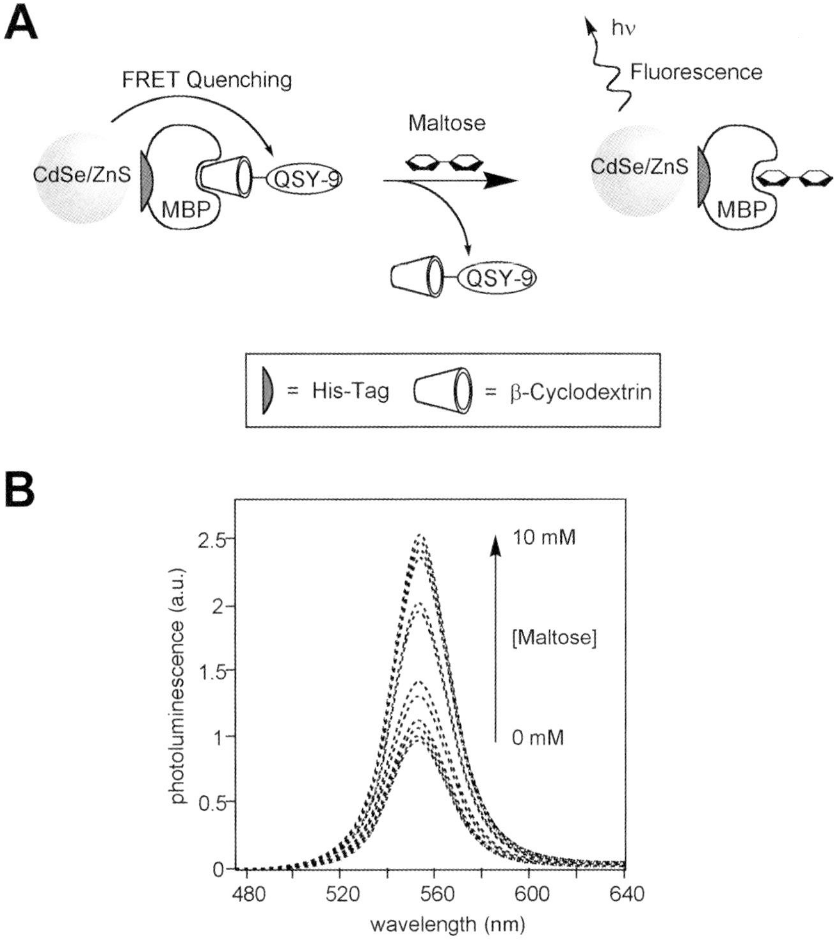

Figure 6.26 (A) Application of CdSe/ZnS QDs for the competitive assay of maltose using the maltose binding protein (MBP) as sensing material and β-cyclodextrin–QSY-9 dye conjugate, β-CD–QSY–9, as FRET quencher. (B) Fluorescence changes of the MBP-functionalized QDs upon analyzing increasing amounts of maltose. (Reprinted by permission from Macmillan Publishers Ltd.; Nature Materials [188], Copyright (2003)).

the presence of the quencher–peptide capping layer. The hydrolytic cleavage of the peptide resulted in the removal of the quencher units and this restored the fluorescence of the QDs. For example, collagenase was used to cleave the Rhodamine Red-X dye-labeled peptide (**38**) linked to CdSe/ZnS QDs [Figure 6.27(A)]. While the tethered dye quenched the fluorescence of the QD, hydrolytic scission of the dye and its corresponding removal restored the fluorescence [Figure 6.27(B)].

In a related study, the activity of tyrosinase (TR) was analyzed by CdSe/ZnS QDs [192]. The QDs were capped with tyrosine methyl ester monolayer. The tyrosinase-induced oxidation of the tyrosine groups to the respective dopaquinone units generated active quencher units that suppressed the fluorescence of the QDs

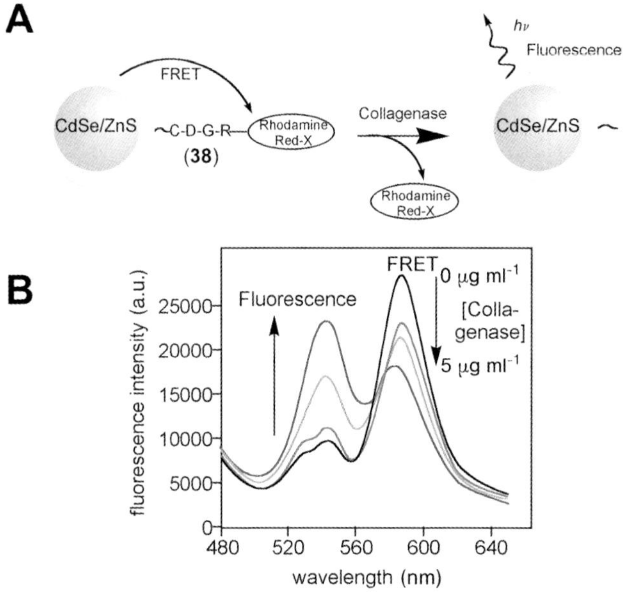

Figure 6.27 (A) Application of CdSe/ZnS QDs for the optical analysis of the protease-mediated hydrolysis of the Rhodamine Red-X-functionalized peptide (**38**). (B) Decrease in the fluorescence of the dye and the corresponding increase in the fluorescence of the QDs upon interaction with different concentrations of collagenase. (Reprinted with permission from [191]. Copyright 2006 American Chemical Society).

[Figure 6.28(A)]. The depletion of the fluorescence of the QDs upon their interaction with different concentrations of tyrosinase (TR) is displayed in Figure 6.28(B). The tyrosinase-stimulated oxidation of phenol residues was further employed to use the QDs to monitor the activity of thrombin [192]. The CdSe/ZnS QDs were functionalized with the peptide **39** that included the specific sequences for cleavage by thrombin and the tyrosine site. The tyrosinase-induced oxidation of tyrosine yields the dopaquinone units that quenched the fluorescence of the QDs [Figure 6.29(A)]. The hydrolytic scission of the peptide by thrombin cleaved off the quinone quencher units and restored the fluorescence of the QDs [Figure 6.29(B)].

The FRET process occurring within a duplex DNA structure consisting of tethered CdSe/ZnS QDs and a dye was applied to probe DNA hybridization and the DNase I cleavage of the DNA [193]. Nucleic acid **40**-functionalized CdSe/ZnS QDs were hybridized with the complementary Texas Red-functionalized nucleic acid **41** [Figure 6.30(A)]. The time-dependent resonance energy transfer from the QDs to the dye units was used to monitor the hybridization process. Treatment of the DNA duplex with DNase I resulted in the cleavage of the DNA and the recovery of the fluorescence properties of the CdSe/ZnS QDs. After cleavage of the double stranded DNA with DNase I, the intensity of the FRET band of the dye decreased and the fluorescence of CdSe/ZnS QDs increased [Figure 6.30(B)]. The luminescence properties of the QDs were only partially recovered due to the nonspecific adsorption of the dye on QDs.

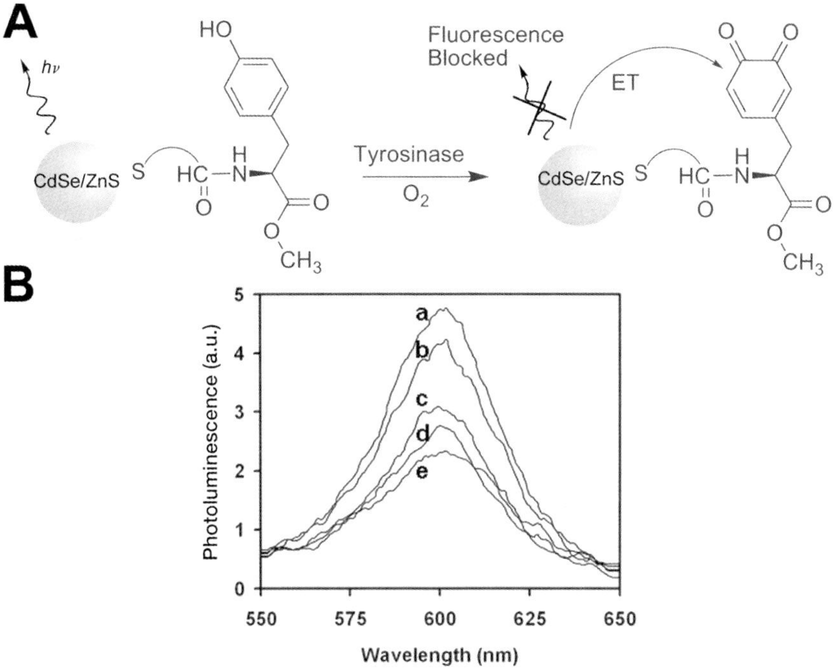

Figure 6.28 (A) Analysis of tyrosinase activity by the biocatalytic oxidation of the methyl ester tyrosine-functionalized CdSe/ZnS QDs to the dopaquinone derivative that results in the electron transfer quenching of the QDs. (B) Time-dependent fluorescence quenching of the QDs upon tyrosinase-induced oxidation of the tyrosine-functionalized QDs: (a) 0, (b) 0.5, (c) 2, (d) 5 and (e) 10 min. (Reprinted with permission from [192]. Copyright 2006 American Chemical Society).

The QDs were also used to probe the formation of aptamer–protein complexes [194]. An anti-thrombin aptamer was coupled to QDs, and the nucleic acid sequence was hybridized with a complementary oligonucleotide–quencher conjugate (Figure 6.31). The fluorescence of the QDs was quenched in the QD–quencher duplex. In the presence of thrombin the duplex was separated and the aptamer underwent a conformational change to the quadruplex structure that binds thrombin. The displacement of quencher units from the blocked aptamer activated the luminescence functions of the QDs, and a about 19-fold increase in their fluorescence was observed.

6.8 Semiconductor Nanoparticles for Photoelectrochemical Applications

The photoexcitation of the semiconductor quantum dots yields the transfer of an electron from the semiconductor valence band to its conduction band, to yield an electron–hole pair. The electron–hole recombination in the QD may lead to either thermal or radiative relaxation of the excited species. The immobilization of QDs

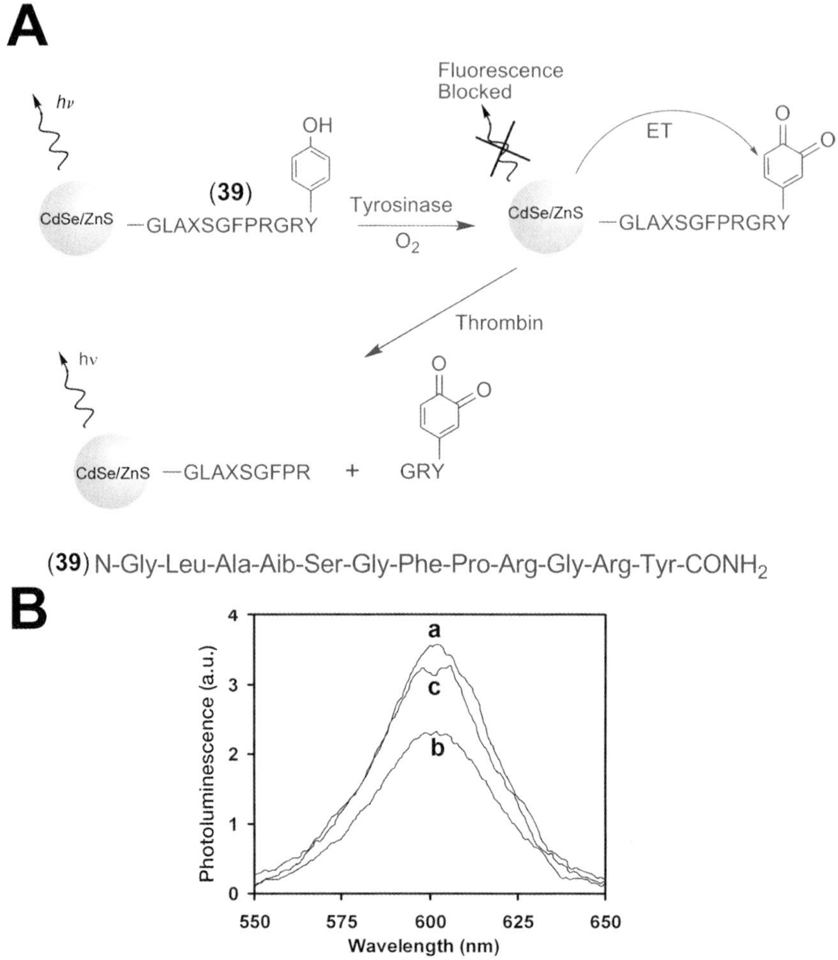

Figure 6.29 (A) Sequential analysis of tyrosinase activity and thrombin activity by the tyrosinase-induced oxidation of the tyrosine-containing peptide (**39**) associated with the CdSe/ZnS QDs that results in the electron transfer quenching of the QDs, followed by the thrombin-induced cleavage of the dopaquinone-modified peptide that restores the fluorescence of the QDs. (B) Fluorescence of: (a) the **39**-modified QDs; (b) after reaction of the QDs with tyrosinase, 10 min, and (c) after treatment of the dopaquinone-functionalized QDs with thrombin, 6 min. (Reprinted with permission from [132]. Copyright 2006 American Chemical Society).

onto electrodes permits the utilization of the photogenerated electron–hole pair for inducing the formation of photocurrents (photoelectrochemical effect). The photocurrent may be formed by the transfer of the conduction band electrons to the bulk electrode and the concomitant transfer of electrons from the electron donor to the valence band holes to yield a steady-state cathodic photocurrent. Alternatively, the photogenerated conduction band electrons may be transferred to a solution-solubilized electron acceptor, with the concomitant transfer of electrons from the

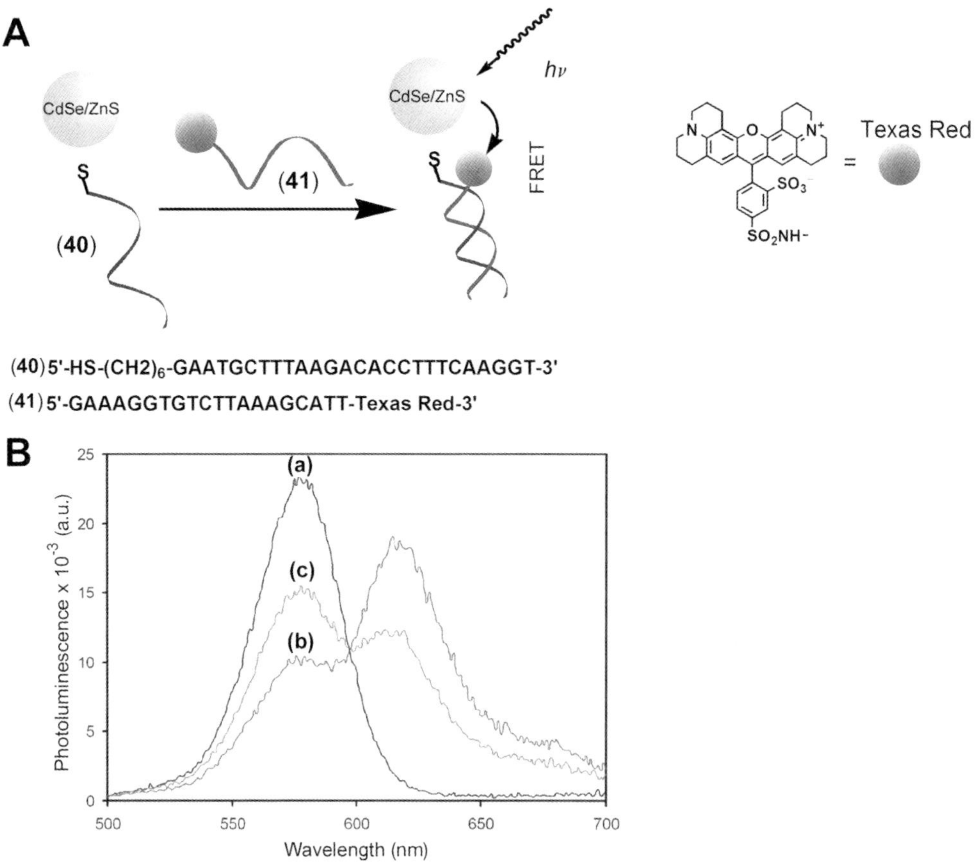

Figure 6.30 (A) Assembly of the CdSe/ZnS and Texas Red-tethered duplex DNA. (B) Fluorescence spectra of: (a) the (**40**)-functionalized CdSe/ZnS QDs; (b) the (**40**)/(**41**) duplex DNA tethered to the QDs and the Texas Red chromophore; (c) after treatment of the duplex DNA tethered to CdSe/ZnS and the dye with DNase I. (Reprinted with permission from [193]. Copyright 2005 American Chemical Society).

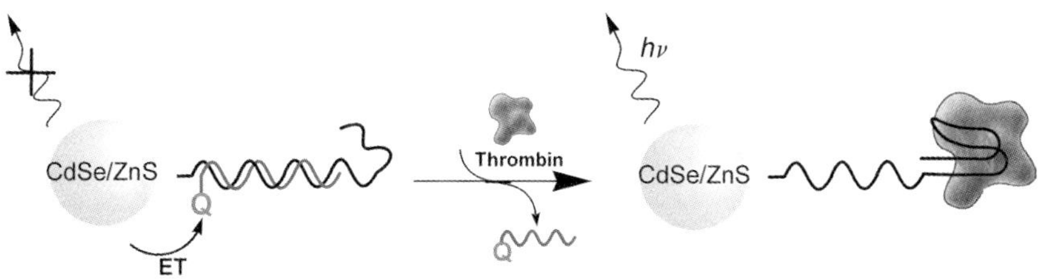

Figure 6.31 Analysis of thrombin by the protein-induced separation of the anti-thrombin aptamer blocked by a quencher-functionalized nucleic acid that restores the fluorescence of the QDs.

electrode to the valence band hole and the formation of an anodic photocurrent. The coupling of biomolecule–QD conjugates to electrode surfaces enables one not only to use the photoelectrochemical effect as a means to transduce biosensing processes, but also to tailor functional nano-architectures on surfaces that perform logic gate operations or act as switching systems. Different QD–DNA hybrid systems [195, 196] or QD–protein conjugates [197, 198] were assembled on electrodes, and the control of the photoelectrochemical properties of the QDs by the biomolecules was demonstrated. CdS nanoparticles were assembled on electrodes by double-stranded nucleic acids acting as bridging units, and the effect of a redox-active intercalator on the resulting photocurrent and its direction was demonstrated [196]. Dithiol-tethered single-stranded ssDNA (**42**) was assembled on an Au electrode and subsequently hybridized with a complementary dithiolated ssDNA (**43**) to yield a double-stranded DNA. The resulting surface was treated with CdS–NPs to yield a semiconductor nanoparticle interface linked to the electrode surface (Figure 6.32). Irradiation of the dsDNA–CdS NP-modified electrode in the presence of triethanolamine (TEOA) as electron donor resulted in an anodic low-intensity photocurrent. The observed generated photocurrent was attributed to an imperfect structure of the CdS NP–DNA assemblies that resulted from direct contact between the NPs and the electrode, rather then from charge transport through the DNA. The DNA duplex structure linking the CdS NPs to the electrode could be employed, however, as a medium to incorporate redox-active intercalators that facilitate electrical contact between the NPs and the electrode, resulting in enhanced photocurrents. Methylene blue (MB) (**44**) was intercalated into the (**42**/**43**)-dsDNA coupled to the CdS NPs [Figure 6.32(A) and (B)]. The cyclic voltammogram of the system implied that at potentials $E > -0.28$ V (vs. SCE) the intercalator exists in its oxidized form (**44**), whereas at potentials $E < -0.28$ V (vs. SCE) the intercalator exists in its reduced leuco form (**45**). Coulometric analysis of the MB redox wave, $E^{\circ} = -0.28$ V (vs. SCE), knowing the surface coverage of the

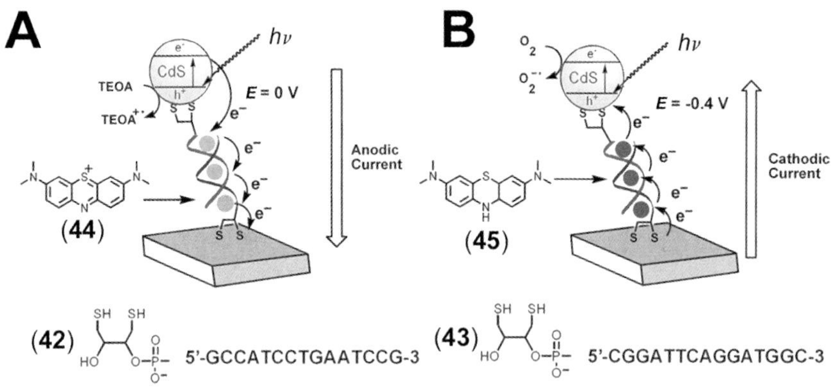

Figure 6.32 Directional electroswitched photocurrents in the CdS NP–ds-DNA–intercalator system. (A) Enhanced generation of anodic photocurrent in the presence of the oxidized methylene blue intercalator (**44**) (applied potential $E = 0$ V). (B) Enhanced generation of cathodic photocurrent in the presence of the reduced methylene blue intercalator (**45**) (applied potential $E = -0.4$ V). (Reproduced with permission from [196]. Copyright 2005 Wiley-VCH).

dsDNA, indicated that about 2–3 intercalator units were associated with the double-stranded DNA. An anodic photocurrent was generated in the system in the presence of TEOA as electron donor and MB intercalated into the dsDNA and while applying a potential of 0 V (vs. SCE) on the electrode. At this potential MB existed in its oxidized state (**44**), which acts as an electron acceptor. The resulting photocurrent was about fourfold higher than that recorded in the absence of the intercalator.

The enhanced photocurrent was attributed to the trapping of conduction band electrons by the intercalator units and their transfer to the electrode that was biased at 0 V, thus retaining the intercalator units in their oxidized form. The oxidation of TEOA by the valence band holes then led to the formation of the steady-state anodic photocurrent. Biasing the electrode at a potential of −0.4 V (vs. SCE), a potential that retained the intercalator units in their reduced state (**45**), led to blocking of the photocurrent in the presence of TEOA and under an inert argon atmosphere. This experiment revealed that the oxidized intercalator moieties with the DNA matrix played a central role in the charge transport of the conduction band electrons and the generation of the photocurrent.

Figure 6.33(A), curve b, shows the photocurrent generated by the (**42**/**43**)–dsDNA linked to the CdS NPs in the presence of the reduced intercalator **45** under conditions where the electrode was biased at −0.4 V (vs. SCE) and the system was exposed to air (oxygen). At a bias potential of −0.4 V (vs. SCE), the intercalator units exist in their reduced *leuco* form (**45**) that exhibits electron-donating properties [Figure 6.32(B)]. Photoexcitation of the CdS NPs yields electron–hole pairs in the conduction band and valence band, respectively. The transport of the conduction band electrons to oxygen with the concomitant transport of electrons from the reduced intercalator units to the

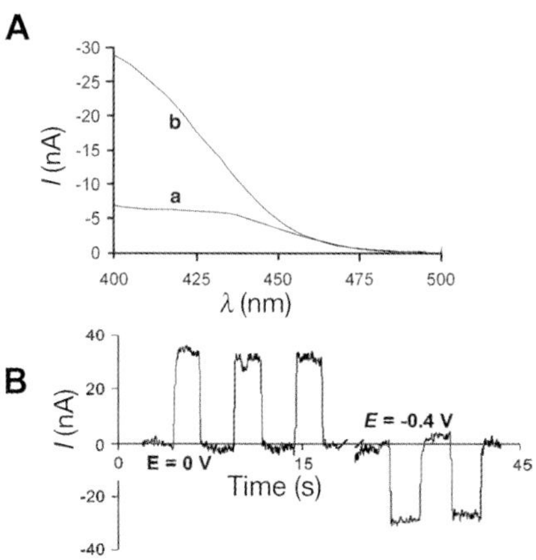

Figure 6.33 (A) Cathodic photocurrents generated in the CdS NP–dsDNA system associated with the electrode: (a) in the absence and (b) in the presence of reduced methylene blue intercalator (**45**). The data were obtained at the applied potential $E = -0.4$ V and in the presence of air. (B) Electrochemically switched anodic and cathodic photocurrents generated by the Cd NP–dsDNA–**44**/**45** systems at 0 and −0.4 V, respectively, in the presence of 20 mM TEOA and air. Photocurrents were generated with irradiation at $\lambda = 420$ nm. (Reproduced with permission from [196]. Copyright 2005 Wiley-VCH).

valence band holes completed the cycle for the generation of the photocurrent. The fact that the electrode potential retained the intercalator units in their reduced state and the infinite availability of the electron acceptor (O_2) yielded the steady-state cathodic photocurrent in the system.

The introduction of TEOA and oxygen to the electrode modified with (**42**/**43**)–DNA duplex and associated CdS NPs allowed the control of the photocurrent direction by switching the bias potential applied on the electrode. Figure 6.33(B) depicts the potential-induced switching of the photocurrent direction upon switching the electrode potential between −0.4 (cathodic photocurrents) and 0 V (anodic photocurrents), respectively. A layer-by-layer deposition of nucleic acid-functionalized CdS QDs on the electrode was followed by the photoelectrochemical transduction of the assembly process [195]. Semiconductor CdS NPs (2.6 ± 0.4 nm) were functionalized with one of the two thiolated nucleic acids **46** and **47** that are complementary to the 5′- and 3′-ends of a target DNA molecule (**48**). An array of CdS NP layers was then constructed on an Au electrode by a layer-by-layer hybridization process with the use of the target DNA **48** as crosslinker of CdS QDs functionalized with nucleic acids (**46** or **47**) complementary to the two ends of the DNA target (Figure 6.34). Illumination of the array in the presence of a sacrificial electron donor resulted in the generation of a photocurrent. The photocurrents increased with the number of generations of CdS NPs associated with the electrode, and the photocurrent action spectra followed the absorbance features of the CdS NPs, which implies that the photocurrents originated from the photoexcitation of the CdS nanoparticles. The ejection of the conduction band electrons into the electrode occurred from the QDs that were in intimate contact with the electrode support. This was supported by the fact that $Ru(NH_3)_6^{3+}$ units ($E° = -0.16$ V vs SCE), which were electrostatically bound to the DNA, enhanced the photocurrent from the DNA–CdS array. The $Ru(NH_3)_6^{3+}$ units acted as charge transfer mediators that facilitated the hopping of conduction band electrons from CdS particles, which lacked contact with the electrode, due to their separation by the DNA tethers.

Enzymes or redox proteins were also linked to semiconductor QDs, and the resulting photocurrents were employed to assay the enzyme activities and to develop different biosensors. Cytochrome *c*-mediated biocatalytic transformations were coupled to CdS NPs, and the direction of the resulting photocurrent was controlled by the oxidation state of the cytochrome c mediator [198]. The CdS NPs were immobilized on an Au electrode through a dithiol linker, and mercaptopyridine units, acting as promoter units that electrically communicate between the cytochrome *c* and the NPs, were linked to the semiconductor NPs (Figure 6.35). In the presence of reduced cytochrome *c*, the photoelectrocatalytic activation of the oxidation of lactate by lactate dehydrogenase (LDH) proceeded, while generating an anodic photocurrent [Figure 6.35(A)]. Photoexcitation of the NPs resulted in the ejection of the conduction band electrons into the electrode and the concomitant oxidation of the reduced cytochrome *c* by the valence band holes. The resulting oxidized cytochrome *c* subsequently mediated the LDH-biocatalyzed oxidation of lactate. Similarly, cytochrome *c* in its oxidized form was used to stimulate the bioelectrocatalytic reduction of NO_3^- to NO_2^- in the presence of nitrate reductase (NR), while generating a cathodic

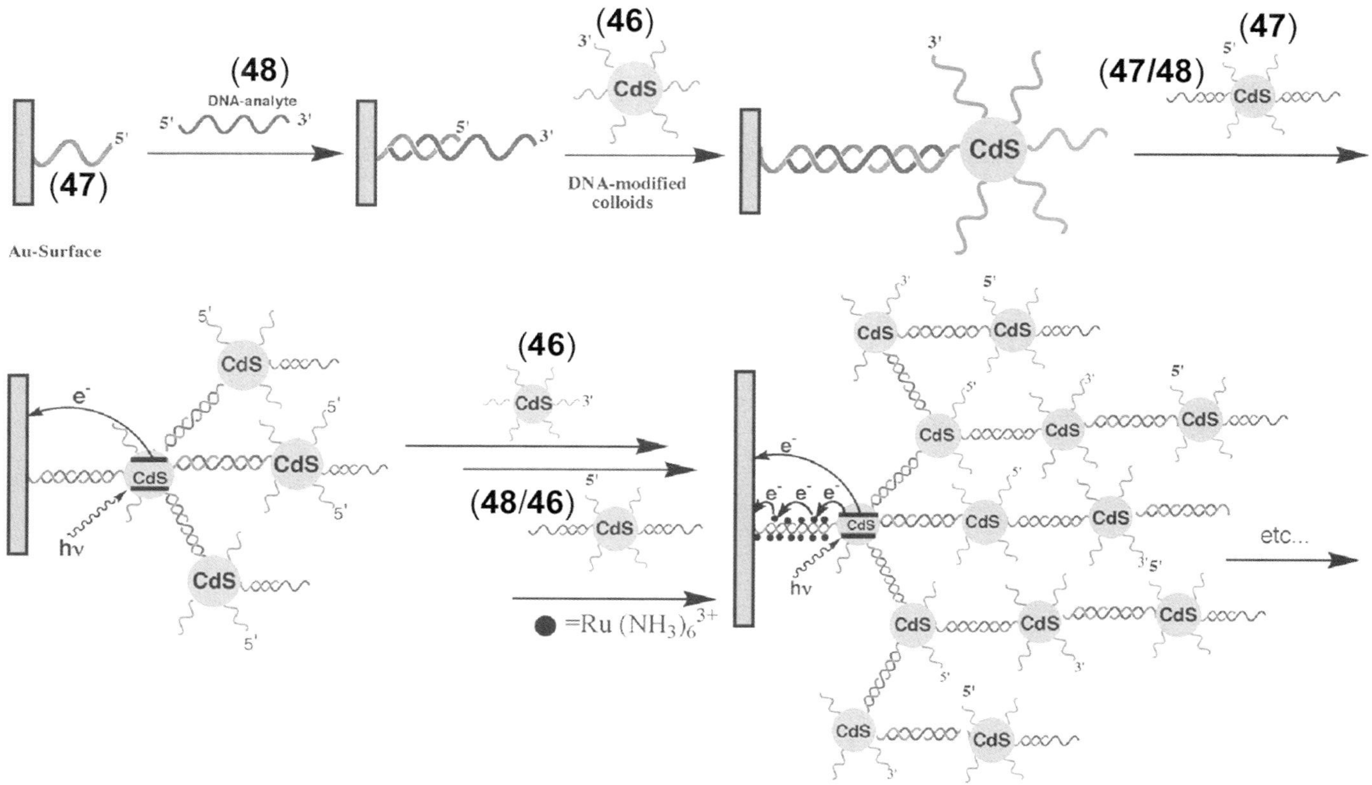

Figure 6.34 Layer-by-layer deposition of CdS NPs using **46**- and **47**-functionalized NPs and **48** as crosslinker. The association of $Ru(NH_3)_6^{3+}$ with the DNA array facilitates charge transport and enhances the resulting photocurrent. (Reproduced with permission from [195]. Copyright 2001 Wiley-VCH).

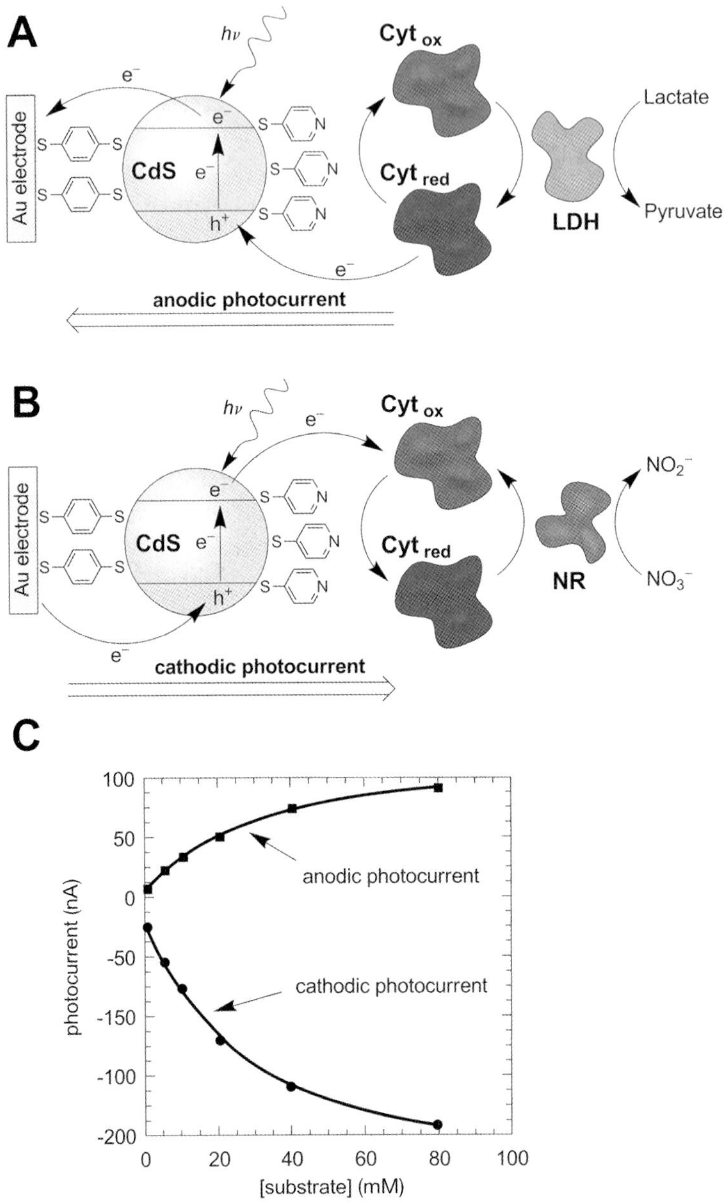

Figure 6.35 Generation of photocurrents by the photochemically induced activation of enzyme cascades by CdS NPs. (A) Photochemical activation of the cytochrome *c*-mediated oxidation of lactate in the presence of LDH. (B) Photochemical activation of the cytochrome *c*-mediated reduction of nitrate (NO_3^-) by nitrate reductase (NR). (C) Photocurrents generated by the biocatalytic cascades in the presence of various concentrations of the substrates (lactate/nitrate). (Reproduced from [198] by permission of The Royel Society of Chemistry).

photocurrent [Figure 6.35(B)]. The transfer of conduction band electrons to the oxidized, heme-containing cofactor generated the reduced cytochrome *c*, while the transfer of electrons from the electrode to the valence band holes of the NPs restored the ground state of the NPs. The cytochrome *c*-mediated biocatalyzed reduction of NO_3^- to nitrite then allowed the formation of the cathodic photocurrent, while biasing the electrode potential at 0 V vs. SCE. The photocurrents generated by the biocatalytic cascades at various concentrations of the different substrates are depicted in Figure 6.35 (C). These results demonstrated that photoelectrochemical functions of semiconductor NPs could be used to develop sensors for biocatalytic transformations. A related study employed CdSe/ZnS QDs capped with mercaptosuccinic acid as a protecting layer for the generation of photocurrents in the presence of cytochrome *c* [199].

In a different study [197], CdS NPs were assembled on an Au electrode, and the NPs were further modified with acetylcholinesterase (Figure 6.36). The biocatalyzed

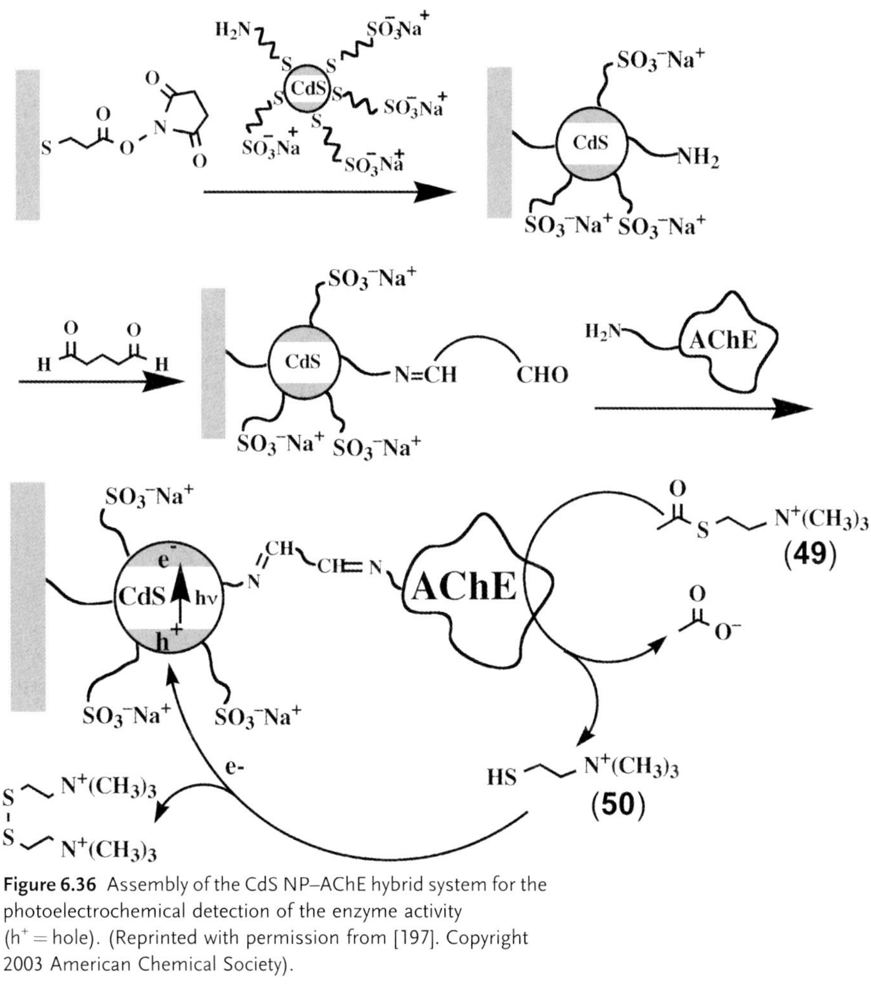

Figure 6.36 Assembly of the CdS NP–AChE hybrid system for the photoelectrochemical detection of the enzyme activity (h^+ = hole). (Reprinted with permission from [197]. Copyright 2003 American Chemical Society).

hydrolysis of acetylthiocholine (**49**) by acetylcholinesterase generated thiocholine (**50**), which acted as an electron donor for the photogenerated holes in the valence band of the CdS NPs. The resulting photocurrent was controlled by the concentration of the substrate and was depleted in the presence of inhibitors of acetylcholinesterase. The system was suggested as a potential sensor for chemical warfare agents that act as inhibitors of acetylcholinesterase.

6.9 Biomolecules as Catalysts for the Synthesis of Nanoparticles

It is well established that living organisms may synthesize NPs and even shaped metallic NPs [200–205]. For example, triangular gold nanoparticles were synthesized by using *Aloe vera* and lemongrass plant (*Cymbopogon flexuosus*) extracts as reducing agents [205, 206]. Although the mechanism of growth of the NPs is not clear, the resulting nanostructures originate from one or more products that are generated by the cell metabolism. This suggests that biomolecules might be active components for synthesizing NPs.

Indeed, a new emerging area in nanobiotechnology involves the use of biomaterials and, specifically, enzymes as active components for the synthesis and growth of particles [207]. As the enlargement of the NPs dominates their spectral properties (extinction coefficient, plasmon excitation wavelength), the biocatalytic reactions that yield the nanoparticles may be sensed by the optical properties of the generated NPs. Furthermore, enzyme–metal NP conjugates may provide biocatalytic hybrid systems that act as amplifying units for biosensing processes.

Different oxidases generate H_2O_2 upon biocatalyzed oxidation of the corresponding substrates by molecular oxygen. The H_2O_2 generated was found to reduce $AuCl_4^-$ in the presence of Au NP seeds that act as catalysts. This observation led to the development of an optical system for the detection of glucose oxidase activity and for the sensing of glucose [208] [Figure 6.37(A)]. A glass support was modified with an aminopropylsiloxane film to which negatively charged Au NPs were linked. Glucose oxidase (GOx) biocatalyzed the oxidation of glucose, and this led to the formation of H_2O_2 that acted as a reducing agent for the catalytic deposition of Au on the Au NPs associated with the glass support. The enlargement of the particles was then followed spectroscopically [Figure 6.37(B)]. Since the amount of the H_2O_2 formed is controlled by the concentration of glucose, the absorbance intensities of the resulting NPs are dominated by the concentration of glucose and the respective calibration curve [Figure 6.37(C)] was extracted. Other enzymes demonstrated, similarly, the biocatalytic generation of reducing products that grow NPs. For example, alkaline phosphatase hydrolyses *p*-aminophenol phosphate to yield *p*-aminophenol and the latter product reduces Ag^+ on Au NP seeds that act as catalytic sites [209].

Different bioactive *o*-hydroquinone derivatives such as the neurotransmitters dopamine (**51**), L-DOPA (**52**), adrenaline (**53**) and noradrenaline (**54**), were found to act as effective reducing agents of metal salts to the respective metal NPs, for example, the reduction of $AuCl_4^-$ to Au NPs without catalytic seeds [210]. For

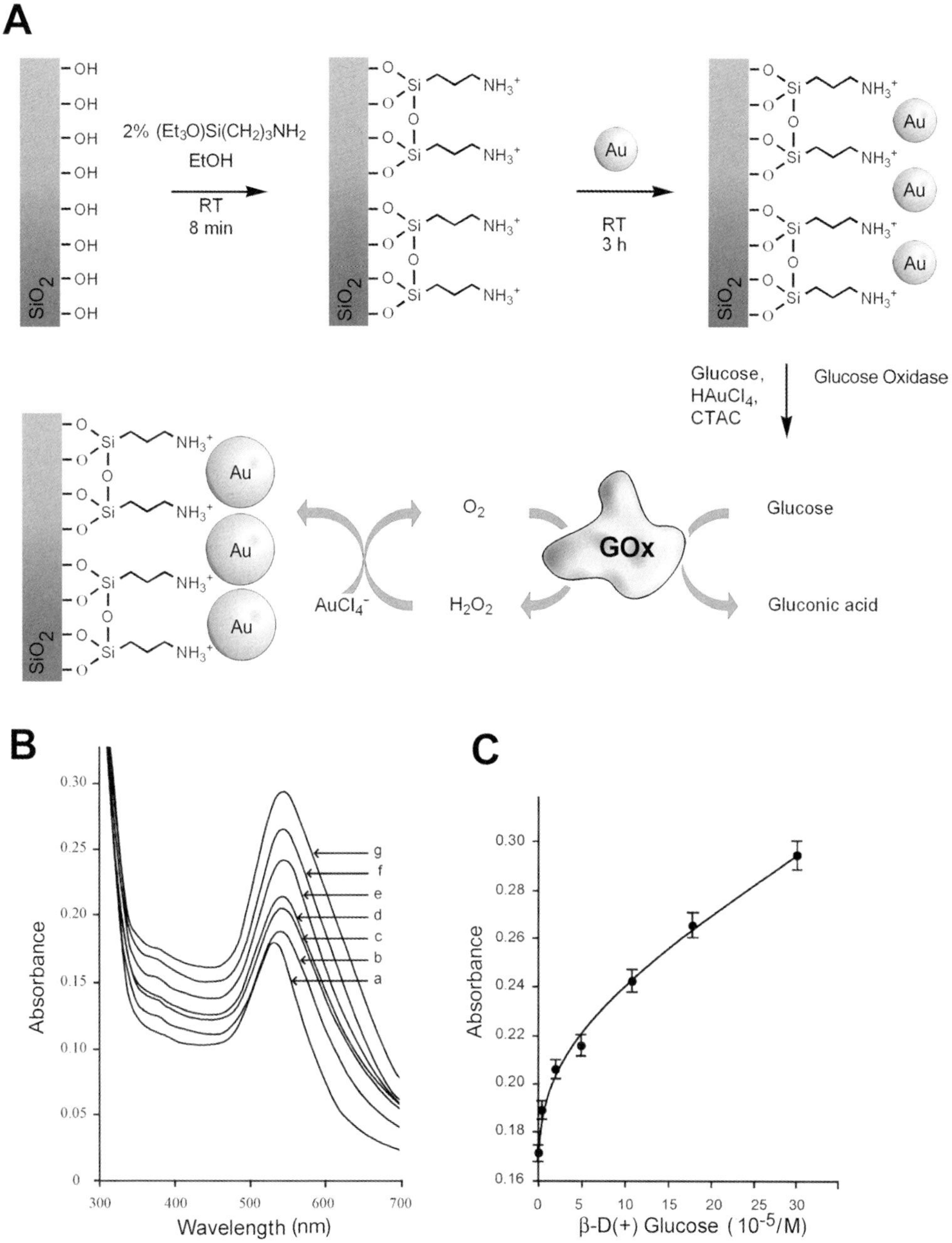

Figure 6.37 (A) Biocatalytic enlargement of Au NPs in the presence of glucose and glucose oxidase (GOx). RT, room temperature; CTAC, cetyltrimethylammonium chloride. (B) Absorbance spectra of Au NP-functionalized glass supports upon reaction with 2×10^{-4} M $HAuCl_4$ and 47 $\mu g\,mL^{-1}$ GOx in 0.01 M phosphate buffer that includes CTAC (2×10^{-3} M) and different concentrations of β-D-(+)-glucose: (a) 0, (b) 5×10^{-6}, (c) 2×10^{-5}, (d) 5×10^{-5}, (e) 1.1×10^{-4}, (f) 1.8×10^{-4} and (g) 3.0×10^{-4} M. For all experiments, the reaction time was 10 min and the temperature was 30 °C. (C) Calibration curve corresponding to the absorbance at $\lambda = 542$ nm of the Au NP-functionalized glass supports upon analyzing different concentrations of glucose. (Reprinted with permission from [208]. Copyright 2005 American Chemical Society).

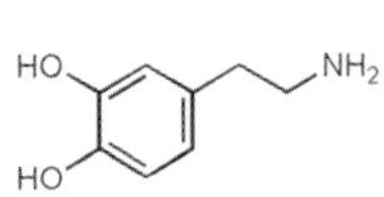

Dopamine (**51**)

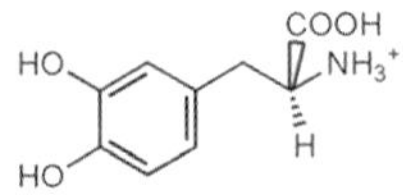

L-DOPA (**52**)

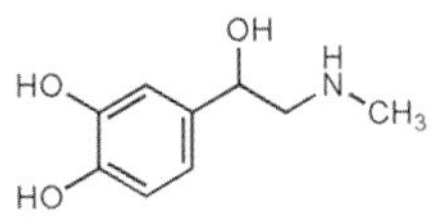

Adrenaline (**53**)

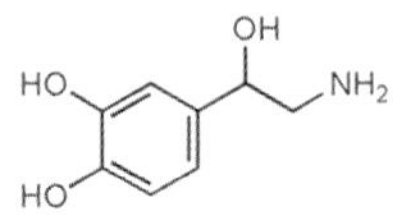

Noradrenaline (**54**)

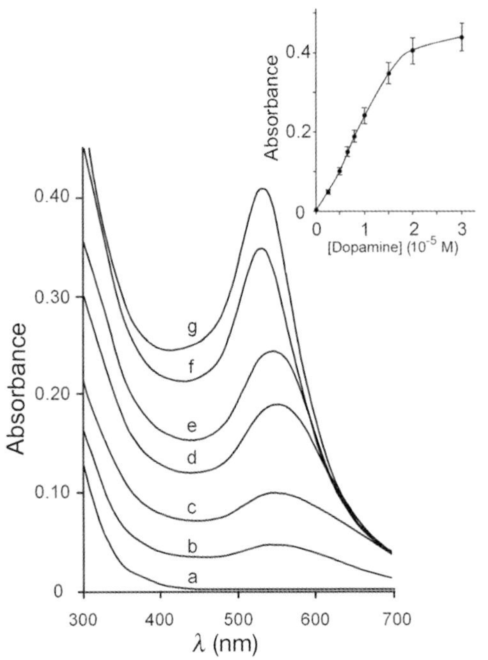

Figure 6.38 Absorbance spectra of Au NPs formed in the presence of different concentrations of dopamine (**51**): (a) 0, (b) 2.5×10^{-6}, (c) 5×10^{-6}, (d) 8×10^{-6}, (e) 1×10^{-5}, (f) 1.5×10^{-5} and (g) 2×10^{-5} M. All systems include $HAuCl_4$ (2×10^{-4} M) and CTAC (2×10^{-3} M) in 0.01 M phosphate buffer. Spectra were recorded after a fixed time interval of 2 min. Inset: calibration curve corresponding to the absorbance of the Au NP solution at $\lambda = 540$ nm formed in the presence of various concentrations of dopamine. (Reprinted with permission from [210]. Copyright 2005 American Chemical Society).

example, Figure 6.38 depicts the absorbance features of the Au NPs generated by different concentrations of dopamine (**51**). As the concentration of dopamine increased, the plasmon absorbance of the Au NPs was intensified. Although the Au NPs were enlarged, as the concentration of dopamine was elevated, the maximum absorbance of the plasmon band was blue-shifted. A detailed TEM analysis revealed that the increase in the concentration of dopamine indeed enhanced the growth of the

Au NP. Nonetheless, the enlargement process generated on the Au NP surface small (1–2 nm) Au clusters that were detached to the solution, and these acted as additional seeds for enlargement. As a result, the absorbance spectra corresponded to the enlarged particles and to the numerous small clusters/NPs, and these led to the blue shift in the plasmon absorbance. The spectral features permitted the extraction of a calibration curve corresponding to the optical detection of dopamine (Figure 6.38, inset). Analogous results were observed for the detection of the other neurotransmitters (**52–54**) [210].

The optical detection of L-DOPA (**52**) by means of the Au NPs enabled the following of the activity of tyrosinase. The enzyme tyrosinase is specifically expressed by melanocytes and melanoma cells and is viewed as a specific marker for these cells [211]. Tyrosinase hydroxylates tyrosine to L-DOPA using O_2 as the oxygen source [Figure 6.39(A)]. Consequently, the biocatalytically generated L-DOPA stimulated the synthesis of the Au NPs, and these acted as optical labels for the activity of the enzyme. Figure 6.39(B) shows the absorbance spectra of the Au NPs upon analyzing different amounts of tyrosinase. It was found that tyrosinase could be detected by this method with a sensitivity limit of 10 units.

The catalytic growth of metallic NPs by enzymes and the optical monitoring of the biocatalytic transformations were extended to probe enzyme inhibition [212]. The biocatalyzed hydrolysis of acetylthiocholine (**49**) (an analog of acetylcholine) yields thiocholine (**50**), and the thiol product was found to act as a reducing agent that reduces $AuCl_4^-$ on Au NP seeds. The enlargement of the NPs was used to follow the AChE

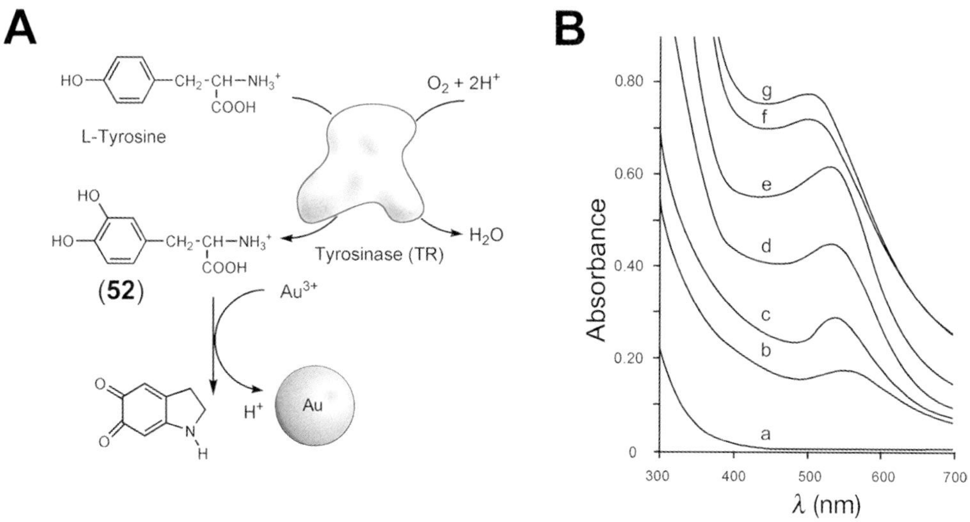

Figure 6.39 (A) Assay of tyrosinase activity through the biocatalyzed oxidation of tyrosine and the L-DOPA (**52**)-mediated formation of Au NPs. (B) Absorbance spectra of the Au NPs formed by various concentrations of tyrosinase: (a) 0, (b) 10, (c) 20, (d) 30, (e) 35, (f) 40 and (g) 60 U mL^{-1}. All systems include tyrosine (2×10^{-4} M), $HAuCl_4$ (2×10^{-4} M) and CTAC (2×10^{-3} M) in 0.01 M phosphate buffer. Spectra were recorded after a fixed time interval of 10 min. (Reprinted with permission from [210]. Copyright 2005 American Chemical Society).

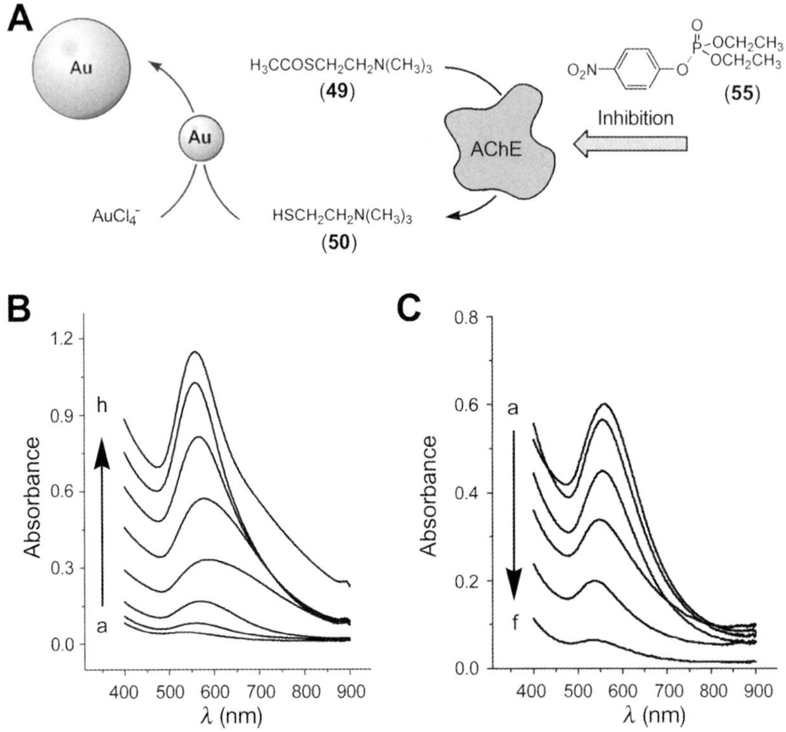

Figure 6.40 (A) AChE-mediated growth of Au NPs and its inhibition by a nerve gas analog. (B) Absorbance spectra of the Au NPs formed in the presence of AChE (0.13 U mL^{-1}), $HAuCl_4$ (1.1×10^{-3} M), Au NP seeds (3.6×10^{-8} M) and various concentrations of acetylthiocholine (**49**): (a) 0, (b) 2.4×10^{-5}, (c) 4.8×10^{-5}, (d) 9.5×10^{-5}, (e) 1.4×10^{-4}, (f) 1.9×10^{-4}, (g) 2.4×10^{-4} and (h) 3.8×10^{-4} M. Spectra were recorded after 5 min of particle growth. (C) Absorbance spectra of the Au NPs formed in the presence of AChE and acetylthiocholine in the presence of different concentrations of the inhibitor (**55**): (a) 0, (b) 1.0×10^{-8}, (c) 5.0×10^{-8}, (d) 1.0×10^{-7}, (e) 2.0×10^{-7} and (f) 4×10^{-7} M. All systems include AChE (0.13 U mL^{-1}), $HAuCl_4$ (1.1×10^{-3} M), Au NP seeds (3.6×10^{-8} M) and acetylthiocholine (1.4×10^{-3} M). In all experiments, AChE was incubated with the respective concentration of the inhibitor for a time interval of 20 min, and the absorbance spectra of the enlarged Au NPs were recorded after 5 min of biocatalytic growth. (Reprinted with permission from [212]. Copyright 2005 American Chemical Society).

activity [Figure 6.40(A)]. The fact that the enzyme controls the absorbance of the Au NPs and thus their degree of enlargement [Figure 6.40(B)] suggested that upon inhibition of AChE the growth of the particles would be blocked. Indeed, the addition of paraoxon (**55**), a well-established AChE irreversible inhibitor that mimics the functions of organophosphate nerve gases, to the biocatalytic system that synthesizes the nanoparticles resulted in the inhibition of the growth of the Au NPs [Figure 6.40(C)]. Also, an electron-transfer mediator, Os(II) bispyridine-4-picolinic acid, was used for the enlargement of Au NP seeds through the biocatalytic oxidation of glucose by GOx [213]. A similar mediated redox mechanism was used to follow the inhibition of AChE through blocking the enlargement of the Au NPs [213].

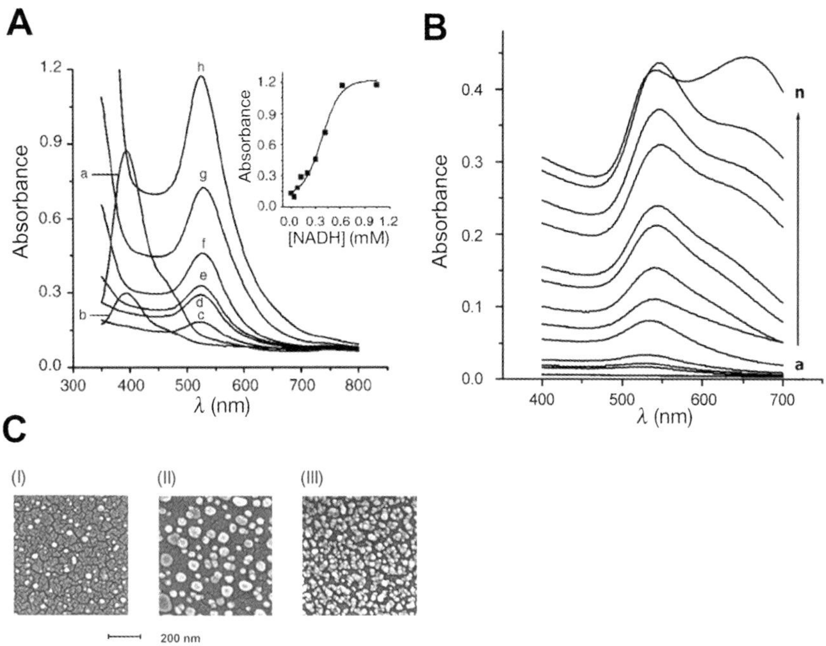

Figure 6.41 (A) Absorption spectra corresponding to the growth of Au NPs (4.0×10^{-10} M) in solution in the presence of increasing concentrations of NADH: (a) 0, (b) 4.2×10^{-5}, (c) 8.4×10^{-5}, (d) 12.6×10^{-5}, (e) 21×10^{-5}, (f) 30×10^{-5}, (g) 42×10^{-5} and (h) 63×10^{-5} M. Inset: calibration curve for the analysis of NADH by the resulting Au NPs at $\lambda = 524$ nm. (B) Absorbance changes of the Au NP–aminopropylsiloxane-functionalized glass slides upon enlargement with different concentrations of NADH: (a) 0, (b) 14×10^{-5}, (c) 27×10^{-5}, (d) 34×10^{-5}, (e) 41×10^{-5}, (f) 44×10^{-5}, (g) 48×10^{-5}, (h) 51×10^{-5}, (i) 54×10^{-5}, (j) 58×10^{-5}, (k) 61×10^{-5}, (l) 65×10^{-5}, (m) 68×10^{-5} and (n) 1.36×10^{-3} M. All systems included $HAuCl_4$ (1.8×10^{-4} M) and CTAB (7.4×10^{-2} M). (C) SEM images of NADH-enlarged Au particles generated on an Au NP-functionalized glass support using $HAuCl_4$ (1.8×10^{-4} M), CTAB (7.4×10^{-2} M) and NADH: (I) 2.7×10^{-4}; (II) 5.4×10^{-4} and (III) 1.36×10^{-3} M. (Reproduced with permission from [75]. Copyright 2004 Wiley-VCH).

1,4-Dihydronicotinamide adenine dinucleotide (phosphate) (NADH or NADPH) cofactors were found to enlarge Au NP seeds by the reduction of $AuCl_4^-$. The enzyme/cofactor-mediated enlargement or synthesis of metal NPs thus provides a means to follow the activities of enzymes or their substrates. Numerous biocatalyzed transformations involve NAD^+-dependent enzymes and the coupling of these biocatalysts to the growth of Au NPs allowed the development of different biosensors [75]. Figure 6.41(A) shows the absorbance spectra of a system consisting of Au NP seeds, $AuCl^{4-}$ and cetyltrimethylammonium bromide (CTAB) upon addition of various concentrations of NADH. The plasmon absorbance of the generated Au NPs is intensified as the concentration of NADH increases. Accordingly, the growth of the Au NPs allowed the quantitative optical analysis of NADH [see the calibration curve in Figure 6.41(A), inset]. The enlargement of the Au NP seeds was also examined on surfaces. Citrate-stabilized Au NP seeds were immobilized on an aminopropylsiloxane film associated with a glass support, and the functionalized surface was treated with different concentrations of NADH in the presence of $AuCl_4^-$. The growth of the Au NPs was followed spectroscopically [Figure 6.41(B)]. The plasmon absorbance of the Au NPs increased as the concentration of NADH was elevated, consistent with the growth of the particles. The SEM images of the particles confirmed the growth of the NPs on the glass support, [Figure 6.41(C)]. The particles generated by 2.7×10^{-4}, 5.4×10^{-4} and 1.36×10^{-3} M NADH are shown in images (I), (II) and (III), which reveal the generation of Au NPs with an average diameter of 13 ± 1, 40 ± 8 and 20 ± 5 nm, respectively. The NPs generated by the high concentration of NADH (III) reveals a 2D array of enlarged particles that touch each other, consistent with the spectral features of the surface.

The quantitative growth of Au NPs by NADH allowed the assay of NAD^+-dependent enzymes and their substrates. This was demonstrated by the application of the enzyme lactate dehydrogenase (LDH) and lactate as substrate [75] [Figure 6.42(A)]. The absorbance spectra of the NPs enlarged in the presence of different concentrations of lactate, are shown in Figure 6.42(B). From the derived calibration curve, lactate

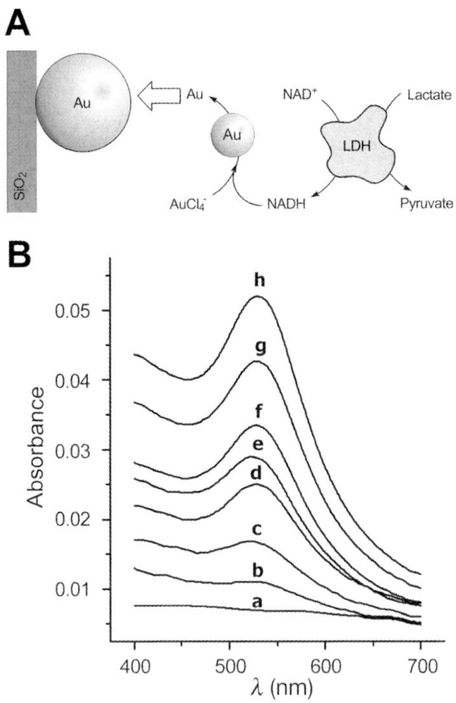

Figure 6.42 (A) Biocatalytic enlargement of Au NPs by the NAD^+–LDH–lactate system. (B) Spectral changes of the Au NP-functionalized glass supports upon interaction with the growth solution consisting of $HAuCl_4$ (1.8×10^{-4} M), CTAB (7.4×10^{-2} M) and lactate–LDH-generated NADH formed within 30 min in the presence of various concentrations of lactate: (a) 0, (b) 2.9×10^{-3}, (c) 3.6×10^{-3}, (d) 5.1×10^{-3}, (e) 5.8×10^{-3}, (f) 6.6×10^{-3}, (g) 7.3×10^{-3}, (h) 9.8×10^{-3} M. (Reproduced with permission from [75]. Copyright 2004 Wiley-VCH).

could be analyzed with a sensitivity limit corresponding to 3×10^{-3} M. Also, the reduced cofactor NADH was also employed to reduce Cu^{2+} ions to Cu^0 metal deposited on core Au NPs [214].

The NADH-induced growth of Au NPs was further developed by demonstrating that the reduced cofactor was able to synthesize, under controlled conditions, shaped NPs (in the form of dipods, tripods and tetrapods). These shaped particles exhibit unique optical properties (color) as a result of their high aspect ratio and the existence of a longitudinal plasmonic exciton [215]. It was demonstrated that in basic aqueous solution (pH = 11) that included ascorbate and NADH as reducing agents and CTAB as surfactant, the rapid formation of the shaped Au NPs was observed. The Au NP shapes consisted of dipods (12%), tripods (45%), tetrapods (13%) and spherical particles (30%) (Figure 6.43). The shaped particles were made of "arms" about 20–25 nm long and with a width of 2–5 nm (Figure 6.43). The development of the shaped particles was controlled by the concentration of NADH. At low NADH concentrations, "embryonic-type" shapes were observed, and at high cofactor concentrations, well-shaped particles were detected [Figure 6.43(A)–(C)]. A series of experiments indicated that the particles were formed in two steps. In the primary step, ascorbate reduced, at pH 11, $AuCl_4^-$ to Au NP seeds. These particles acted as catalysts for the rapid NADH-mediated reduction of $AuCl_4^-$ to Au^0 that was deposited on the seeds. High-resolution transmission electron microscopy (HRTEM) allowed

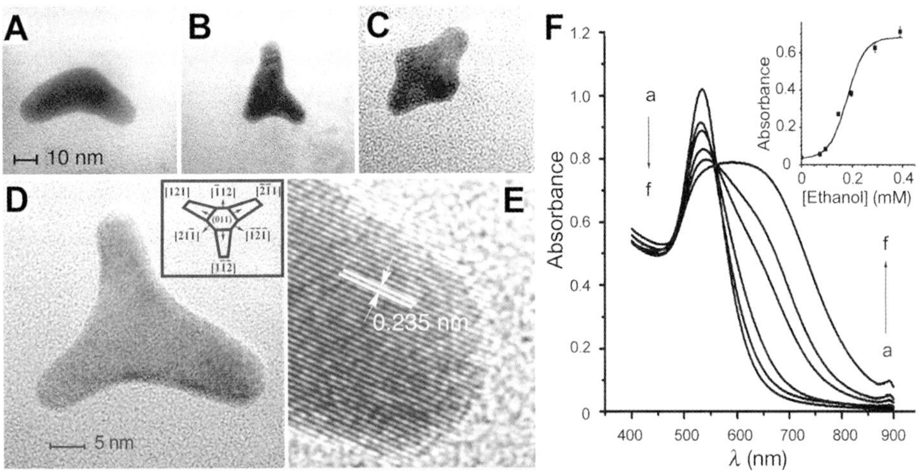

Figure 6.43 (A–C) Typical TEM images of dipod-, tripod- and tetrapod-shaped Au NPs formed in the presence of 4.0×10^{-6} M NADH, respectively. (D) HRTEM image of a representative tripod-shaped Au NP formed in the presence of, 4.0×10^{-6} M NADH. (E) HRTEM analysis of the tripod Au NP. The inset in (D) shows the crystal planes and the tripod growth directions extracted from the HRTEM analysis. (F) Absorbance spectra corresponding to the Au NPs formed by the biocatalytic generation of NADH in the presence of alcohol dehydrogenase (AlcDH) and variable concentrations of ethanol: (a) 0, (b) 7.3×10^{-5}, (c) 9.4×10^{-5}, (d) 1.5×10^{-4}, (e) 2.0×10^{-4} and (f) 2.9×10^{-4} M. Inset: calibration curve corresponding to the absorbance of the shaped Au NPs at $\lambda = 680$ nm formed in the presence of different concentrations of ethanol. (Reprinted with permission from [215]. Copyright 2005 American Chemical Society).

the detailed analysis of the growth directions of the shaped NPs, and this enabled the growth mechanism to be proposed. For example, HRTEM images of tripod particles are given, Figure 6.43(D) and (E). The lattice planes exhibit an inter-planar distance of 0.235 nm that corresponds to the {1 1 1} type planes of crystalline gold. The pods, separated by 120°, revealed a crystallite orientation of [0 1 1] with growth directions of the pods of type <211>; namely the pods are extended the direction $[1\bar{1}\bar{2}]$, $[\bar{2}\bar{1}1]$ and [1 2 1] [Figure 6.43(D), inset]. The shaped Au NPs revealed a red-shifted plasmon absorbance band at $\lambda = 680$ nm, consistent with a longitudinal plasmon exciton in the "rod-like" structures. The blue color of the shaped Au NPs distinctly differs from the red spherical Au NPs.

The fact that the degree of shaping and the absorbance spectra of the resulting Au NPs were controlled by the concentration of NADH allowed the development of an ethanol biosensor based on the shape-controlled synthesis of NPs. The biocatalyzed oxidation of ethanol by AlcDH yields NADH, and the biocatalytically generated NADH acted as an active reducing agent for the formation of blue shaped Au NPs. As the amount of NADH is controlled by the concentration of ethanol, the extent of the structurally developed shaped nanoparticles exhibiting the red-shifted longitudinal plasmon was then controlled by the substrate concentration. Figure 6.43(F) shows the absorbance spectra resulting from the gradual development of the shaped Au NPs formed in the presence of different concentrations of ethanol. The derived calibration curve corresponding to the optical analysis of ethanol by the shaped Au NPs is depicted in Figure 6.43(F), inset.

6.10
Biomolecule Growth of Metal Nanowires

The synthesis of nanowires is one of the challenging topics in nanobiotechnology [216, 217]. The construction of objects at the molecular or supramolecular level could be used to generate templates or "seeds" for nanometer-sized wires. Nanowires are considered as building blocks for the self-assembly of logic and memory circuits in future nanoelectronic devices [218]. Thus, the development of methods to assemble metal or semiconductor nanowires is a basic requirement for the construction of nanocircuits and nanodevices. Furthermore, the nanowires should include functional sites that allow their incorporation into structures of higher complexity and hierarchical functionality. The use of biomaterials as templates for the generation of nanowires and nanocircuitry is particularly attractive. Biomolecules exhibit dimensions that are comparable to those of the envisaged nano-objects. In addition to the fascinating structures of biomaterials that may lead to new inorganic or organic materials, templates of biological origin may act as "factories" for the production of "molds" for nanotechnology. The replication of DNA, the synthesis of proteins and the self-assembly of protein monomers into sheets, tubules and filaments all represent biological processes for the high-throughput synthesis of biomolecular templates for nanotechnology. Specifically, the coupling of NPs to biomolecules might yield new materials that combine the self-assembly properties of the biomolecules with the catalytic functions of the NPs. That is, the catalytic enlargement of the NPs associated with the biomolecules might generate the metallic nanowires.

Among the different biomaterials, DNA is of specific interest as a template for the construction of nanowires [219–221]. The ease of synthesis of DNA of controlled lengths and predesigned shapes, together with the information stored in the base sequence, introduce rich structural properties and addressable structural domains for the binding of the reactants that form nanowires. Also, numerous biocatalysts such as endonucleases, ligase, telomerase and polymerase can "cut", "paste", elongate or replicate DNA and may be considered as nanotools for shaping the desired DNA and, eventually, for the generation of nanocircuits. In addition, the intercalation of molecular components into DNA and the binding of cationic species such as metal ions to the phosphate units of nucleic acids allow the assembly of chemically active functional complexes, which may be used as precursors for the formation the nanowires.

One of the early examples that demonstrated the synthesis of Ag nanowires [222] is depicted in Figure 6.44. Two microelectrodes, which were positioned opposite one

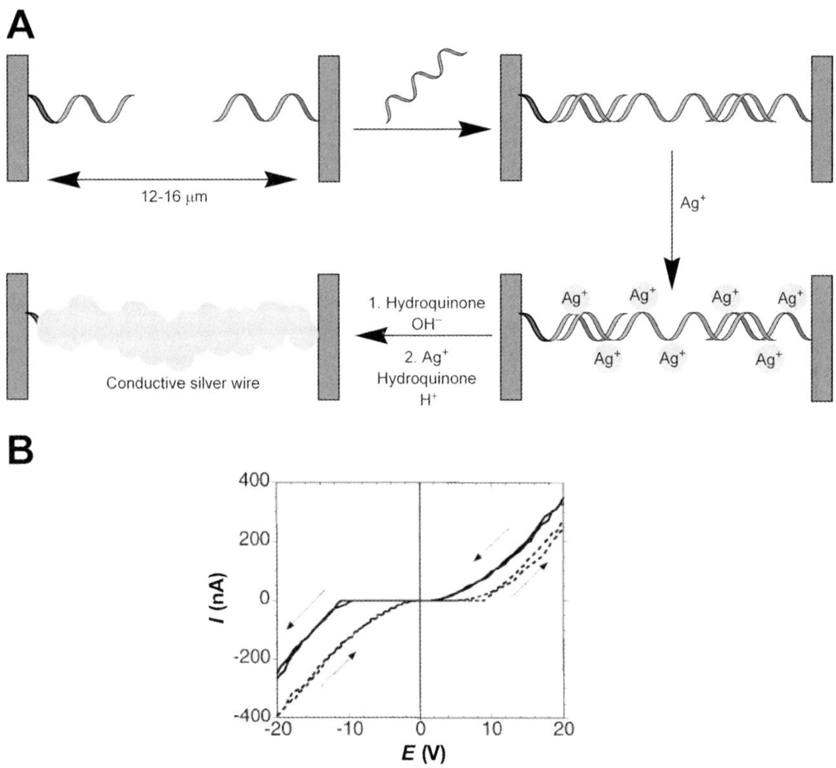

Figure 6.44 (A) Construction of a nanowire that bridges two microelectrodes by the deposition of Ag^+ ions on a bridging DNA strand followed by the chemical reduction of the Ag^+ ions to the metallic agglomerate. (B) Current versus voltage (I–V) curves obtained with the structures produced. (Reprinted with permission from Macmillan Publishers Ltd: Nature [222]. Copyright 1998).

another with a 12–16-μm separation gap, were functionalized with 12-base oligonucleotides that were then bridged with a 16-μm long λ-DNA [Figure 6.44(A)]. The resulting phosphate units of the DNA bridge were loaded with Ag^+ ions by ion-exchange and the bound Ag^+ ions were reduced to Ag metal with hydroquinone. The resulting small Ag aggregates along the DNA backbone were further reduced by reducing Ag^+ under acidic conditions and catalytic deposition of Ag on the Ag aggregates formed metallic nanowires. The resulting Ag nanowires exhibited dimensions corresponding to micrometer long and about 100 nm wide nanowires. The conduction properties of the nanowires revealed non-ohmic behavior and threshold potentials were needed to activate electron transport through the wires [Figure 6.44 (B)]. The potential gap in which no current passes through the nanowires was attributed to the existence of structural defects in the nanowires. The "hopping" of electrons across these barriers requires an overpotential that is reflected by the break voltage. This study was extended with the synthesis of many other metallic nanowires on DNA templates, and Cu [223], Pt [224] and Pd [225] nanowires were generated on DNA backbones.

The use of DNA and metallic NPs as templates and NPs as building blocks of nanowires was further demonstrated by using telomers synthesized by HeLa cancer cell extracts [226]. The constant repeat units that exist in the telomers provided addressable domains for the self-assembly of the NPs and the subsequent synthesis of the nanowires. By one approach, the primer **56** was telomerized in the presence of the dNTPs nucleotide mixture, which contained (aminoallyl)-dNTPs (**57**) [Figure 6.45(A)]. The resulting amine-containing telomers were treated with Au NPs (1.4 nm) (**58**), which were functionalized with the single *N*-hydroxysuccinimidyl ester groups, to yield Au NP-modified telomers. The Au NP-decorated DNA wires were then enlarged by electroless gold deposition to generate metallic nanowires [Figure 6.45(B) and (C)]. A second approach involved the telomerase-induced generation of telomers that included constant repeat units. The hybridization of the Au NPs functionalized with nucleic acids complementary to the telomer repeat domains, followed by electroless enlargement of the NPs, yielded Au nanowires.

The binding of proteins, such as RecA, to DNA has been used as a means for the patterning of nanoscale DNA-based metal wires with nonconductive or semiconductive gaps [227]. The metallization of the free, non-protein-coated DNA segments permitted the sequence-specific biomolecular lithography of DNA-based nanowires [228]. A single-stranded nucleic acid sequence was complexed with RecA and was carried to a double-stranded duplex DNA, a process that led to the nanolithographic patterned insulation of the DNA template by the protein [Figure 6.46(A)]. A carbon nanotube was then specifically attached to the protein patch using a series of antigen–antibody recognition processes: The anti-RecA antibody (Ab) was bound to the protein linked to the DNA duplex and the biotinylated anti-antibody was then linked to the RecA Ab. The latter Ab was used to bind specifically the streptavidin-coated carbon nanotubes on the protein patch. Ag^+ ions were bound to the free DNA domains and these were reduced to Ag clusters that were associated with the DNA assembly. The dsDNA was modified with aldehyde groups prior to this process to allow the chemical reduction of the electrostatically bound Ag^+ ions to Ag clusters. The

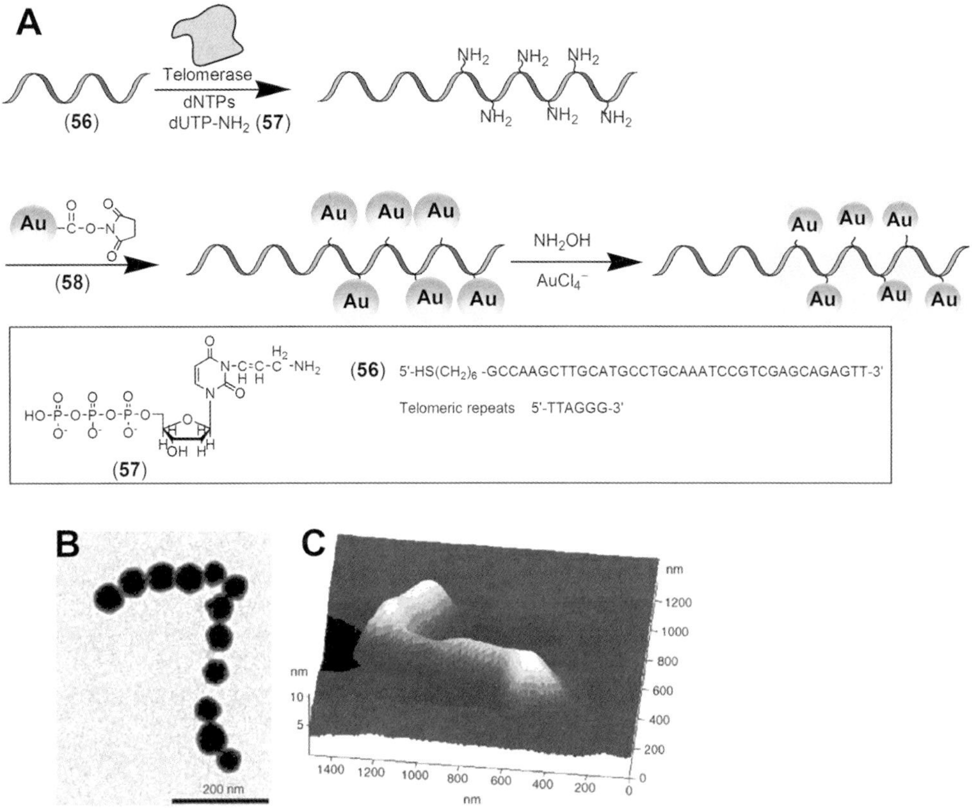

Figure 6.45 Assembly of Au nanowires on a telomer template. (A) Covalent attachment of Au NPs to amine groups, which were introduced into the telomer structure during the telomerization step, followed by catalytic enlargement of the NPs. (B) TEM and (C) AFM images of an Au nanowire that was generated according to the procedure outlined in (A). (Reprinted with permission from [226]. Copyright 2004 American Chemical Society).

subsequent catalytic electroless enlargement of the Ag^0 nanoclusters with gold generated the device where the two Au contacts were bridged by the carbon nanotubes. A representative SEM image of the device is depicted in Figure 6.46(B). The resulting device deposited on a Si substrate acted as a field-effect nanotransistor, where the gold contacts bridged by the CNT acted as the nanoscale source and drain electrodes, and the current flow through the CNT was gated by the applied potential on the Si support [Figure 6.46(C)]. Other DNA–protein arrays based on streptavidin and biotinylated DNA were used to organize tailored NP nanostructures that were subsequently metallized to composite architectures of nanowires [229].

Proteins may be applied as templates for the deposition of metal nanoparticles and for the formation of metal nanowires. For example, aromatic short-chain peptides, such as the Alzheimer's diphenylalanine β-amyloid (**59**), form well-ordered nano-

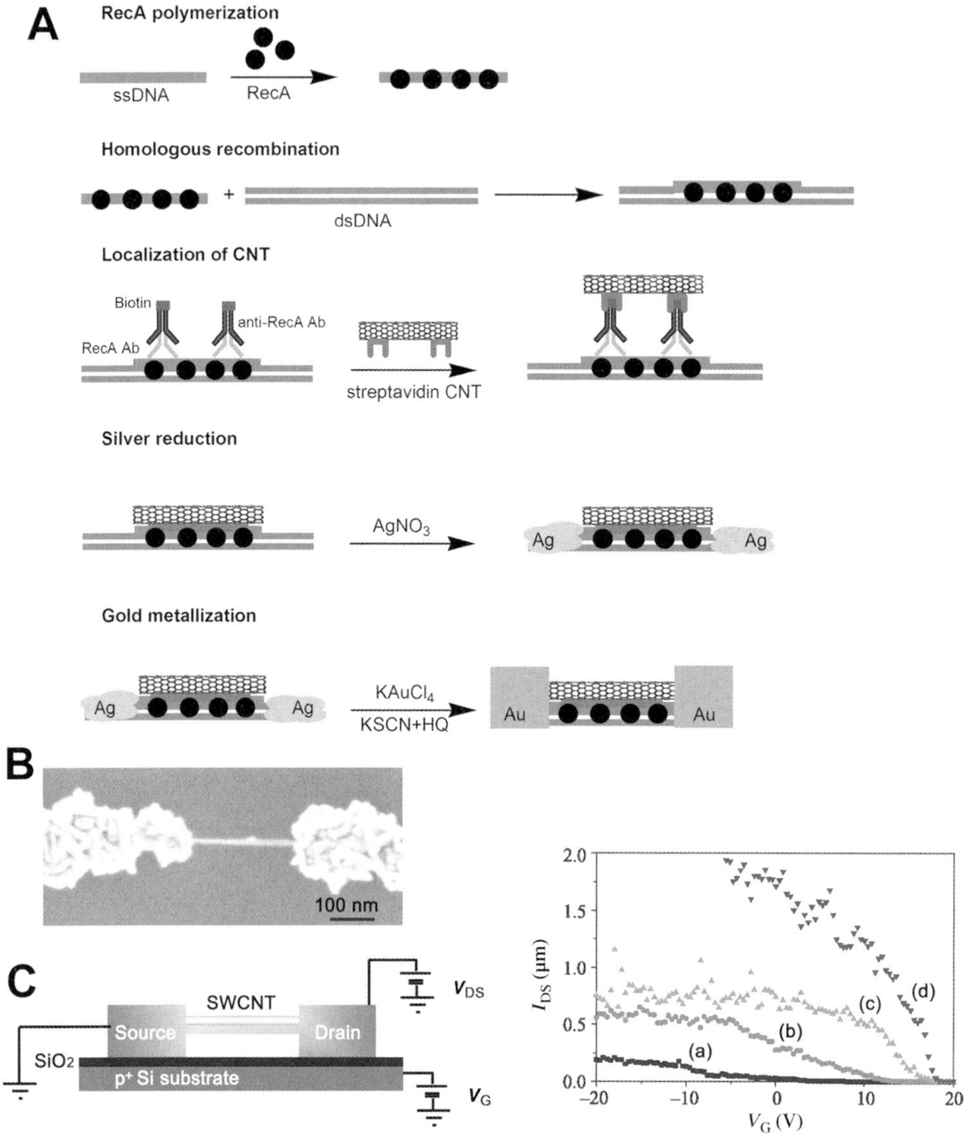

Figure 6.46 (A) Construction of a DNA-templated CNT FET. (B) SEM image of a rope of CNTs and metallic wires contacting it. (C) Electrical circuit and electrical characterization of the DNA-templated CNT FET. The drain–source current is given versus gate voltage for different values of drain–source bias: V_{DS} = (a) 0.5, (b) 1, (c) 1.5 and (d) 2 V. (Parts B and C reproduced from [227] with permission from AAAS).

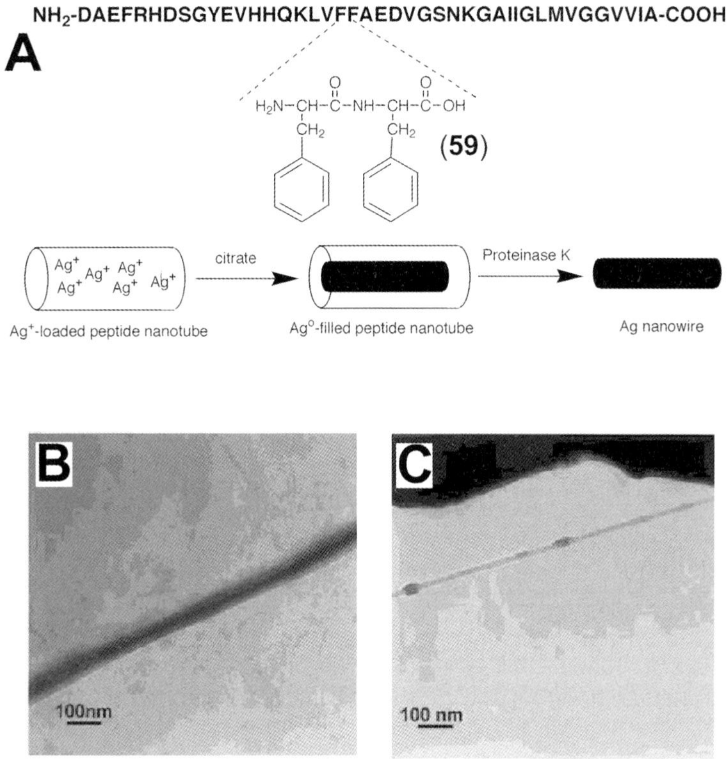

Figure 6.47 (A) Formation of a silver nanowire inside a channel of a short-chain diphenylalanine peptide tube. (B) TEM image of the peptide template, which is filled with a metallic silver nanowire. (C) TEM image of the silver nanowire after the degradation of the peptide template in the presence of proteinase K. (Parts B and C reproduced from [230] with permission from AAAS).

tubes that were used as templates for growing Ag nanowires [230] [Figure 6.47(A)]. The peptide nanotubes were loaded with Ag^+ ions, which were reduced with citrate to yield metallic silver nanowires inside the peptide nanotubes [Figure 6.47(B)]. The peptide coating was then removed by enzymatic degradation in the presence of proteinase K to yield micrometer-long Ag nanowires with a diameter of 20 nm [Figure 6.47(C)]. Upon application of D-phenylalanine-based peptide fibrils, which are resistant to proteinase K, the peptide coating of the Ag nanowires was preserved. Peptide nanotubes were also used to generate coaxial metal–insulator–metal nanotubes [231] [Figure 6.48(A)]. The D-phenylalanine peptide **59** assembled nanotube was interacted with Ag^+ ions and the resulting intra-tube-associated ions were reduced to form the Ag^0 nanowire. The resulting composite was further modified by tethering of a thiol-functionalized peptide to the nanowire peptide coating. The association of Au NPs with the thiol groups followed by deposition of gold on the Au NP seeds generated the resulting coaxial metal nanowires [Figure 6.48(B)].

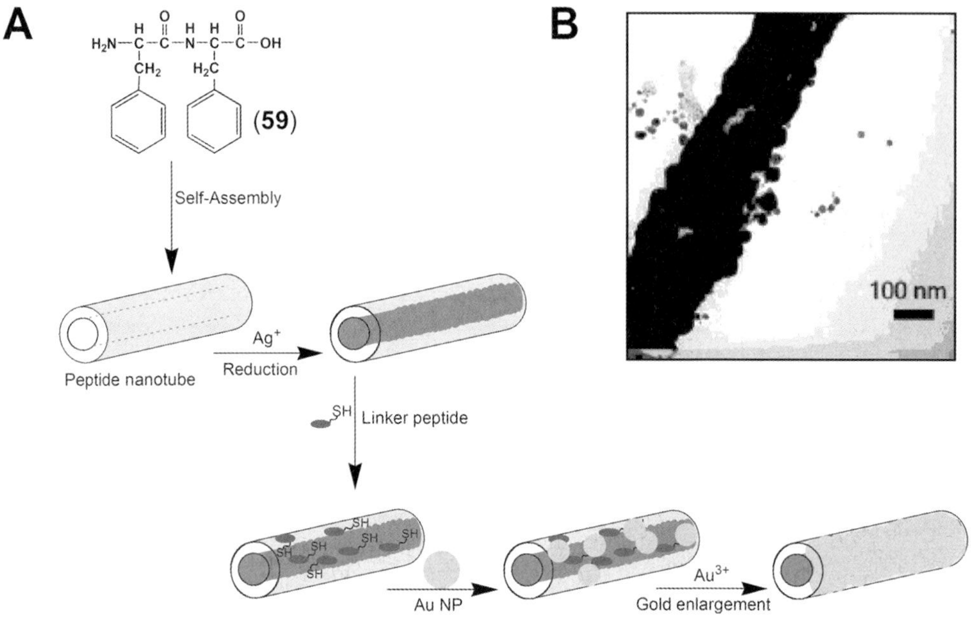

Figure 6.48 (A) Stepwise synthesis of coaxial Ag/Au nanowires in a peptide template. (B) TEM image of the coaxial wire formed in the peptide template. (Reproduced with permission from [231]. Copyright 2006 American Chemical Society).

The specific assembly of protein subunits into polymeric structures could provide a means for the patterning of the generated metal nanowires. The f-actin filament provides specific binding for the biomolecular motor protein myosin, which forms a complex with the filament, where the motility of myosin along the filament is driven by ATP [232, 233]. The f-actin filament is formed by the reversible association of g-actin subunits in presence of ATP, Mg^{2+} and Na^{+} ions. Accordingly, the f-actin filament was used as a template for the formation of metallic nanowires [234]. The f-actin filament was covalently modified with Au NPs (1.4 nm) that were functionalized with single *N*-hydroxysuccinimidyl ester groups, and the Au NP-functionalized g-actin subunits were then separated and used as versatile building blocks for the formation of the nanostructure. The Mg^{2+}/Na^{+}/ATP-induced polymerization of the functionalized monomers yielded the Au NP-functionalized filaments [Figure 6.49(A)], and electroless catalytic deposition of gold on the Au NP-functionalized f-actin filament yielded 1–3-μm long gold wires of height 80–150 nm. The gold wires revealed metallic conductivity with a resistance similar to that of bulk gold. By the sequential polymerization of naked actin filament units on the preformed Au NP–actin wire and the subsequent electroless catalytic deposition of gold on the Au NPs, patterned actin–Au wire–actin filaments were generated [Figure 6.49(C)]. A related approach was applied to yield the inverse Au wire–actin–Au wire patterned filaments [Figure 6.49(B)]. The nanostructure consisting of actin–Au nanowire–actin was used

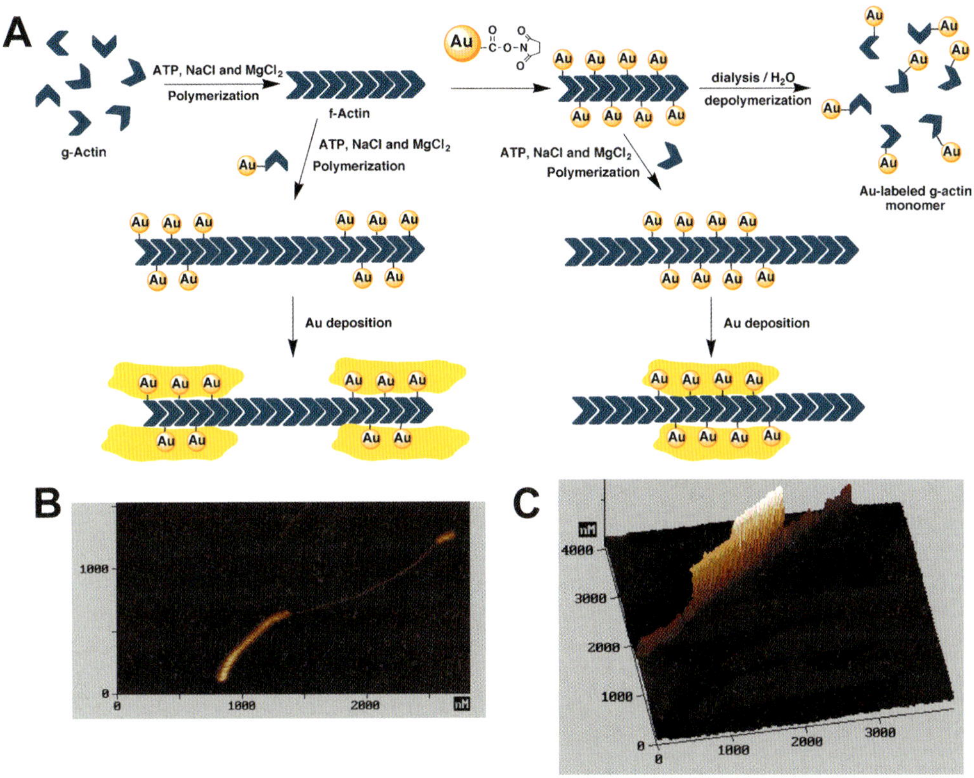

Figure 6.49 (A) Assembly of patterned actin-based Au nanowires: Left: an Au wire–actin–Au wire filament; Right: an actin–Au wire–actin filament. (B) AFM image of the Au wire–actin–Au wire filament. (C) AFM image of the actin–Au wire–actin filament. (Reprinted by permission from Macmillan Publishers Ltd: Nature Materials 234, Copyright (2004)).

as a nanotransporter driven by an external fuel. The actin–Au nanowire–actin nanostructure was deposited on a myosin-modified glass surface. Addition of ATP resulted in the motility of the nanostructures on the surface of $250 \pm 50\,\text{nm s}^{-1}$. Thus, such metallic nano-objects conjugated to motor proteins were suggested as potential nanotransporters, where chemicals deposited on the Au cargo are carried by the proteins.

In contrast with the use of proteins as passive templates for the growth of nanowires, one can use enzymes and NPs as active hybrid systems for the synthesis of nanocircuitry and for the preparation of patterned nanostructures. The participation of enzymes in growing metal NPs and particularly the biocatalytic enlargement of metal NP seeds suggests that metal NP–enzyme hybrids could be used as active components for the synthesis of metallic nanowires. The flavoenzyme glucose oxidase was functionalized with Au NPs (1.4 nm, average loading of 12 NPs per

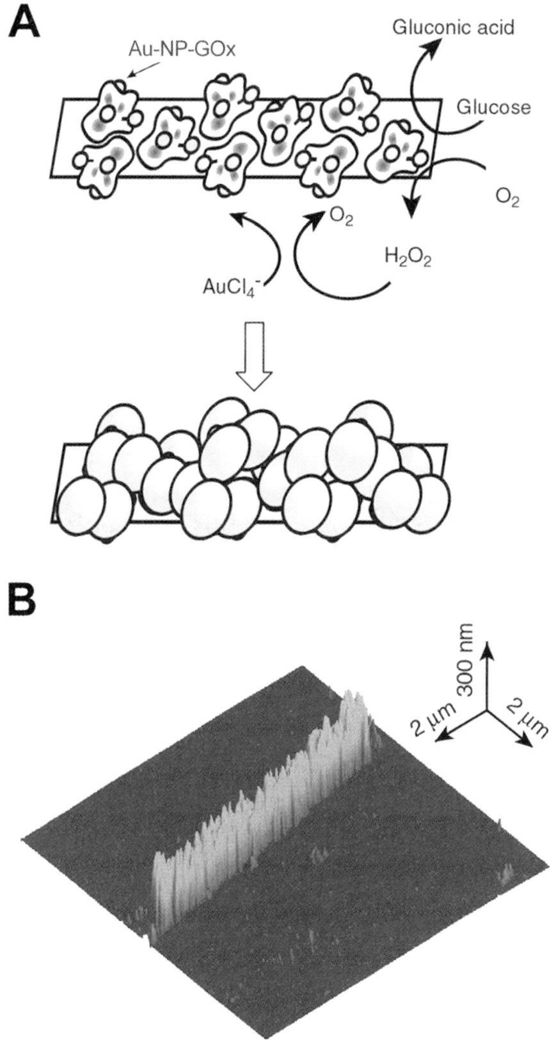

Figure 6.50 (A) Generation of an Au nanowire by the biocatalytic enlargement of an Au NP functionalized GOx line deposited on a silicon support by DPN. (B) Atomic force microscopy (AFM) image of the Au nanowire generated by the Au NP-functionalized GOx "biocatalytic ink". (Adapted with permission from [209]. Copyright 2006 Wiley-VCH).

enzyme unit) and the biocatalytic–NP hybrid material was used as a template for the stepwise synthesis of metallic nanowires [209] [Figure 6.50(A)]. The enzyme–Au NP hybrid was used as a "biocatalytic ink" for the pattering of Si surfaces using dip-pen nanolithography (DPN). The subsequent glucose-mediated generation of H_2O_2 and the catalytic enlargement of the NPs resulted in the catalytic growth of the particles, and this yielded micrometer-long Au metallic wires exhibiting heights and widths in the region of 150–250 nm, depending on the biocatalytic "development" time interval [Figure 6.50(B)].

This process is not limited to redox enzymes, and NP-functionalized biocatalysts that transform a substrate to an appropriate reducing agent may be similarly used as "biocatalytic inks" for the generation of metallic nanowires. This was exemplified with the application of Au NP-functionalized alkaline phosphatase, AlkPh (average loading of 10 NPs per enzyme unit) as "biocatalytic ink" for the generation of Ag nanowires [209] [Figure 6.51(A)]. The alkaline phosphatase-mediated hydrolysis of *p*-aminophenol phosphate (**60**) yielded *p*-aminophenol (**61**), which reduced Ag^+ to Ag^0 on the Au NP seeds associated with the enzyme. This allowed the enlargement of the Au particles acting as a core for the formation of a continuous silver wires exhibiting height of 30–40 nm, on the protein template [Figure 6.51 (B)]. The method allowed the stepwise, orthogonal formation of metal nanowires composed of different metals and controlled dimensions [Figure 6.51(C)]. Furthermore, the biocatalytic growth of the nanowires exhibited a self-inhibition mechanism and upon coating of the protein with the metal no further enlargement occurred. This self-inhibition mechanism is specifically important since the dimensions of the resulting nanowires are controlled by the sizes of the biocatalytic templates.

6.11 Conclusions and Perspectives

Nanobiotechnology, and particularly, biomolecule–NP hybrid systems, that represent the functional elements of nanobiotechnology, provide an emerging interdisciplinary scientific discipline. This chapter has addressed these scientific areas by discussing numerous biomolecule–NP conjugates as functional units for biosensing, building elements of nanocircuitry and devices and composite materials for intracellular diagnostics or medical therapeutics. Biomolecule–NP hybrid systems combine the evolution-optimized recognition and reactivity properties of biomolecules with the unique electronic, catalytic and optical properties of metallic or semiconductor NPs. The biomolecule–NP hybrids provide new materials that reveal unique properties and functions as a result of the composite structures and the quantum size effects accompanying the different nano-assemblies.

The analytical applications of biomolecule–NP systems have seen tremendously advances in the last decade. The electronic properties of metallic NPs were used to electrically communicate redox proteins with electrodes and to develop amperometric biosensors. The catalytic functions of metallic NPs were broadly applied to develop amplification methods for biosensing events, and the localized plasmonic features of NPs were extensively used to develop new optical methods that probe biorecognition events and design novel biosensor configurations. Optical phenomena such as the surface-enhanced fluorescence or Raman signals by dye-modified metallic NPs, the coupling of the localized plasmon of metallic NPs with surface plasmonic waves,

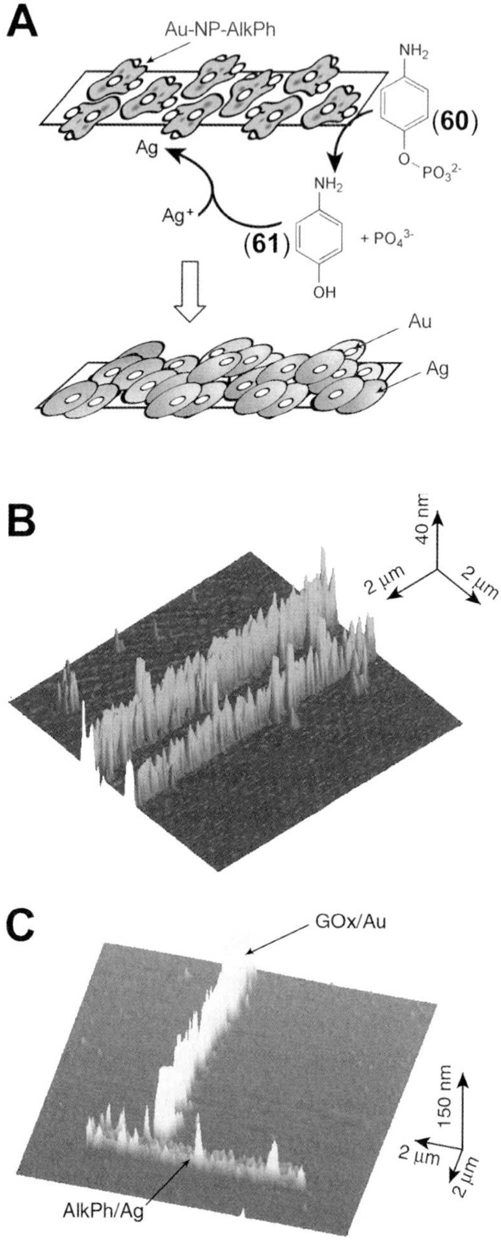

Figure 6.51 (A) Biocatalytic enlargement of Au NP-modified AlkPh with silver and fabrication of silver lines on a support using DPN and the "biocatalytic ink". (B) AFM image of Ag nanowires generated on the Au NP–AlkPh template deposited on the silicon support by DPN after 40 min of enlargement in the Ag growth solution. (C) AFM image of Au and Ag nanowires generated by deposition and enlargement of the Au NP–GOx template, followed by the passivation of the Au nanowire with mercaptoundecanoic acid and the subsequent deposition of Au NP–AlkPh template and its enlargement to the Ag nanowire. (Adapted with permission from [209]. Copyright 2006 Wiley-VCH).

the interparticle plasmon coupling in metallic NP aggregates and the reflectance of NPs were creatively used to image biorecognition events and to develop optical biosensors. Similarly, the coupling of biomolecules to semiconductor NPs (quantum dots) demonstrated the utility of these systems to assemble new optical and photoelectrochemical biosensor systems. The size-controlled emission properties of semiconductor QDs, as a result of the quantum confinement of the electronic levels in the nanoparticles, allow the multiplexed analysis of different targets and the design of high-throughput analyses in array formats. The development of photoelectrochemically based biosensors by the use of biomolecule–semiconductor hybrid systems highlights the bridge between the optical and electronic applications of biomolecule–NP hybrids for biosensing.

The development of biomolecule–NP-based sensors reached the level of practical applicability, and various analytical systems for clinical diagnostics, the analysis of food products, environmental pollutants and homeland security biosensors are expected to emerge from these hybrid nanocomposites. Some new challenging research directions in the analytical applications of biomolecule–NP hybrid systems can be identified, however. The unique optical properties of metal or semiconductor nanorods [235, 236] pave the way to new optical applications of these nano-objects. Composite nanorods may act as optical barcodes for biosensing events and could provide a rich library of labels for the parallel analysis of complex mixtures. Similarly, metal NP–semiconductor quantum dot (dumbbell) nanostructures [237] may provide new composites to assemble biomolecule hybrid systems of tailored electronic and optical functions.

Significant progress in the use of biomolecules as templates for the synthesis of nanostructures and nanocircuitry has been achieved in recent years. DNA and proteins act as efficient nanoscale templates for the synthesis of metallic or semiconductor nanowires. Although the scientific approaches to construct the nanowires are innovative, important challenges in this field are still ahead of us. The non-ohmic behavior of the conductance through nanowires and the difficulties in electrically wiring the nanowires to micro-contacts requires a fundamental experimental and theoretical understanding of charge transport phenomena across these nanostructures. The use of enzymes as catalysts that grow nanowires of predesigned structures [209] represents an important path to the further development of nanocircuitry and devices.

Numerous other applications of biomolecule–NP hybrid systems may be envisaged. The use of such systems in nanomedicine [238] has attracted increasing interest, and different applications such as photodynamic anticancer therapy [239], targeted delivery of radioisotopes [240], drug delivery [241] and gene therapy [242] have been demonstrated.

Physicists, chemists, biologists and material scientists have already recognized the limitless scientific opportunities and challenges in the applications of biomolecule–NP hybrid systems. The rapid progress in the field suggests that these interdisciplinary efforts will lead to exciting new science of immense practical and technological utility.

References

1 Mulvaney, P. (1996) *Langmuir*, **12**, 788–800.

2 Alvarez, M.M., Khoury, J.T., Schaaff, T.G., Shafigullin, M.N., Vezmar, I. and Whetten, R.L. (1997) *The Journal of Physical Chemistry B*, **101**, 3706–3712.

3 Mirkin, C.A., Letsinger, R.L., Mucic, R.C. and Storhoff, J.J. (1996) *Nature*, **382**, 607–609.

4 Aslan, K., Luhrs, C.C. and Perez-Luna, V.H. (2004) *The Journal of Physical Chemistry B*, **108**, 15631–15639.

5 Brus, L.E. (1991) *Applied Physics A*, **53**, 465–474.

6 Alivisatos, A.P. (1996) *Science*, **271**, 933–937.

7 Grieve, K., Mulvaney, P. and Grieser, F. (2000) *Current Opinion in Colloid and Interface Science*, **5**, 168–172.

8 Chan, W.C.W., Maxwell, D.J., Gao, X., Bailey, R.E., Han, M. and Nie, S. (2002) *Current Opinion in Biotechnology*, **13**, 40–46.

9 Sapsford, K.E., Pons, T., Medintz, I.L. and Mattoussi, H. (2006) *Sensors*, **6**, 925–953.

10 Katz, E., Willner, I. and Wang, J. (2004) *Electroanalysis*, **16**, 19–44.

11 Wang, J. (2003) *Analytica Chimica Acta*, **500**, 247–257.

12 Patolsky, F., Weizmann, Y., Lioubashevski, O. and Willner, I. (2002) *Angewandte Chemie-International Edition*, **41**, 2323–2327.

13 Daniel, M.-C. and Astruc, D. (2004) *Chemical Reviews*, **104**, 293–346.

14 Katz, E. and Willner, I. (2004) *Angewandte Chemie-International Edition*, **43**, 6042–6108.

15 Rosi, N.L. and Mirkin, C.A. (2005) *Chemical Reviews*, **105**, 1547–1562.

16 Wang, J. (2005) *Small*, **1**, 1036–1043.

17 Medintz, I.L., Uyeda, H.T., Goldman, E.R. and Mattoussi, H. (2005) *Nature Materials*, **4**, 435–446.

18 Pellegrino, T., Kudera, S., Liedl, T., Javier, A.M., Manna, L. and Parak, W.J. (2005) *Small*, **1**, 48–63.

19 Willner, I. (2002) *Science*, **298**, 2407–2408.

20 Bartlett, P.N., Tebbutt, P. and Whitaker, R.C. (1991) *Progress in Reaction Kinetics*, **16**, 55–155.

21 Degani, Y. and Heller, A. (1987) *Journal of Physical Chemistry*, **91**, 1285–1289.

22 Schuhmann, W., Ohara, T.J., Schmidt, H.-L. and Heller, A. (1991) *Journal of the American Chemical Society*, **113**, 1394–1397.

23 Willner, I., Riklin, A., Shoham, B., Rivenzon, D. and Katz, E. (1993) *Advanced Materials*, **5**, 912–915.

24 Heller, A. (1992) *Journal of Physical Chemistry*, **96**, 3579–3587.

25 Bu, H., Mikkelsen, S.R. and English, A.M. (1995) *Analytical Chemistry*, **67**, 4071–4076.

26 Willner, I., Katz, E., Lapidot, N. and Baüerle, P. (1992) *Bioelectrochemistry and Bioenergetics*, **29**, 29–45.

27 Willner, I., Heleg-Shabtai, V., Blonder, R., Katz, E., Tao, G., Bückmann, A.F. and Heller, A. (1996) *Journal of the American Chemical Society*, **118**, 10321–10322.

28 Zayats, M., Katz, E. and Willner, I. (2002) *Journal of the American Chemical Society*, **124**, 2120–2121.

29 Raitman, O.A., Katz, E., Bückmann, A.F. and Willner, I. (2002) *Journal of the American Chemical Society*, **124**, 6487–6496.

30 Raitman, O.A., Patolsky, F., Katz, E. and Willner, I. (2002) *Chemical Communications*, 1936–1937.

31 Willner, I. and Katz, E. (2000) *Angewandte Chemie-International Edition*, **39**, 1180–1218.

32 Murphy, L. (2006) *Current Opinion in Chemical Biology*, **10**, 177–184.

33 Heller, A. (2004) *Physical Chemistry Chemical Physics*, **6**, 209–216.

34 Willner, B., Katz, E. and Willner, I. (2006) *Current Opinion in Biotechnology*, **17**, 589–596.

35 Niemeyer, C.M. (2003) *Angewandte Chemie-International Edition*, **42**, 5796–5800.

36 Xiao, Y., Patolsky, F., Katz, E., Hainfeld, J.F. and Willner, I. (2003) *Science*, **299**, 1877–1881.

37 Zayats, M., Katz, E., Baron, R. and Willner, I. (2005) *Journal of the American Chemical Society*, **127**, 12400–12406.

38 Park, S.-J., Lazarides, A.A., Mirkin, C.A., Brazis, P.W., Kannewurf, C.R. and Letsinger, R.L. (2000) *Angewandte Chemie-International Edition*, **39**, 3845–3848.

39 Parak, W.J., Pellegrino, T., Micheel, C.M., Gerion, D., Williams, S.C. and Alivisatos, A.P. (2003) *Nano Letters*, **3**, 33–36.

40 Ghosh, S.S., Kao, P.M., McCue, A.W. and Chappelle, H.L. (1990) *Bioconjugate Chemistry*, **1**, 71–76.

41 Dubertret, B., Calame, M. and Libchaber, A.J. (2001) *Nature Biotechnology*, **19**, 365–370.

42 Niemeyer, C.M. (2001) *Chemistry – A European Journal*, **7**, 3188–3195.

43 Merkoçi, A., Aldavert, M., Marín, S. and Alegret, S. (2005) *Trends in Analytical Chemistry*, **24**, 341–349.

44 Merkoçi, A. (2007) *FEBS Journal*, **274**, 310–316.

45 Wang, J. (1985) *Stripping Analysis*, VCH, Weinheim.

46 Gonzáles-García, M.B. and Costa-García, A. (1995) *Bioelectrochemistry and Bioenergetics*, **38**, 389–395.

47 González-García, M.B., Fernández-Sánchez, C. and Costa-García, A. (2000) *Biosensors & Bioelectronics*, **15**, 315.

48 Ozsoz, M., Erdem, A., Kerman, K., Ozkan, D., Tugrul, B., Topcuoglu, N., Ekren, H. and Taylan, M. (2003) *Analytical Chemistry*, **75**, 2181–2187.

49 Wang, J., Xu, D., Kawde, A.-N. and Polsky, R. (2001) *Analytical Chemistry*, **73**, 5576–5581.

50 Authier, L., Grossirod, C., Brossier, P. and Limoges, B. (2001) *Analytical Chemistry*, **73**, 4450–4456.

51 Cai, H., Xu, Y., Zhu, N., He, P. and Fang, Y. (2002) *Analyst*, **127**, 803–808.

52 Dequire, M., Degrand, C. and Limoges, B. (2000) *Analytical Chemistry*, **72**, 5521–5528.

53 Wang, J., Polsky, R. and Xu, D. (2001) *Langmuir*, **17**, 5739–5741.

54 Lee, T.M.-H., Li, L.-L. and Hsing, I.-M. (2003) *Langmuir*, **19**, 4338–4343.

55 Li, L.-L., Cai, H., Lee, M.-H., Barford, J. and Hsing, I.-M. (2004) *Electroanalysis*, **16**, 81–87.

56 Wang, J. and Kawde, A.-N. (2002) *Electrochemistry Communications*, **4**, 349–352.

57 Wang, J., Xu, D. and Polsky, R. (2002) *Journal of the American Chemical Society*, **124**, 4208–4209.

58 Pumera, M., Castañeda, M.T., Pividori, M.I., Eritja, R., Merkoçi, A. and Alegret, S. (2005) *Langmuir*, **21**, 9625–9629.

59 Kawde, A. and Wang, J. (2004) *Electroanalysis*, **16**, 101–107.

60 Wang, J., Li, J., Baca, A.J., Hu, J., Zhou, F., Yan, W. and Pang, D.W. (2003) *Analytical Chemistry*, **75**, 3941–3945.

61 Wang, J., Rincón, O., Polsky, R. and Dominguez, E. (2003) *Electrochemistry Communications*, **5**, 83–86.

62 Park, S.-J., Taton, T.A. and Mirkin, C.A. (2002) *Science*, **295**, 1503–1506.

63 Velev, O.D. and Kaler, E.W. (1999) *Langmuir*, **15**, 3693–3698.

64 Polsky, R., Gill, R., Kaganovsky, L. and Willner, I. (2006) *Analytical Chemistry*, **78**, 2268–2271.

65 Buttry, D.A. and Ward, M.D. (1992) *Chemical Reviews*, **92**, 1355–1379.

66 Zhou, X.C., O'Shea, S.J. and Li, S.F.Y. (2000) *Chemical Communications*, 953–954.

67 Patolsky, F., Ranjit, K.T., Lichtenstein, A. and Willner, I. (2000) *Chemical Communications*, 1025–1026.

68 Liu, T., Tang, J. and Jiang, L. (2004) *Biochemical and Biophysical Research Communications*, **313**, 3–7.

69 Han, S., Lin, J., Satjapipat, M., Baca, A.J. and Zhou, F. (2001) *Chemical Communications*, 609–610.

70 Willner, I., Patolsky, F., Weizmann, Y. and Willner, B. (2002) *Talanta*, **56**, 847–856.

71 Weizmann, Y., Patolsky, F. and Willner, I. (2001) *Analyst*, **126**, 1502–1504.

72 Pavlov, V., Xiao, Y., Shlyahovsky, B. and Willner, I. (2004) *Journal of the American Chemical Society*, **126**, 11768–11769.

73 Bock, L.C., Griffin, L.C., Latham, J.A., Vermass, E.H. and Toole, J.J. (1992) *Nature*, **355**, 564–566.

74 Tasset, D.M., Kubik, M.F. and Steiner, W. (1997) *Journal of Molecular Biology*, **272**, 688–698.

75 Xiao, Y., Pavlov, V., Levine, S., Niazov, T., Markovitch, G. and Willner, I. (2004) *Angewandte Chemie-International Edition*, **43**, 4519–4522.

76 Wang, J., Liu, G., Polsky, R. and Merkoçi, A. (2002) *Electrochemistry Communications*, **4**, 722–726.

77 Wang, J., Liu, G. and Merkoçi, A. (2003) *Journal of the American Chemical Society*, **125**, 3214–3215.

78 Liu, G., Wang, J., Kim, J., Jan, M. and Collins, G. (2004) *Analytical Chemistry*, **76**, 7126–7130.

79 Wang, J., Liu, G. and Zhu, Q. (2003) *Analytical Chemistry*, **75**, 6218–6222.

80 Hansen, J.A., Wang, J., Kawde, A.-N., Xiang, Y., Gothelf, K.V. and Collins, G. (2006) *Journal of the American Chemical Society*, **128**, 2228–2229.

81 Wang, J., Lee, T. and Liu, G. (2005) *Journal of the American Chemical Society*, **127**, 38–39.

82 Hutter, E. and Fendler, J.H. (2004) *Advanced Materials*, **16**, 1685–1706.

83 Karlsson, R. (2004) *Journal of Molecular Recognition*, **17**, 151–161.

84 Englebienne, P., Hoonacker, A.V. and Verhas, M. (2003) *Spectroscopy*, **17**, 255–273.

85 Shankaran, D.R., Gobi, K.V. and Miura, N. (2007) *Sensors and Actuators B*, **121**, 158–177.

86 Schultz, D.A. (2003) *Current Opinion in Biotechnology*, **14**, 13–22.

87 Elghanian, R., Storhoff, J.J., Mucic, R.C., Letsinger, R.L. and Mirkin, C.A. (1997) *Science*, **277**, 1078–1081.

88 Storhoff, J.J., Elghanian, R., Mucic, R.C., Mirkin, C.A. and Letsinger, R.L. (1998) *Journal of the American Chemical Society*, **120**, 1959–1964.

89 Reynolds, R.A., III, Mirkin, C.A. and Letsinger, R.L. (2000) *Journal of the American Chemical Society*, **122**, 3795–3796.

90 Souza, G.R. and Miller, J.H. (2001) *Journal of the American Chemical Society*, **123**, 6734–6735.

91 Jin, R., Wu, G., Li, Z., Mirkin, C.A. and Schatz, G.C. (2003) *Journal of the American Chemical Society*, **125**, 1643–1654.

92 Beissenhirtz, M.K. and Willner, I. (2006) *Organic & Biomolecular Chemistry*, **4**, 3392–3401.

93 Weizmann, Y., Beissenhirtz, M.K., Cheglakov, Z., Nowarski, R., Kotler, M. and Willner, I. (2006) *Angewandte Chemie-International Edition*, **45**, 7384–7388.

94 Beissenhirtz, M.K., Elnathan, R., Weizmann, Y. and Willner, I. (2007) *Small*, **3**, 375–379.

95 Sato, K., Hosokawa, K. and Maeda, M. (2003) *Journal of the American Chemical Society*, **125**, 8102–8103.

96 Sato, K., Hosokawa, K. and Maeda, M. (2007) *Analytical Sciences*, **23**, 17–20.

97 Li, H. and Rothberg, L. (2004) *Proceedings of the National Academy of Sciences of the United States of America*, **101**, 14036–14039.

98 Li, H. and Rothberg, L. (2004) *Journal of the American Chemical Society*, **126**, 10958–10961.

99 Ellington, A.D. and Szostak, J.W. (1990) *Nature*, **346**, 818–822.

100 Tuerk, C. and Gold, L. (1990) *Science*, **249**, 505–510.

101 Wilson, D.S. and Szostak, J.W. (1999) *Annual Review of Biochemistry*, **68**, 611–647.

102 Schlatterer, J.C., Stuhlmann, F. and Jäschke, A. (2003) *ChemBioChem*, **4**, 1089–1092.
103 Breaker, R.R. and Joyce, G.F. (1994) *Chemistry & Biology*, **1**, 223–229.
104 Cuenoud, B. and Szostak, J.W. (1995) *Nature*, **375**, 611–614.
105 Santoro, S.W., Joyce, G.F., Sakthivel, K., Gramatikova, S. and Barbas, C.F. III (2000) *Journal of the American Chemical Society*, **122**, 2433–2439.
106 Travascio, P., Witting, P.K., Mauk, A.G. and Sen, D. (2001) *Journal of the American Chemical Society*, **123**, 1337–1348.
107 Travascio, P., Li, Y. and Sen, D. (1998) *Chemistry & Biology*, **5**, 505–517.
108 Santoro, S.W. and Joyce, G.F. (1997) *Proceedings of the National Academy of Sciences of the United States of America*, **94**, 4262–4266.
109 Faulhammer, D. and Famulok, M. (1996) *Angewandte Chemie-International Edition in English*, **35**, 2837–2841.
110 Liu, J. and Lu, Y. (2003) *Journal of the American Chemical Society*, **125**, 6642–6643.
111 Liu, J. and Lu, Y. (2004) *Analytical Chemistry*, **76**, 1627–1632.
112 Liu, J. and Lu, Y. (2006) *Angewandte Chemie-International Edition*, **45**, 90–94.
113 Guarise, C., Pasquato, L., De Filillis, V. and Scrimin, P. (2006) *Proceedings of the National Academy of Sciences of the United States of America*, **103**, 3978–3982.
114 Laromaine, A., Koh, L., Murugesan, M., Ulijn, R.V. and Stevens, M.M. (2007) *Journal of the American Chemical Society*, **129**, 4156–4157.
115 Choi, Y., Ho, N.-H. and Tung, C.-H. (2007) *Angewandte Chemie-International Edition*, **46**, 707–709.
116 Schofield, C.L., Field, R.A. and Russell, D.A. (2007) *Analytical Chemistry*, **79**, 1356–1361.
117 Huang, C.-C., Huang, Y.-F., Cao, Z., Tan, W. and Chang, H.-T. (2005) *Analytical Chemistry*, **77**, 5735–5741.
118 Taton, T.A., Mirkin, C.A. and Letsinger, R.L. (2000) *Science*, **289**, 1757–1760.
119 Taton, T.A., Lu, G.L. and Mirkin, C.A. (2001) *Journal of the American Chemical Society*, **123**, 5164–5165.
120 Storhoff, J.J., Marla, S.S., Bao, P., Hagenow, S., Mehta, H., Lucas, A., Garimella, V., Patno, T., Buckingham, W., Cork, W. and Müller, U.R. (2004) *Biosensors & Bioelectronics*, **19**, 875–883.
121 Bao, P., Huber, M., Wei, T.-F., Marla, S.S., Storhoff, J.J. and Müller, U.R. (2005) *Nucleic Acids Research*, **33**, e15.
122 Nam, J.-M., Thaxton, C.S. and Mirkin, C.A. (2003) *Science*, **301**, 1884–1886.
123 Nam, J.-M., Stoeva, S.I. and Mirkin, C.A. (2004) *Journal of the American Chemical Society*, **126**, 5932–5933.
124 Stoeva, S.I., Lee, J.-S., Smith, J.E., Rosen, S.T. and Mirkin, C.A. (2006) *Journal of the American Chemical Society*, **128**, 8378–8379.
125 Georganopoulou, D.G., Chang, L., Nam, J.M., Thaxton, C.S., Mufson, E.J., Klein, W.L. and Mirkin, C.A. (2005) *Proceedings of the National Academy of Sciences of the United States of America*, **102**, 2273–2276.
126 Stoeva, S.I., Lee, J.-S., Thaxton, C.S. and Mirkin, C.A. (2006) *Angewandte Chemie-International Edition*, **45**, 3303–3306.
127 Oh, B.-K., Nam, J.-M., Lee, S.W. and Mirkin, C.A. (2006) *Small*, **2**, 103–108.
128 Nam, J.-M., Wise, A.R. and Groves, J.T. (2005) *Analytical Chemistry*, **77**, 6985–6988.
129 Storhoff, J.J., Lucas, A.D., Garimella, V., Bao, Y.P. and Müller, U.R. (2004) *Nature Biotechnology*, **22**, 883–887.
130 Du, B.-A., Li, Z.-P. and Liu, C.-H. (2006) *Angewandte Chemie-International Edition*, **45**, 8022–8025.
131 Ray, P.C. (2006) *Angewandte Chemie-International Edition*, **45**, 1151–1154.
132 Kneipp, K., Kneipp, H., Itzkan, I., Dasari, R.R. and Feld, M.S. (1999) *Chemical Reviews*, **99**, 2957–2975.
133 Moskovits, M. (2005) *Journal of Raman Spectroscopy*, **36**, 485–496.
134 Haynes, C.L., McFarland, A.D. and Van Duyne, R.P. (2005) *Analytical Chemistry*, **77**, 338A–346A.

135 Faulds, K., Smith, W.E. and Graham, D. (2004) *Analytical Chemistry*, **76**, 412–417.

136 Graham, D., Faulds, K. and Smith, W.E. (2006) *Chemical Communications*, 4363–4371.

137 Faulds, K., Stewart, L., Smith, W.E. and Graham, D. (2005) *Talanta*, **67**, 667–671.

138 Faulds, K., Fruk, L., Robson, D.C., Thompson, D.G., Enright, A., Smith, W.E. and Graham, D. (2006) *Faraday Discussions*, **132**, 261–268.

139 Cao, Y.C., Jin, R. and Mirkin, C.A. (2002) *Science*, **297**, 1536–1540.

140 Cao, Y.C., Jin, R., Nam, J.-M., Thaxton, C.S. and Mirkin, C.A. (2003) *Journal of the American Chemical Society*, **125**, 14676–14677.

141 Grubisha, D.S., Lipert, R.J., Park, H.-Y., Driskell, J. and Porter, M.D. (2003) *Analytical Chemistry*, **75**, 5936–5943.

142 Driskell, J.D., Kwarta, K.M., Lipert, R.J. and Porter, M.D. (2005) *Analytical Chemistry*, **77**, 6147–6154.

143 Xu, S., Ji, X., Xu, W., Li, X., Wang, L., Bai, Y., Zhao, B. and Ozaki, Y. (2004) *Analyst*, **129**, 63–68.

144 Lyon, L.A., Musick, M.D. and Natan, M.J. (1998) *Analytical Chemistry*, **70**, 5177–5183.

145 Englebienne, P., Hoonacker, A.V. and Verhas, M. (2001) *Analyst*, **126**, 1645–1651.

146 Huang, L., Reekmans, G., Saerens, D., Friedt, J.-M., Frederix, F., Francis, L., Muyldermans, S., Campitelli, A. and van Hoof, C. (2005) *Biosensors & Bioelectronics*, **21**, 483–490.

147 He, L., Musick, M.D., Nicewarner, S.R., Salinas, F.G., Benkovic, S.J., Natan, M.J. and Keating, C.D. (2000) *Journal of the American Chemical Society*, **122**, 9071–9077.

148 Yao, X., Li, X., Toledo, F., Zurita-Lopez, C., Gutova, M., Momand, J. and Zhou, F. (2006) *Analytical Biochemistry*, **354**, 220–228.

149 Sato, Y., Sato, K., Hosokawa, K. and Maeda, M. (2006) *Analytical Biochemistry*, **355**, 125–131.

150 Yang, X., Wang, Q., Wang, K., Tan, W. and Li, H. (2007) *Biosensors & Bioelectronics*, **22**, 1106–1110.

151 Lioubashevski, O., Chegel, V., Patolsky, F., Katz, E. and Willner, I. (2004) *Journal of the American Chemical Society*, **126**, 7133–7143.

152 El-Sayed, M.A. (2001) *Accounts of Chemical Research*, **34**, 257–264.

153 Yonzon, C.R., Zhang, X., Zhao, J. and Van Duyne, R.P. (2007) *Spectroscopy*, **22**, 42–56.

154 Moores, A. and Goettmann, F. (2006) *New Journal of Chemistry*, **30**, 1121–1132.

155 Haes, A.J. and Van Duyne, R.P. (2004) *Analytical and Bioanalytical Chemistry*, **379**, 920–930.

156 Zhao, J., Zhang, X., Yonzon, C.R., Haes, A.J. and Van Duyne, R.P. (2006) *Nanomedicine*, **1**, 219–228.

157 Haes, A.J. and Van Duyne, R.P. (2002) *Journal of the American Chemical Society*, **124**, 10596–10604.

158 Riboh, J.C., Haes, A.J., McFarland, A.D., Yonzon, C.R. and Van Duyne, R.P. (2003) *The Journal of Physical Chemistry B*, **107**, 1772–1780.

159 Haes, A.J., Hall, W.P., Chang, L., Klein, W.L. and Van Duyne, R.P. (2004) *Nano Letters*, **4**, 1029–1034.

160 Haes, A.J., Chang, L., Klein, W.L. and Van Duyne, R.P. (2005) *Journal of the American Chemical Society*, **127**, 2264–2271.

161 Endo, T., Kerman, K., Nagatani, N., Takamura, Y. and Tamiya, E. (2005) *Analytical Chemistry*, **77**, 6976–6984.

162 Endo, T., Kerman, K., Nagatani, N., Hiepa, H.M., Kim, D.-K., Yonezawa, Y. and Tamiya, E. (2006) *Analytical Chemistry*, **78**, 6465–6475.

163 Alivisatos, A.P. (2004) *Nature Biotechnology*, **22**, 47–52.

164 Nirmal, M. and Brus, L.E. (1999) *Accounts of Chemical Research*, **32**, 407–414.

165 Niemeyer, C.M. (2001) *Angewandte Chemie-International Edition*, **40**, 4128–4158.

166 Katz, E., Shipway, A.N., and Willner, I. (2003) in *Nanoscale Materials*, (eds L.M.

Liz-Marzan and P. Kamat), Kluwer, Dordrecht, Chapter 2, pp. 5–78.

167 Michalet, X., Pinaud, F.F., Bentolila, L.A., Tsay, J.M., Doose, S., Li, J.J., Sundaresan, G., Wu, A.M., Gambhir, S.S. and Weiss, S. (2005) *Science*, **307**, 538–544.

168 Costa-Fernandez, J.M., Pereiro, R. and Sanz-Medel, A. (2006) *Trends in Analytical Chemistry*, **25**, 207–218.

169 Klostranec, J.M. and Chan, W.C.W. (2006) *Advanced Materials*, **18**, 1953–1964.

170 Murphy, C.J. (2002) *Analytical Chemistry*, **74**, 520A–526A.

171 Goldman, E.R., Balighian, E.D., Mattoussi, H., Kuno, M.K., Mauro, J.M., Tran, P.T. and Anderson, G.P. (2002) *Journal of the American Chemical Society*, **124**, 6378–6382.

172 Goldman, E.R., Anderson, G.P., Tran, P.T., Mattoussi, H., Charles, P.T. and Mauro, J.M. (2002) *Analytical Chemistry*, **74**, 841–847.

173 Goldman, E.R., Clapp, A.R., Anderson, G.P., Uyeda, H.T., Mauro, J.M., Medintz, I.L. and Mattoussi, H. (2004) *Analytical Chemistry*, **76**, 684–688.

174 Yang, L. and Li, Y. (2006) *Analyst*, **131**, 394–401.

175 Hoshino, A., Fujioka, K., Manabe, N., Yamay, S., Goto, Y., Yasuhara, M. and Yamamoto, K. (2005) *Microbiology and Immunology*, **49**, 461–470.

176 Gerion, D., Chen, F., Kannan, B., Fu, A., Parak, W.J., Chen, D.J., Majumdar, A. and Alivisatos, A.P. (2003) *Analytical Chemistry*, **75**, 4766–4772.

177 Pathak, S., Choi, S.-K., Arnheim, N. and Thompson, M.E. (2001) *Journal of the American Chemical Society*, **123** 4103–4104.

178 Xiao, Y. and Barker, P.E. (2004) *Nucleic Acids Research*, **32**, e28.

179 Chan, P.M., Yuen, T., Ruf, F., Gonzalez-Maeso, J. and Sealfon, S.C. (2005) *Nucleic Acids Research*, **33**, e161.

180 Mattoussi, H., Mauro, J.M., Goldman, E.R., Anderson, G.P., Sunder, V.C., Mikulec, F.V. and Bawendi, M.G. (2000) *Journal of the American Chemical Society*, **122**, 12142–12150.

181 Warren, C.W. and Nie, S. (1998) *Science*, **281**, 2016–2018.

182 Gerion, D., Pinaud, F., Willimas, S.C., Parak, W.J., Zanchet, D., Weiss, S. and Alivisatos, A.P. (2001) *The Journal of Physical Chemistry B*, **105**, 8861–8871.

183 Bruchez, M., Jr. Moronne, M., Gin, P., Weiss, S. and Alivisatos, A.P. (1998) *Science*, **281**, 2013–2016.

184 Gao, X., Cui, Y., Levenson, R.M., Chung, L.W.K. and Nie, S. (2004) *Nature Biotechnology*, **22**, 969–976.

185 Clapp, A.R., Medintz, I.L. and Mattoussi, H. (2006) *ChemPhysChem*, **7**, 47–57.

186 Sapsford, K.E., Berti, L. and Medintz, I.L. (2006) *Angewandte Chemie-International Edition*, **45**, 4562–4588.

187 Patolsky, F., Gill, R., Weizmann, Y., Mokari, T., Banin, U. and Willner, I. (2003) *Journal of the American Chemical Society*, **125**, 13918–13919.

188 Medintz, I.L., Clapp, A.R., Mattoussi, H., Goldman, E.R., Fisher, B. and Mauro, J.M. (2003) *Nature Materials*, **2**, 630–638.

189 Goldman, E.R., Medintz, I.L., Whitley, J.L., Hayhurst, A., Clapp, A.R., Uyeda, H.T., Deschamps, J.R., Lassman, M.E. and Mattoussi, H. (2005) *Journal of the American Chemical Society*, **127**, 6744–6751.

190 Medintz, I.L., Clapp, A.R., Brunel, F.M., Tiefenbrunn, T., Uyeda, H.T., Chang, E.L., Deschamps, J.R., Dawson, P.E. and Mattoussi, H. (2006) *Nature Materials*, **5**, 581–589.

191 Shi, L., De Paoli, V., Rosenzweig, N. and Rosenzweig, Z. (2006) *Journal of the American Chemical Society*, **128**, 10378–10379.

192 Gill, R., Freeman, R., Xu, J., Willner, I., Winograd, S., Shweky, I. and Banin, U. (2006) *Journal of the American Chemical Society*, **128**, 15376–15377.

193 Gill, R., Willner, I., Shweky, I. and Banin, U. (2005) *The Journal of Physical Chemistry B*, **109**, 23715–23719.

194 Levy, M., Cater, S.F. and Ellington, A.D. (2005) *ChemBioChem*, **6**, 2163–2166.

195 Willner, I., Patolsky, F. and Wasserman, J. (2001) *Angewandte Chemie-International Edition*, **40**, 1861–1864.

196 Gill, R., Patolsky, F., Katz, E. and Willner, I. (2005) *Angewandte Chemie-International Edition*, **44**, 4554–4557.

197 Pardo-Yissar, V., Katz, E., Wasserman, J. and Willner, I. (2003) *Journal of the American Chemical Society*, **125**, 622–623.

198 Katz, E., Zayats, M., Willner, I. and Lisdat, F. (2006) *Chemical Communications*, 1395–1397.

199 Stoll, C., Kudera, S., Parak, W.J. and Lisdat, F. (2006) *Small*, **2**, 741–743.

200 Klaus, T., Joerger, R., Olsson, E. and Granqvist, C.G. (1999) *Proceedings of the National Academy of Sciences of the United States of America*, **96**, 13611–13614.

201 Gardea-Torresdey, J.L., Parsons, J.G., Gomez, E., Peralta-Videa, J., Troiani, H.E., Santiago, P. and Yacaman, M.J. (2002) *Nano Letters*, **2**, 397–401.

202 Ahmad, A., Senapati, S., Khan, M., Kumar, R. and Sastry, M. (2003) *Langmuir*, **19**, 3550–3553.

203 Shankar, S.S., Ahmad, A. and Sastry, M. (2003) *Biotechnology Progress*, **13**, 1627–1631.

204 Ahmad, A., Mukherjee, P., Mandal, D., Senapati, S., Khan, M.I., Kumar, R. and Sastry, M. (2002) *Journal of the American Chemical Society*, **124**, 12108–12109.

205 Rai, A., Singh, A., Ahmad, A. and Sastry, M. (2006) *Langmuir*, **22**, 736–741.

206 Chandran, S.P., Chaudhary, M., Pasricha, R., Ahmad, A. and Sastry, M. (2006) *Biotechnology Progress*, **22**, 577–583.

207 Willner, I., Baron, R. and Willner, B. (2006) *Advanced Materials*, **18**, 1109–1120.

208 Zayats, M., Baron, R., Popov, I. and Willner, I. (2005) *Nano Letters*, **5**, 21–25.

209 Basnar, B., Weizmann, Y., Cheglakov, Z. and Willner, I. (2006) *Advanced Materials*, **18**, 713–718.

210 Baron, R., Zayats, M. and Willner, I. (2005) *Analytical Chemistry*, **77**, 1566–1571.

211 Angeletti, C., Khomitch, V., Halaban, R. and Rimm, D.L. (2004) *Diagnostic Cytopathology*, **31**, 33–37.

212 Pavlov, V., Xiao, Y. and Willner, I. (2005) *Nano Letters*, **5**, 649–653.

213 Xiao, Y., Pavlov, V., Shlyahovsky, B. and Willner, I. (2005) *Chemistry – A European Journal*, **11**, 2698–2704.

214 Shlyahovsky, B., Katz, E., Xiao, Y., Pavlov, V. and Willner, I. (2005) *Small*, **1**, 213–216.

215 Xiao, Y., Shlyahovsky, B., Popov, I., Pavlov, V. and Willner, I. (2005) *Langmuir*, **21**, 5659–5662.

216 Gazit, E. (2007) *FEBS Journal*, **274**, 317–322.

217 Baron, R., Willner, B. and Willner, I. (2007) *Chemical Communications*, 323–332.

218 Kovtyukhova, N.I. and Mallouk, T.E. (2002) *Chemistry – A European Journal*, **8**, 4354–4363.

219 Seeman, N.C. (1998) *Annual Review of Biophysics and Biomolecular Structure*, **27**, 225–248.

220 Seeman, N.C. (2005) *Trends in Biochemical Sciences*, **30**, 119–125.

221 Gu, Q., Cheng, C., Gonela, R., Suryanarayanan, S., Anabathula, S., Dai, K. and Haynie, D.T. (2006) *Nanotechnology*, **17**, R14–R25.

222 Braun, E., Eichen, Y., Sivan, U. and Ben-Yoseph, G. (1998) *Nature*, **391**, 775–778.

223 Monson, C.F. and Wooley, A.T. (2003) *Nano Letters*, **3**, 359–363.

224 Merting, M., Colombi Ciacchi, L., Seidel, R., Pompe, W. and De Vita, A. (2002) *Nano Letters*, **2**, 841–844.

225 Richter, J., Mertig, M., Pompe, W., Mönch, I. and Schackert, H.K. (2001) *Applied Physics Letters*, **78**, 536–538.

226 Weizmann, Y., Patolsky, F., Popov, I. and Willner, I. (2004) *Nano Letters*, **4**, 787–792.

227 Keren, K., Berman, R.S., Buchstab, E., Sivan, U. and Braun, E. (2003) *Science*, **302**, 1380–1382.

228 Keren, K., Krueger, M., Gilad, R., Ben-Yoseph, G., Sivan, U. and Braun, E. (2002) *Science*, **297**, 72–75.

229 Yan, H., Park, S.H., Finkelstein, G., Reif, J.H. and LaBean, T.H. (2003) *Science*, **301**, 1882–1884.

230 Reches, M. and Gazit, E. (2003) *Science*, **300**, 625–627.

231 Carny, O., Shalev, D.E. and Gazit, E. (2006) *Nano Letters*, **6**, 1594–1597.

232 Vale, R.D. (2003) *Journal of Cell Biology*, **163**, 445–450.

233 dos Remedios, C.G. and Moens, P.D.J. (1995) *Biochimica et Biophysica Acta*, **1228**, 99–124.

234 Patolsky, F., Weizmann, Y. and Willner, I. (2004) *Nature Materials*, **3**, 692–695.

235 Murphy, C.J., San, T.K., Gole, A.M., Orendorff, C.J., Gao, J.X., Gou, L., Hunyadi, S.E. and Li, T. (2005) *The Journal of Physical Chemistry B*, **109**, 13857–13870.

236 El-Sayed, M.A. (2004) *Accounts of Chemical Research*, **37**, 326–333.

237 Mokari, T., Rothenberg, E., Popov, I., Costi, R. and Banin, U. (2004) *Science*, **304**, 1787–1790.

238 Moghimi, S.M., Hunter, A.C. and Murray, J.C. (2005) *FASEB Journal*, **19**, 311–330.

239 Samia, A.C.S., Chen, X. and Burda, C. (2003) *Journal of the American Chemical Society*, **125**, 15736–15737.

240 Lockman, P.R., Oyewumi, M.O., Koziara, J.M., Roder, K.E., Mumper, R.J. and Allen, D.D. (2003) *Journal of Controlled Release*, **93**, 271–282.

241 Allen, T.M. and Cullis, P.R. (2004) *Science*, **303**, 1818–1822.

242 Miller, A.D. (2004) *ChemBioChem*, **5**, 53–54.

7
Philosophy of Nanotechnoscience

Alfred Nordmann

7.1
Introduction: Philosophy of Science and of Technoscience

One way or another, the philosophy of science always informs and reflects the development of science and technology. It appears in the midst of disputes over theories and methods, in the reflective thought of scientists and since the late nineteenth century also in the analyses of so-called philosophers of science. Four philosophical questions, in particular, are answered implicitly or contested explicitly by any scientific endeavor:

- How is a particular science to be defined and what are the objects and problems in its domain of interest?
- What is the methodologically proper or specifically scientific way of approaching these objects and problems?
- What kind of knowledge is produced and communicated, how does it attain objectivity, if not certainty, and how does it balance the competing demands of universal generality and local specificity?
- What is its place in relation to other sciences, where do its instruments and methods, its concepts and theories come from and should its findings be explained on a deeper level by more fundamental investigations?

When researchers publish their results, when they review and critique their peers, argue for research funds or train graduate students, they offer examples of what they consider good scientific practice and thereby adopt a stance on all four questions. When, for example, there is a call for more basic research on some scientific question, one can look at the argument that is advanced and discover how it is informed by a particular conception of science and the relation of science and technology. Frequently it involves the idea that basic science identifies rather general laws of causal relations. These laws can then be applied in a variety of contexts and the deliberate control of causes and effects can give rise to new technical devices. If one encounters such an argument for basic science, one might ask, of course, whether this picture of

Nanotechnology. Volume 1: Principles and Fundamentals. Edited by Günter Schmid

ISBN: 978-3-527-31732-5

basic versus applied science is accurate. While it may hold here and there, especially in theoretical physics, it is perhaps altogether inadequate for chemistry. And thus one may find that the implicit assumptions agree less with the practice and history of science and more with a particular self-understanding of science. According to this self-understanding, basic science disinterestedly studies the world as it is, whereas the engineering sciences apply this knowledge to change the world in accordance with human purposes.

Science and scientific practice are always changing as new instruments are invented, new problems arise, new disciplines emerge. Also, the somewhat idealized self-understandings of scientists can change. The relation of science and technology provides a case point. Is molecular electronics a basic science? Is nanotechnology applied nanoscience? Are the optical properties of carbon nanotubes part of the world as it is or do they appear only in the midst of a large-scale engineering pursuit that is changing the world according to human purposes? There are no easy or straightforward answers to these questions and this is perhaps due to the fact that the traditional ways of distinguishing science and technology, and basic and applied research, do not work any longer. As many authors are suggesting, we should speak of "technoscience" [1, 2] which is defined primarily by the interdependence of theoretical observation and technical intervention [3].[1)] Accordingly, the designation "nanotechnoscience" is more than shorthand for "nanoscience and nanotechnologies" but signifies a mode of research other than traditional science and engineering. Peter Galison, for example, notes that "[n]anoscientists aim to build – not to demonstrate existence. They are after an engineering way of being in science" [5]. Others appeal to the idea of a "general purpose technology" and thus suggest that nanotechnoscience is fundamental research to enable a new technological development at large. Richard Jones sharpens this when he succinctly labels at least some nanotechnoscientific research as "basic gizmology."[2)]

Often, nanoscience is defined as an investigation of scale-dependently discontinuous properties or phenomena [6]. This definition of nanoscience produces in its wake an ill-defined conception of nanotechnologies – these encompass all possible technical uses of these properties and phenomena. In its 2004 report on nanoscience and nanotechnologies, the Royal Society and Royal Academy of Engineering defines these terms as follows:

1) This is in reference to Ian Hacking's distinc-tion of "representing" and "intervening" [4]: In technoscientific research, the business of theoretical representation cannot be dissociated, even in principle, from the material conditions of knowledge production and thus from the interventions that are required to make and stabilize the phenomena. In other words, technoscience knows only one way of gaining new knowledge and that is by first making a new world. If the business of science is the theoretical representation of an eternal and immutably given nature and if the business of technology is to control the world, to intervene and change the "natural" course of events, "technoscience" is a hybrid where theoretical representation becomes entangled with technical intervention.

2) Jones used this term in conversation (and on his web site, www.softmachines.org) and referred, for example, to Nadrian Seeman's systematic exploration of DNA as a building block or component of future technical systems.

> *"Nanoscience is the study of phenomena and manipulation of materials at atomic, molecular and macromolecular scales, where properties differ significantly from those at a larger scale. Nanotechnologies are the design, characterisation, production and application of structures, devices and systems by controlling shape and size at the nanometre scale" [7].*

The notion of "nanotechnoscience" does not contradict such definitions but assumes a different perspective – it looks from within the organization of research where fundamental capabilities are typically acquired in the context of funded projects with a more or less concretely imagined technical goal. This is what Galison means by an engineering way of being in science. Even though a great deal of scientific knowledge and experience goes into the acquisition of such capabilities and the investigation of novel phenomena, it is not quite "science" because the point of this investigation is not normally to question received conceptions and to establish new truths, nor is it to produce and test hypotheses or to develop theories that close important gaps in our understanding of the world. And even though nanoscale research practice involves a good bit of tinkering and pursues technological challenges and promises, it is also not "engineering" because most researchers are not in the business of building devices for more or less immediate use. At best, they lay the groundwork for concrete engineering projects in the future.

For this "engineering way of being in science", a philosophy of technoscience is needed that asks for nanotechnological, biomedical or semiconductor research the four questions that were identified above: what is the role of theory and theory-development in nanoscale research and what kinds of theories are needed for nanotechnological development?; what are the preferred methods and tools and the associated modes of reasoning in nanoscientific research?; what is nanotechnoscience and how are its objects constituted?; and what kind of knowledge do technoscientific researchers typically produce and communicate? The four main sections of this chapter will address these questions – and in all four cases, strictly philosophical considerations will shade into societal dimensions and questions of value. That this is so is due to the fact that there may have been "pure science" but that there is no such thing as "pure technoscience." Indeed, one way of characterizing technoscience is by noting that academic laboratory research is no longer answerable just to standards of peer researchers but has entered the "ethical space" of engineering with its accountability also to patrons and clients, to developers and users [8, 9].

7.2
From "Closed Theories" to Limits of Understanding and Control

7.2.1
Closed Relative to the Nanoscale

In the late 1940s, the physicist Werner Heisenberg introduced the notion of "closed theories". In particular, he referred to four closed theories: "Newtonian mechanics,

Maxwell's theory with the special theory of relativity, thermodynamics and statistical mechanics, non-relativistic quantum mechanics with atomic physics and chemistry". These theories he considered to be closed in four complementary respects:

1. Their historical development has come to an end, they are finished or have reached their final form.
2. They constitute a hermetically closed domain in that the theory defines conditions of applicability such that the theory will be true wherever its concepts can be applied.
3. They are immune to criticism; problems that arise in contexts of application are deflected to auxiliary theories and hypotheses or to the specifics of the set-up, the instrumentation, and so on.
4. They are forever valid: wherever and whenever experience can be described with the concepts of such a theory, the laws postulated by this theory will be proven correct [10].[3)]

All this holds for nanotechnoscience: It draws on an available repertoire of theories that are closed or considered closed in respect to the nanoscale, but it is concerned neither with the critique or further elaboration of these theories, nor with the construction of theories of its own.[4)] This is not to say, however, that closed theories are simply "applied" in nanotechnoscience.

When Heisenberg refers to the hermetically closed character of closed theories (in condition 2 above), he states merely that the theory will be true where its concepts *can* be applied and leaves quite open how big or small the domain of its actual applicability is. Indeed, he suggests that this domain is so small that a "closed theory does not contain any absolutely certain statement about the world of experience" [10]. Even for a closed theory, then, it remains to be determined how and to what extent its concepts can be applied to the world of experience.[5)] Thus, there is no pre-existing domain of phenomena to which a closed theory is applied. Instead, it is "a question of success",

3) Heisenberg's notion of closed theories influenced Thomas Kuhn's conception of a paradigm [11]. It also informed the so-called finalization thesis, one of the first systematic accounts of technoscience [12]. Heisenberg also emphasized a fifth and especially contentious aspect of closed theories: an expansion of their domain of application will not introduce a change to the theory. This aspect and Heisenberg's particular list of closed theories plays no part in the following discussion.

4) In the case of nanotechnoscience, this repertoire includes far more than the four theories singled out by Heisenberg. It is a bold claim, to be sure, that nanotechnoscience is not concerned with the construction of theories of its own. One counterexample might be the discovery and subsequent theoretical work on the giant magnetoresistance effect [13]. Also, there are certain isolated voices who call for the development of theory specifically suited to the complexities of the nanocosm [14, 15]. These voices are isolated, indeed, and the consensus appears to be that the development of nanotechnologies can do without such theories – which might be hard to come by anyway [16].

5) Here, Heisenberg might have been inspired by Heinrich Hertz, who formulated the *Principles of Mechanics* as a closed theory [17]. He defined as mechanical problems all those phenomena of motion that can be accounted for by his fundamental law, albeit with the help of additional assumptions. Phenomena that cannot be accounted for in such a way, are not mechanical problems and simply outside the domain of mechanics (for example, the problems of life).

that is, of calibration, tuning or mutual adjustment to what extent phenomena of experience can be assimilated to the theory such that its concepts can be applied to them.

7.2.2
Applying Theory to the Nanoscale: Fitting Versus Stretching

This notion of "application" has been the topic of many recent discussions on modeling[6] – but it does not capture the case of nanotechnoscience. For in this case, researchers are not trying to bring nanoscale phenomena into the domain of quantum chemistry or fluid dynamics or the like. They are not using models to extend the domain of application of a closed theory or general law. They are not engaged in fitting the theory to reality and *vice versa*. Instead, they take nanoscale phenomena as parts of a highly complex mesocosm between classical and quantum regimes. They have no theories that are especially suited to account for this complexity, no theories, for example, of structure–property relations at the nanoscale.[7] Nanoscale researchers understand, in particular, that the various closed theories have been formulated for far better-behaved phenomena in far more easily controlled laboratory settings. Rather than claim that the complex phenomena of the nanoscale can be described such that the concepts of the closed theory now apply to them, they draw on closed theories eclectically, stretching them beyond their intended scope of application to do some partial explanatory work at the nanoscale.[8] A certain measurement of a current through an organic–inorganic molecular complex, for example, might be reconstructed quantum-chemically or in the classical terms of electrical engineering – and yet, the two accounts do not

6) See the work, in particular, of Nancy Cartwright, Margaret Morrison and Mary Morgan [18, 19].

7) Note that the term "complexity" is used here in a deliberately nontechnical manner. It does not refer to phenomena that fit the constraints of nonlinear complex dynamics, "complexity theory" or the like. The complexity at the nanoscale is one of great messiness, too many relevant variables and properties and multiple complicated interactions. This becomes apparent especially in contrast to the comparatively neat world of the laboratory phenomena that underwrite classical and quantum physics. In its complexity, the "real-world situation" of the nanoworld is precariously situated between classical and quantum regimes.

8) Here is another way to characterize this contrast between "applying" theory by fitting and by stretching: In the standard case of fitting theory to reality and vice versa, the problem concerns ways to compensate for the idealizations or abstractions that are involved in formulating a theory and constructing a model. However, classical theories do not abstract from nanoscale properties and processes, nor do they refer to idealizations of nanoscale phenomena. In this case, the challenge is that of crossing from the intended domain of a classical theory into quite another domain. – like all attempts to distinguish systematically the new nanotechnoscience from old-fashioned "science and engineering," this one is vulnerable to the critique that the two notions of "application" (bringing phenomena into the domain of application, stretching the domain of application to areas for which the theory has not been made) are not categorically distinct but differ only by degree. I thank Eric Winsberg for suggesting this line of thought.

compete against each other for offering a better or best explanation [20]. Armed with theories that are closed relative to the nanoscale, researchers are well equipped to handle phenomena in need of explanation, but they are also aware that they bring crude instruments that are not made specifically for the task and that these instruments therefore have to work in concert. Indeed, nanoscale research is characterized by a tacit consensus according to which the following three propositions hold true simultaneously:

1. There is a fundamental difference between quantum and classical regimes such that classical theories cannot describe quantum phenomena and such that quantum theories are inappropriate for describing classical phenomena.
2. The nanoscale holds intellectual and technical interest because it is an "exotic territory" [14] where classical properties such as color and conductivity emerge when one moves up from quantum levels and where phenomena such as quantized conductance emerge as one moves down to the quantum regime.
3. Nanoscale researchers can eclectically draw upon a large toolkit of theories from the quantum and classical regimes to construct explanations of novel properties, behaviors or processes.

Taken together, these three statements express a characteristic tension concerning nanotechnology, namely that it is thought to be strange, novel and surprising on the one hand, familiar and manageable on the other. More significantly for the present purposes, however, they express an analogous tension regarding available theories: they are thought to be inadequate on the one hand but quite sufficient on the other. The profound difference between classical and quantum regimes highlights what makes the nanocosm special and interesting – but this difference melts down to a matter of expediency and taste when it comes to choosing tools from classical or quantum physics. Put yet another way: what makes nanoscale phenomena scientifically interesting is that they cannot be adequately described from either perspective, but what makes nanotechnologies possible is that the two perspectives make do when it comes to accounting for these phenomena.

Available theories need to be stretched in order to manage the tension between these three propositions. How this stretching actually takes place in research practice needs to be shown with the help of detailed case studies. One might look, for example, at the way in which theory is occasionally "stuck in" to satisfy an extraneous explanatory demand.[9] A more prominent case is the construction of simulation

9) See, for example, a publication in Science on observed effects (large on–off ratios, negative differential resistance) in a molecular device. Asked by peer reviewers to offer an explanation of the observed effect, the authors suggest a somewhat arbitrary but plausible candidate mechanism and call for theories and future experimental work to "elucidate the transport mechanisms" [21]. This discussion was introduced reluctantly since it is clearly unnecessary for the point they wish to make (namely that they can consistently pass a current where no one had done so before) and because it is obviously easy to come up with a sufficiently credible explanation from the toolkit of available theory. The authors implicitly acknowledge that another explanation could easily substitute for theirs.

models where integrations of different levels of theoretical description are tuned to the actual behavior of a nanoscale system or process [22, 23]. This implies also that the very meaning of theories is stretched, especially where they account for the causal structure behind the observed phenomena: as these theories are applied in situations that are taken to be far more complex than the one for which the theories were developed, the causal story offered by them takes a backseat to the contributions they can make towards a description of the phenomena. In other words, algorithms descriptive of a certain dynamics become detached from the causal explanation they originally helped to provide, since it is the initial or structural conditions precisely that are not thought to hold continuously from macro to nano to quantum regimes.[10)]

There is quite another symptom of the ways in which theories and concepts are stretched as they are applied to the nanoscale. The nanoworld is taken to be complex, self-organizing, full of surprises – a world characterized by chemical and biological activity. The aspirations of nanotechnologies therefore emphasize the construction of active rather than merely passive devices.[11)] The so-called first generation of nanotechnological achievements was limited to the generation of new materials (passive structures), the second generation is supposed to incorporate molecular activity into nanotechnical systems.[12)] However, from the point of view of theories that are closed relative to the nanoscale, one cannot "see" any of that novel activity and liveliness but only what has become stabilized in the formulations and formalisms of those theories. Several descriptive or programmatic terms for nanoscale phenomena therefore strain to reach beyond their actual meanings. A prime example of this is the term "selective surface" which attributes agency to something that remains quite passive: Cells may attach to a given surface differentially, but the surface is not therefore doing anything to favor or disfavor certain cells; the selection is entirely on the side of the engineer who selects that surface in order to achieve some functionality. The same holds for

10) At the panel "Ontologies of Technoscience" at the October 2006 Bielefeld Conference ("Science in the Context of Application"), Bernadette Bensaude-Vincent showed that in the development of materials science and nanotechnologies the focus on structure–function relations (Crick's dogma that scientific understanding requires that function is referred to underlying structure) gives way to analyses of dynamic patterns in the observed functions and properties. Davis Baird offered as an example of this a particular nanotechnological detection device that physically instantiates factor analysis and therefore statistically infers underlying causes from observed properties (in other words, it does not perform a physical or chemical analysis to identify the presence of what is measured). Nicole Karafyllis finally suggested that (nano)technologies are now entering into a novel relation to biology as they design function not through construction (e.g. from structural principles) but by way of growth (e.g. by way of harnessing of self-organization).

11) Regarding the prestige of the device *vis-à-vis* the material, see [23a] on Herbert Gleiter as a pioneer of nanotechnology.

12) The notion of first/second generation passive/active devices was established and promoted especially by Mihail Roco of the National Nanotechnology Initiative. This paper is agnostic as to whether the second generation will ever be attained.

"self-cleaning surfaces", "smart materials", "autonomous (self-propelled) movement", the different conceptions of "self-assembly" or "soft machines".[13] All these terms have a specific meaning and at the same time refer to something more visionary, more genuinely "nano" that goes beyond their origin in theories that come from outside the nanoworld.[14]

7.2.3 Mute Complexity

So far, the notion of stretching what we know in one regime to phenomena in another one has been taken descriptively to characterize nanoscale research. Here, however, arises an occasion for critical questioning by scientists, citizens, concerned policy makers. To the extent that one cannot see the specific complexity from the point of view of theories closed relative to the nanoscale, we may find that the difficulties of understanding and controlling nanoscale phenomena are not adequately expressed. By stretching closed theories one recovers partial explanations of phenomena and thereby partial stories only of success. In other words, the assurance that much is amenable to explanation from the large toolkit of available theories finds ample expression, but there is no theoretical framework for the actual struggle of taming and controlling nanoscale phenomena – this part of the story remains untold, locked up in the laboratory.[15] Put bluntly, one might be doing years and years of interesting research only to discover that most of the phenomena one is tinkering with, that one is stabilizing and probing in the laboratory, will never be robust enough to serve as components in nanotechnological devices. There is no language, in other words, to identify specific limits of knowledge and control.

Having arrived at this point in a rather roundabout manner, one might ask whether the limited ways to speak of limits of understanding and control can be shown more straightforwardly. A telling illustration or example is provided by

13) Since the publication of Richard Jones's book on Soft Machines, that concept has been the subject of an emerging discussion [25, 26]. It concerns the question of whether the term "machine" retains any meaning in the notion of a "soft machine" when this is thought of as a non-mechanical, biological machine (while it clearly does retain meaning if thought of as a "concrete" machine in the sense of Simondon).

14) This is especially true, perhaps, for the concept "self-assembly", which has been cautiously delimited, for example, by George Whitesides [27], but which keeps escaping the box and harks backwards and forwards to far more ambitious notions of order out of chaos, spontaneous configurations at higher levels, etc.

15) To be sure, it is a commonplace that laboratory practice is more complex than the stories told in scientific papers. Traditional scientific research often seeks to isolate particular causal relations by shielding them against interferences from the complex macroscopic world of the laboratory. Whether it is easy or difficult to isolate these relations, whether they are stable or evanescent, is of little importance for the scientific stories to be told. The situation changes in respect to nanotechnoscience: its mission is to ground future technologies under conditions of complexity. In this situation, it is more troubling that scientific publications tell stories only of success.

nanotoxicology. It is finding out the hard way that physico-chemical characterization does not go very far, and that even the best methods for evaluating chemical substances do not REACh all the way to the nanoscale [28].[16] In other words, the methods of chemical toxicology go only so far and tell only a small part of the toxicological story – though regarding chemical composition, at least, there are general principles, even laws that can be drawn upon. With regard to the surface characteristics and shape of particles of a certain size, one has to rely mostly on anecdotes from very different contexts, such as the story of asbestos. For lack of better approaches, therefore, one begins from the vantage point of chemical toxicology and confidently stretches available theories and methods as far as they will go – while the complexities of hazard identification, let alone risk assessment (one partially characterized nanoparticle or nanosystem at a time?!) tend to be muted.[17]

There is yet another, again more general, way to make this point. Theories that are closed relative to the nanoscale can only introduce nonspecific constraints. The prospects and aspirations of nanotechnologies are only negatively defined: Everything is thought to be possible at the nanoscale that is not ruled out by those closed theories or the known laws of nature. This, however, forces upon us a notion of technical possibility that is hardly more substantial than that of logical possibility. Clearly, the mere fact that something does not contradict known laws is not sufficient to establish that it can be realized technically under the complex conditions of the nanoregime. Yet once again, there is no theoretical framework or language available to make a distinction here and to acknowledge the specificities and difficulties of the nanoworld – since all we have are theories that were developed elsewhere and that are now stretched to accommodate phenomena from the nanosphere.[18] However, failure to develop an understanding also of limits of understanding and control at the nanoscale has tremendous cost as it misdirects expectations, public debate and possibly also research funding.

16) The pun is intended: REACh refers to the new style of regulating chemical substances in and by the EU. It is widely acknowledged that it does not apply where properties depend not only on chemical composition but also on surface characteristics, size, shape, perhaps also engineered functionality and the specificities of their environments. (Along similar lines, Joachim Schummer [29] has argued that REACh does not even reach the products of conventional synthetic chemistry.)

17) Sabine Maasen and Monika Kurath have shown that this difficulty for chemical toxicology creates interesting new opportunities for nanotoxicology [31]. For another illustration of the predicament, one might recall that carbon nanotubes were "discovered" in the 1980s, that for a good number of years they have been being commercially manufactured and that researchers are still complaining that no two batches are alike.

18) I have been urging that more attention should be paid to limits of understanding and control at the nanoscale. If I am right in this section, I have been asking for something that cannot be done (as of now) in a straightforward way.

7.3 From Successful Methods to the Power of Images

7.3.1 (Techno)scientific Methodology: Quantitative Versus Qualitative

As was shown above, Heisenberg considered "a question of success" the extent to which phenomena of experience can be fitted to closed theories [10]. This suggests the question what "success" amounts to in nanoscale research, that is, what it takes to satisfy oneself that one has reached a sufficiently good understanding or control of the phenomena under investigation.

For Heisenberg and any philosopher of science who is oriented towards theoretical physics, this question boils down to the predictive success of a quantitative science. Here, "quantitative" means more than the employment of numbers and even of precision measurements. The characteristics of quantitative approaches include the following. First, predicted numerical values are compared with values obtained by measurement. The reasonably close agreement between two numbers thus serves to establish the agreement of theory and reality. Second, this quantitative agreement emphatically makes do without any appeal to a likeness or similarity between theoretical models and the real-world systems they are said to represent. Quantitative science rests content if it reliably leads from initial conditions to accurate predictions, it does not require that all the details of its conceptual apparatus (every term in its algorithms) has a counterpart in reality. Both characteristics of quantitative science are familiar especially from twentieth century theoretical physics – but do they serve to characterize also nanotechnoscience [31]?

In the light of the extremely heterogeneous research practices under the general heading of "nanoscience and nanotechnologies" there may not be a general answer to this question. Yet it is fair to say that much nanotechnoscientific research is qualitative. Its epistemic success consists in constructions of likeness.[19)]

The shift sounds innocent enough but may have significant consequences: the agreement of predicted and measured quantities is being displaced by an agreement of calculated and experimental images. The latter qualitative agreement consists

19) Here, a case study of Jan Hendrik Schön might show that he was caught between quantitative and qualitative methodologies. He was "caught cheating", after all, when it was discovered that for different experiments he included an exactly identical plot of current flow. This diagram is supposed to be generated from a series of measurements but the characteristic shape of the curve is also a qualitative short-hand expression for "current is flowing." In a culture of research that is moving increasingly to produce effects, Schön may well have "written" this diagram as it is generally "read" – without regard to the particular values but as a symbol for a certain type of event. Overall, Schön's case is less innocent and more complicated than this [32]. But perhaps in other regards, too, it is symptomatic of the ambivalence that results from the transdisciplinary qualitative orientation of nanotechnoscience even as nanoscale research continues to be informed mostly by rigorously quantitative disciplinary traditions.

primarily in the absence, even deliberate suppression of visual clues by which to hold calculated and experimental images apart. Indeed, the (nano)technoscientific researcher frequently compares two displays or computer screens. One display offers a visual interpretation of the data that were obtained through a series of measurements (e.g. by an electron or scanning probe microscope), the other presents a dynamic simulation of the process he might have been observing – and for this simulation to be readable as such, the simulation software produces a visual output that looks like the output for an electron or scanning probe microscope. Agreement and disagreement between the two images then allows the researchers to draw inferences about probable causal processes and to what extent they have understood them. Here, the likeness of the images appears to warrant the inference from the mechanism modeled in the simulation to the mechanism that is probably responsible for the data that were obtained experimentally. Accordingly (and this cannot be done here), one would need to show how nanoscale researchers construct mutually reinforcing likenesses, how they calibrate not only simulations to observations and visual representations to physical systems but also their own work to that of others, current findings to long-term visions. This kind of study would show that unifying theories play little role in this, unless the common availability of a large tool-kit of theories can be said to unify the research community. Instead of theories, it is instruments (STM, AFM, etc.), their associated software, techniques and exemplary artefacts (buckyballs, carbon nanotubes, gold nanoshells, molecular wires) that provide relevant common referents [33–35].

7.3.2
"Ontological Indifference": Representation Versus Substitution

This is also not the place to subject this qualitative methodology to a sustained critique. Such a critique is easy, in fact, from the point of view of rigorous and methodologically self-aware quantitative science [31]. Far more interesting is the question of why, despite this critique, a qualitative approach appears to be good enough for the purposes of nanoscale research. As Peter Galison has pointed out, these purposes are not to represent the nanoscale accurately and, in particular, not to decide what exists and what does not exist, what is more fundamental and what is derivative. He refers to this as the "ontological indifference" of nanotechnoscience [5]. Why is it, then, that nanotechnological research can afford this indifference? For example, molecular electronics researchers may invoke more or less simplistic pictures of electron transport but they do not need to establish the existence of electrons. Indeed, electrons are so familiar to them that they might think of them as ordinary macroscopic things that pass through a molecule as if it were another material thing with a tunnel going through it [20]. Some physicists and most philosophers of physics strongly object to such blatant disregard for the strangely immaterial and probabilistic character of the quantum world that is the home of electrons, orbitals, standing electron waves [36, 37]. And indeed, to achieve a practical understanding of electron transport, it may be necessary to entertain more subtle accounts. However, it is the privilege of ontologically indifferent technoscience that it can always develop more complicated accounts as the need arises. For the time being, it can see how far it gets with rather more simplistic pictures.[20)]

Ontological indifference amounts to a disinterest in questions of representation and an interest, instead, in substitution.[21)] Instead of using sparse modeling tools to represent only the salient causal features of real systems, nanoresearchers produce in the laboratory and in their models a rich, indeed oversaturated substitute reality such that they begin by applying alternative techniques of data reduction not to "nature out there" but to some domesticated chunk of reality in the laboratory. These data reduction and modeling techniques, in turn, are informed by algorithms which are concentrated forms of previously studied real systems, they are tried and true components of substitute realities that manage to emulate real physical systems [38].[22)] In other words, there is so much reality in the simulations or constructed experimental systems before them, that nanotechnology researchers can take them for reality itself [39]. They study these substitute systems and, of course, have with these systems faint prototypes for technical devices or applications. While the public is still awaiting significant nanotechnological products to come out of the laboratories, the researchers in the laboratories are already using nanotechnological tools to detach and manipulate more or less self-sufficient nanotechnological systems which "only" require further development before they

20) A particularly interesting and challenging example of this is Don Eigler's famous picture of a quantum corral that confines a standing electron wave. The picture's seemingly photographic realism suggests that the quantum corral is just as thing-like as a macroscopic pond. It brazenly bypasses all discussions regarding the interpretation of quantum mechanics and thus displays its ontological indifference. Nevertheless, it is an icon of nanotechnoscience, testimony to new capabilities of manipulation and visualization and a down-payment of sorts on the promise that technical control does not stop at the threshold to quantum effects.

21) Compare Peter Galison's suggestion above that the relevant contrast is that between demonstrating existence and building things. Yet, as will be shown in Section 7.5 below, "building" is too narrow and too "technical" a notion. It does not do justice to the intellectual engagement, even passion for the challenges encountered at the nanoscale.

22) Rom Harré contrasts scientific instruments that serve as probes into causal processes and modeling apparatus (including simulations) that domesticates or produces phenomena. It is this modeling apparatus that underwrites epistemic success in constructions of likeness: Instruments typically obtain measurements that can be traced back down a causal chain to some physical state, property or process. As such, the instruments are detached from nature – measurements tell us something about the world. Physical models, in contrast, are part of nature and exhibit phenomena such that the relevant causal relations obtain within the apparatus and the larger apparatus–world complex. Whether it domesticates a known phenomenon such as the rainbow or elicits an entity or process that does not occur "naturally", it does not allow for straightforward causal inference to the world within which the apparatus is nested [38]. As the metaphor of domestication and Harré's conception of an apparatus–world complex suggest, causal inference from the apparatus to the world may be required only for special theoretical purposes that are characterized by a specific concern for reality (for example, when something goes wrong and one wants to explore the reasons for this). At the same time, the very fact that the apparatus is nested in the world delivers an (unarticulated) continuity of principles and powers and the affordance of ontological indifference.

can exist as useful devices outside the laboratory, devices that not only substitute for but improve upon something in nature.

7.3.3
Images as the Beginning and End of Nanotechnologies

Again, it may have appeared like a cumbersome path that led from qualitative methodology and its constructions of likeness to the notion that models of nanoscale phenomena do not represent but substitute chunks of reality and that they thereby involve the kind of constructive work that is required also for the development of nanotechnological systems and devices. For a more immediate illustration of this point, we need to consider only the role of visualization technologies in the history of nanotechnological research.[23)] Many would maintain, after all, that it all began for real when Don Eigler and Erhard Schweizer created an image with the help of 35 xenon atoms. By arranging the atoms to spell "IBM" they did not represent a given reality but created an image that replaces a random array of atoms by a technically ordered proto-nanosystem. Since then, the ability to create images and to spell words has served as a vanguard in attempts to assert technical control in the nano-regime – the progress of nanotechnological research cannot be dissociated from the development of imaging techniques that are often at the same time techniques for intervention. Indeed, Eigler and Schweizer's image has been considered proof of concept for moving atoms at will. It is on exhibit in the STM web gallery of IBMs Almaden laboratory and is there appropriately entitled "The Beginning" – a beginning that anticipates the end or final purpose of nanotechnologies, namely to directly and arbitrarily inscribe human intentions on the atomic or molecular scale.

Images from the nanocosm are at this point (early 2008) still the most impressive as well as popular nanotechnological products. By shifting from quantitative coordinations of numerical values to the construction of qualitative likeness, from the conventional representation of reality to the symbolic substitution of one reality by another, nanotechnoscience has become beholden to the power of images. Art historians and theorists like William Mitchell and Hans Belting, in particular, have emphasized the difference between conventional signs that serve the purpose of representation and pictures or images that embody visions and desires, that cannot be controlled in that they are not mere vehicles of information but produce an excess of meaning that is not contained in a conventional message [40, 41].

The power of images poses some of the most serious problems of and for nanoscience and nanotechnologies. This is readily apparent already for "The Beginning". As mentioned above, it is taken to signify that for the first time in history humans have manipulated atoms at will and thus as proof of concept for the most daring nanotechnological visions and by the most controversial nanotechnological visionaries such as Eric Drexler. This was not, of course, what Eigler and

23) It is no accident that this is perhaps the best-studied and most deeply explored aspect of nanotechnologies.

Schweizer wanted to say. Their image is testimony also to the difficulty, perhaps the limits of control of individual atoms. But the power of their image overwhelms any such testimony.

Here arises a problem similar to the one encountered in Section 7.2.3. The specificity, complexity and difficulty of work at the nanoscale do not have a language and do not find expression. The theories imported from other size regimes can only carve out an unbounded space of unlimited potential, novelty, possibility. And the pictures from the nanocosm show us a world that has already been accommodated to our visual expectations and technical practice.[24] Ontologically indifferent, nanotechnoscience may work with simplistic conceptions of electron transport and it produces simplistic pictures of atoms, molecules, standing electron waves which contradict textbook knowledge of these things. For example, it is commonly maintained that nanosized things consist only of surface and have no bulk. This is what makes them intellectually and technically interesting. But pictures of the nanocosm invariably show objects with very familiar bulk-surface proportions, a world that looks perfectly suited for conventional technical constructions. And thus, again, we might be facing the predicament of not being told or shown what the limits of nanotechnical constructions and control might be.

The power of images also holds another problem, however. In the opposition of conventional sign and embodied image the totemistic, fetishistic, magical character of pictures comes to the fore. To the extent that the image invokes a presence and substitutes for an absence, its kinship to voodoo-dolls, for example, becomes apparent. This is not the place to explore the analogy between simulations and voodoo-dolls [31], but it should be pointed out that nanotechnologies in a variety of ways cultivate a magical relation to technology – and their imagery reinforces this. Indeed, in the history of humankind we might have begun with an enchanted and uncanny nature that needed to be soothed with prayer to the spirits that dwelled in rocks and trees. Science and technology began as we wondered at nature, became aware of our limits of understanding and yet tamed and rationalized nature in a piece-meal fashion. Technology represents the extent to which we managed to defeat a spirited, enchanted world and subjected it to our control. We technologized nature. Now, however, visitors to science museums are invited to marvel at nanotechnologies, to imagine technological agency well beyond human thresholds of perception, experience and imagination and to pin societal hopes for technological innovation not on intellectual understanding but on a substitutive emulation that harnesses the self-organizing powers of nature. We thus naturalize technology, replace rational control over brute environments by a magical dependency on smart environments and we may end up rendering technology just as uncanny as nature used to be with its earthquakes, diseases and thunderstorms [42, 43].[25]

24) Compare footnote 20 above.

25) This is a strong indictment not of particular nanotechnologies but of certain ways of propagating our nanotechnological future. Considered another way, it is simply an engineering challenge to design nanotechnology for the human scale.

7.4 From Definitions to Visions

7.4.1 Wieldy and Unwieldy Conceptions

The first two sections gave rise to the same complaint. After surveying the role of theories and methodologies for the construction of technical systems that can substitute for reality, it was noted that this tells us nothing about the specificity, complexity and difficulty of control at the nanoscale. The nanocosm appears merely as that place from where nanotechnological innovations emanate and so far it appears that it can be described only in vaguely promising terms: the domain of interest to nanoscience and nanotechnologies is an exotic territory that comprises all that lies in the borderland of quantum and classical regimes, all that is unpredictable (but explicable) by available theories and all that is scale-dependently discontinuous, complex, full of novelty and surprise.[26)]

However, as one attempts a positive definition of nanotechnoscience and its domain of phenomena or applications, one quickly learns how much is at stake. In particular, definitions of "nanotechnology" suggest the unity of a program so heterogeneous and diverse that we cannot intellectually handle or manage the concept any more. By systematically overtaxing the understanding, such definitions leave a credulous public and policy makers in awe and unable to engage with "nanotechnology" in a meaningful manner. The search for a conceptually manageable definition is thus guided by an interest in specificity but also by a political value – it is to facilitate informed engagement on clearly delimited issues. In purely public contexts, therefore, it is best not to speak of nanotechnology in the singular at all but only of specific nanotechnologies or nanotechnological research programs [44]. In the present context, however, an effort is made to circumscribe the scope or domain of nanotechnoscience, that is, to consider the range of phenomena that are encountered by nanoscience and nanotechnologies. This proves to be a formidable challenge.

7.4.2 Unlimited Potential

There is an easy way to turn the negative description of the domain into a positive one. One might say that nanoscience and nanotechnologies are concerned with everything molecular or, slightly more precisely, with the investigation and manipulation of molecular architecture (as well as the properties or functionalities that depend on molecular architecture).

26) Tellingly, the most sophisticated definition of nanoscience is quite deliberate in saying nothing about the "nanocosm" at all. Indeed, this definition is not limited to nanoscale phenomena or effects but intends a more general nanoscience of scale-dependently discontinuous behaviors at all scales: nanoscience is everywhere where one encounters a specific kind of novelty or surprise [6].

Everything that consists of atoms is thus an object of study and a possible design target of nanoscale research. This posits a homogeneous and unbounded space of possibility, giving rise, for example, to the notion of an all-powerful nanotechnology as a combinatorial exercise that produces the "little BANG" [45] – since bits, atoms, neurons, genes all consist of atoms, since all of them are molecular, they all look alike to nanoscientists and engineers who can recombine them at will. And thus comes with the notion of an unlimited space of combinatorial possibilities the transgressive character of nanotechnoscience: categorial distinctions of living and inanimate, organic and inorganic, biological and technical things, of nature and culture appear to become meaningless. Although hardly any scientist believes literally in the infinite plasticity of everything molecular, the molecular point of view proves transgressive in many nanotechnological research programs. It is particularly apparent where biological cells are redescribed as factories with molecular nanomachinery. Aside from challenging cultural sensibilities and systematic attempts to capture the special character of living beings and processes, nanotechnoscience here appears naively reductionist. In particular, it appears to claim that context holds no sway or, in other words, that there is no top–down causation such that properties and functionalities of the physical environment partially determine the properties and behaviors of the component molecules.[27)]

This sparsely positive and therefore unbounded view of nanoscale objects and their combinatorial possibilities thus fuels also the notion of unlimited technical potential along with visions of a nanotechnological transgression of traditional boundaries. Accordingly, this conception of the domain of nanoscience and nanotechnologies suffers from the problem of unwieldiness – it can play no role in political discourse other than to appeal to very general predispositions of technophobes and technophiles [46].

Three further problems, at least, come with the conception of the domain as "everything molecular out there." And as before, internally scientific problems are intertwined with matters of public concern. There is first the (by now familiar) "scientific" and "societal" problem that there is no cognizance of limits of understanding and control – as evidenced by a seemingly naive reductionism. There is second the (by now also familiar) problem that technoscientific achievements and conceptions have a surplus of meaning which far exceeds what the research community can take responsibility for – the power of images is dwarfed by the power of visions (positive or negative) that come with the notion of unlimited potential. And there is finally the problem of the relation of technology and nature.

27) I cannot pass judgement on these claims. However, even Richard Jones's Soft Machines [32] with its vivid appreciation of the complexities of "biological nanotechnology" does not reflect the findings of developmental biologists regarding environmental stimuli to gene expression. Recent work on adult stem cells appears to reveal that they can be reverted to earlier states but that they nevertheless "remember" what they were. Such findings complicate immensely the apparently unbounded promise that nanotechnology can solve all problems at the level of molecules.

Martin Heidegger, one of the sharpest critics of modern technology, chastised it for treating all of nature as a mere resource that is "standing in reserve" to be harnessed by science and industry [47]. The power of his argument derives precisely from the fact that he saw all of modern technoscience as one: it is a scaffolding or harness (the German word is *Gestell*) that recruits humans and nature into a universal scheme of production. Rather than accept as a gift what nature, poetry or craft brings forth, it demands the deliverance of what it has learned rationally to expect from the study of nature as a calculable system of forces. Conceived as a unified enterprise with an unbounded domain of "everything molecular", nanotechnology fits the bill of such an all-encompassing modern technology. It does so because it employs what one might call a thin conception of nature. According to this conception, nature is circumscribed by the physical laws of nature. All that accords with these laws is natural. Thus, nanotechnology can quickly and easily claim for itself that it always emulates nature, that it manufactures things nanotechnologically just as nature does when it creates living organisms. This conception, however, is too "thin" or superficial to be credible and it suffers from the defect that the conditions of (human) life on earth have no particular valence in it: from the point of view of physics and the eternal laws of nature, life on Earth is contingent and not at all necessary. The laws predate and will outlive the human species. In contrast, a substantial, richly detailed or "thick" conception of nature takes as a norm the special evolved conditions that sustain life on Earth. Here, any biomimetic research that emulates nature will be characterized by care and respect as it seeks to maintain these special conditions. This involves an appreciation of how these conditions have evolved historically. On this conception, context holds sway and a molecule that occurs in a technical system will not be the same as one in a biological system, even if it had the same chemical composition.

It is an open question and challenge to nanoscience and nanotechnologies, however, whether it can embrace such a thick or substantial conception of nature.

7.4.3
A Formidable Challenge

It was not very difficult to identify at least four major problems with the commonly held view that the domain of nanotechnological research encompasses "everything molecular" It proves quite difficult, in contrast, to avoid those problems. In particular, it appears to defy common sense and the insights of the physical sciences to argue that molecules should have a history or that they should be characterized by the specific environments in which they appear. Is it not the very accomplishment of physical chemistry ever since Lavoisier that it divested substances of their local origins by considering them only in terms of their composition, in terms of analysis and synthesis? [48]. And should one not view nanoscience and nanotechnologies as an extension of traditional physics, physical chemistry and molecular biology as they tackle new levels of complexity? All this appears evident enough, but yet there are grounds on which to tackle the formidable challenge and to differentiate the domain of nanoscientific objects.[28)]

As noted above, bulk chemical substances are registered and assessed on the grounds of a physico–chemical characterization. Once a substance has been approved, it can be used in a variety of contexts of production and consumption. On this traditional model, there appears no need to consider its variability of interactions in different biochemical environments (but see [29, 30]). Although the toolkit of nanotoxicology is still being developed, there is a movement afoot according to which a carbon nanotube is perhaps not a carbon nanotube. What it is depends on its specific context of use: dispersed in water or bound in a surface, coated or uncoated, functionalized or not – all this is toxicologically relevant. Moreover, a comprehensive physico-chemical characterization that includes surface properties, size and shape would require a highly complex taxonomy with too many species of nanoparticles, creating absurdly unmanageable tasks of identification perhaps one particle at a time. Instead, the characterization of nanoparticles might proceed by way of the level of standardization that is actually reached in production and that is required for integration in a particular product – with a smaller or larger degree of variability, error tolerance, sensitivity to environmental conditions, as the case may be for a specific product in its context of use. Nanotoxicology would thus be concerned with product safety rather than the safety of component substances. On this account, the particles would indeed be defined by their history and situation in the world and thus thickly by their place within and their impact upon nature as the specific evolved conditions of human life on Earth.[29)]

There is another, more principled, argument for a thickly differentiated account of the objects that make up the domain of nanoscience and nanotechnologies. The unbounded domain of "everything molecular" includes not only the objects and properties that we now have access to and that we can now measure and control. It also includes those objects and properties that one may gain access to in the future. This way of thinking is indifferent to the problem of actual technical access also in that it does not consider how observational instruments and techniques structure, shape and perhaps alter the objects in the domain. On this account, the domain appears open and unlimited because

28) One might argue that a definition requiring scale-dependent discontinuities already does offer such a differentiation [6]. This is not the case, however. It is a beginning at best. As shown above, this definition excludes certain phenomena and processes from nanoscience and thus claims specificity, but it leaves nanotechnology entirely undetermined: nanoscience tends to certain novel or surprising properties and processes, nanotechnology is whatever one can make of these properties and processes. More significantly, however, the appearance of scale-dependent novel properties can be claimed rather generically. Not every property at the nanoscale is discontinuous in respect of scale. However, for every substance one can claim that it may or will have some such properties simply by virtue of the proportion of atoms in the boundary layers.

29) Nanotoxicology in particular, and nanotechnological research in general, might thus become a "social science of nature" [49].

it refers to an imaginary (future) state of total information and the nonintrusive and presence of observers in the nanoworld. In contrast to this account, the domain could be delimited more concretely and its visionary surplus could be contained more effectively if it did not include all nanoscale objects "out there" but considers how these objects are constituted, how they become accessible to nanoscale research. Accordingly, the domain of objects and processes would consist of just those phenomena and effects that are revealed by scanning tunneling microscopy and other specifically nanotechnological procedures [50, 51].

However, more so than the current attempt to formulate a philosophy of nanotechnoscience, this proposal by Peter Janich ascribes to nanotechnoscience a methodological unity or basis in common practice. He suggests a philosophical program of systematizing the operations by which nanoscale objects become amenable to measurement and observation. Such a systematic reconstruction of the domain of objects of nanotechnological research might begin by looking at length measurement or scanning probe microscopy. However, research practice is not actually unified in this manner. Even scanning probe microscopy – to many a hallmark or point of origin for nanotechnologies – plays a minor role in the work of many nanoscale researchers [52]. Also, the above-mentioned struggles to attain standard measures or to characterize nanomaterials testify to the unruliness of the objects of research. They are not constituted through methodical procedures that individuate objects and make them comparable throughout the scientific community. Instead, it appears that they are constituted through complicated interactions that are difficult to reproduce and that rely on proximate likeness.

Since Janich's approach faces considerable odds, all one can do perhaps is to generalize the previous lesson from nanotoxicology: The objects of nanoscale research are constituted through their specific histories – histories that concern their origin (in a tissue sample, in the soil, in a chemically produced batch), that include nanotechnological interventions as well as their location finally in a technical system. This would promote, of course, the fragmentation of "nanotechnology" into as many "nanotechnologies" as there are nanotechnological devices or applications. A nightmare vision for some, others may consider this an intellectual requirement. If this is so, it becomes impossible to uphold the idea of carbon nanotubes as all-purpose technical components. If they contribute to the performance of some product, then they are individuated or characterized as being carbon-nanotubes-in-that-product and they are as safe or unsafe as that product is. By the same token, they are no longer conceived as molecular objects that are combinable in principle with just about any other. The open space of unlimited potential differentiates into a manifold of specific technological trajectories.

The formidable challenge has not been met by this proposal. It does help dramatize, however, the inherent tension in the commonly held view of nanotechnological objects, as well as the difficulties (once again) of prediction and control at the nanoscale.

7.5
From Epistemic Certainty to Systemic Robustness

7.5.1
What Do Nanoscientists Know?

The previous sections considered research practices of nanotechnosciences – how theories are stretched to the complexities at the nanoscale, how a qualitative methodology serves the construction of likeness and inferences from that likeness, how the research objects are individuated and encountered. All these practices contribute to the generation of knowledge but it remains to be explored in which sense this is "objective knowledge." As in traditional science, the findings of nanotechnoscientific research are published in scientific journals, so the questions is, more concretely, what kind of knowledge is expressed or communicated in a nanoscientific journal article? To answer this question properly, contrasts need to be established and particular publications compared. Here, a summary must suffice.

A typical research article in classical science states a hypothesis, offers an account of the methods, looks at the evidence produced and assesses the hypothesis in the light of the evidence. It participates in a public process of evaluating propositions, of finding certain statements true or false and of seeking certainty even where it is impossible to attain. In contrast, a technoscientific research article provides testimony to an acquired capability. It offers a sign or proof of what has been accomplished in the laboratory and tells a story of what has been done. The telling of the story does not actually teach the capability but it offers a challenge to the reader that they might develop this capability themselves. As opposed to epistemic knowledge (concerned with truth or falsity of propositions), nanoscale research produces skill knowledge. This is not an individualized skill, however, or tacit knowledge. Acquired capabilities can be objective and public, specifically scientific and communicable. They grasp causal relations and establish habits of action. They are assessed or validated not by the application of criteria or norms but by being properly entrenched in a culture of practice. One cannot judge their truth or falsity (skills are not true or false) but one can judge the robustness of demonstrability: If one has acquired a capability, one can more or less consistently do something in the context of an "apparatus–world complex" [38]. As opposed to the truth or falsity, certainty or uncertainty of hypotheses, the hallmarks of technoscientific knowledge are robustness, reliability, resilience of technical systems or systematic action.

7.5.2
The Knowledge Society

This account of skill knowledge presses the question of where the "science" is in "technoscience". The answer to this question can be found in the first section of this chapter: it is in the (closed) theories that are brought as tools to the achievement of partial control and partial understanding. Nanotechnoscience seeks not to improve theory or to change our understanding of the world but primarily to manage complexity and novelty. As such, nanotechnoscience is just technical tinkering, just product development, just an attempt to design solutions to societal problems or to shape and reshape the world. However, the conceptual and physical tools it tinkers with do not come from ordinary experience, from common sense and a craft tradition but concentrate within them the labors of science. So, the "science" of "nanotechnoscience" is what goes into it. What comes out is skill knowledge and this knowledge does not rely on a corresponding scientific understanding. As long as one can produce an effect in a reasonably robust manner, it does not really matter whether scientific understanding catches up. Indeed, the complexities may be such that it cannot fully catch up.[30)]

The standard example of technology being ahead of science is the steam engine, which was developed without a proper understanding of the relation between heat and work [53]. This understanding came much later and, indeed, was prompted by the efficient performance of the steam engine. The steam engine itself was therefore not applied science but the result of technical tinkering. It was made of valves, pumps, gears, and so on, of which there was good nonscientific craft-knowledge – and it worked just fine before the advent of thermodynamics. In a sense, it did not need to be understood.

As opposed to the steam engine, nanotechnological devices (whatever they may be), genetically modified organisms, drug delivery systems are offsprings of the knowledge society. They are not made of valves and pumps but assembled from highly "scientized" components such as algorithms, capabilities acquired by scientifically trained researchers who are using measuring and monitoring devices that have plenty of knowledge built in [39]. The science that goes into the components is well understood, not so the interactions of all the components and their sensitivities in the context of the overall technical system. Still, like the steam-engine, it may work just fine without being fully understood. And although one cannot attain positive knowledge from which to derive or predict its performance, we may learn to assess its robustness.

7.5.3 Social Robustness

The shift from hypotheses that take the form of sentences to actions within techno-cultural systems, from epistemic questions of certainty to systemic probes of robustness has implications also for the "risk society" that looks to government mostly for protection from risk [54].[31)]

30) This diagnosis is not entirely novel or surprising. Technology, writes Heidegger, is always ahead of science and, in a deep sense, science is only applied technology [47]. By this he means not only that laboratory science requires instruments and experimental apparatus for stabilizing the phenomena. He means more generally that a technological attitude informs the scientific way of summoning phenomena to predictably appear once certain initial conditions are met.

31) The precautionary principle refers to the certainty and uncertainty of knowledge regarding risks. Where technology assessment shifts from truth of sentences about risk to the robustness or resilience of emerging technical systems and their interaction with other technical systems. In this case, a different kind of prudential approach is required – for example, Dupuy and Grinbaum's "ongoing normative assessment" [55].

Expectations of certainty and assurances of safety will not be met by nanotechnologies. Other technologies already fail to meet them. Certainty about the safety of a new drug, for example, is produced by the traditional method of a clinical trial that establishes or refutes some proposition about the drug's efficacy and severity of side-effects. A far more complex and integrated mechanism is required where such certainty is unattainable and where robustness needs to be demonstrated. Here, several activities have to work in tandem, ranging from traditional toxicology, occupational health and epidemiology all the way to the deliberate adoption of an unknown risk for the sake of a significant desired benefit. If this integration works, social robustness will be built into the technical system along with the robustness of acquired skills, tried and true algorithms, measuring and monitoring apparatus. The fact that nanoscale researchers demonstrate acquired capabilities and that they thus produce "mere" skill knowledge creates a demand for skill knowledge also in a social arena where nanotechnological innovations are challenged, justified and appropriated.

7.6
What Basic Science Does Nanotechnology Need?

The preceding sections provided a survey of nanotechnoscience in terms of disciplinary questions (a complex field partially disclosed by stretching closed theories), of methodology (constructions and qualitative judgments of likeness), of ontology (a thin conception of nature as unlimited potential) and of epistemology (acquisition and demonstration of capabilities). This does not exhaust a philosophical characterization of the field which would have to include, for example, a sustained investigation of nanotechnology as a conquest of space or a kind of territorial expansion.[32] Also, nothing has been said so far about nanotechnology as an enabling technology that might enable, in particular, a convergence with bio- and information technologies. Finally, it might be important to consider nanotechnoscience as an element or symptom of a larger cultural transition from scientific to technoscientific research.

This survey is limited in other ways. It glossed over the heterogeneity of research questions and research traditions. And it focused exclusively on the way in which nanotechnological research has developed thus far. There is nothing in the preceding account to preclude a profound reorientation of nanoscience and nanotechnologies. Indeed, one reorientation might consist in the whole enterprise breaking apart and continuing in rather more traditional disciplinary settings – with "nano" ceasing to be a funding umbrella but becoming a prefix that designates a certain approach. Thus, under the sectoral funding umbrellas "food and agriculture", "energy", "health", "manufacturing" or "environment" researchers with the "nano" prefix would investigate how problems and solutions can be viewed at the molecular level. Their work

32) One implication of this is that nanotechnology should not be judged as the promise of a future but, instead, as a collective experiment in and with the present [56].

would then have to be integrated into more comprehensive approaches to the problem at hand.

Alternatively, nanotechnological researchers may pursue and promote disciplinary consolidation and unification.[33] In that case, they might be asking the question, "what kind of basic science does nanotechnology need?". From quantum mechanics, hydrodynamics, and so on, derive the (closed) theories that serve as the toolkit on which nanoscale research is drawing. While these are basic sciences, of course, they are not therefore the basis of nanoscience. What, then, is the basic scientific research that needs to be done in order to ground nanotechnologies properly or to establish nanoscience as a field in its own right? There have been no attempts so far to address this question in a systematic way.[34] And obviously, one should not expect any consensus regarding the following list of proposed basic research for nanotechnology.

In terms of empirical grounding or a theoretical paradigm, some call for general theories of (supra-)molecular structure–property relations, others imagine that there will be a future science of molecular and nanotechnical self-organization.[35] Following the suggestion of Peter Janich (see above, Section 7.4.3), one might identify and systematize how nanoscale phenomena are constituted through techniques of observation and measurement – this might render theories of instrumentation basic to nanoscience.[36]

Another kind of basic research entirely would come from so-called *Bildwissenschaft* (image or picture science) that could provide a foundation for image-production and visualization practice in nanotechnoscience. Such investigations might contribute visual clues for distinguishing illustrations from animations, from simulations, from visualizations of microscopically obtained data. They might also turn to image–text relations or develop conventions for reducing the photographic intimations of realism while enhancing informational content.[37]

33) The field of "nanomedicine" appears to be moving in that direction by distinguishing its research questions and paradigms from "medical nanotechnologies." It is not at all clear yet whether nanomedicine will emerge from this with a disciplinary identity of its own, including perhaps a unique body of theory.

34) To be sure, there are piecemeal approaches. One might say, for example, that a theory of electron transport is emerging as a necessary prerequisite for molecular electronics (but see [57]). Also, the giant magnetoresistance effect might be considered a novel nanotechnological phenomenon that prompted "basic" theory development [13].

35) See, for example, [15]. In [14] Michael Roukes calls for the identification of the special laws that govern the nanoscale. To be sure, there is profound skepticism in the scientific community (a) that there can be laws of structure-property relations at the nanoscale and (b) that they are needed in order to pursue nanotechnological research. On this latter view, the account provided in the first four sections of this paper provides sufficient "grounding" of nanotechnology.

36) See, for example, [58] on modeling of measurements at the nanoscale. Can this kind of theory development and modification serve to constitute a nanoscale research community – or does it belong to a special tribe of instrument developers that merely enters into a trade with other nanotechnology researchers? [59].

37) Compare the suggestion by Thomas Staley (at the conference on Imaging Nanospace) that visualizations of data could be constructed like maps with graphic elements even text imposed upon the quasi-photographic image [60]. This might break the spell of the powerful image (see above, Section 7.3.3) and return ownership of the image to the scientific community.

Finally, one might ask whether nanotechnoscience can and should be construed as a "social science of nature".[38] As an enabling, general-purpose or key technology it leaves undetermined what kinds of applications will be enabled by it. This sets it apart from cancer research, the Manhattan project, the arms race, space exploration, artificial intelligence research, and so on. As long as nanotechnoscience has no societal mandate other than to promote innovation, broadly conceived, it remains essentially incomplete, requiring social imagination and public policy to create an intelligent demand for the capabilities it can supply. As research is organized to converge upon particular societal goals [61], nanoscience and nanotechnology might be completed by incorporating social scientists, anthropologists and philosophers in its ambitions to design or shape a world atom by atom.

Nanotechnologies are frequently touted for their transformative potential, for bringing about the next scientific or industrial revolution. This chapter did not survey a revolutionary development, but pragmatic and problematic integrations of pre-existing scientific knowledge with the novel discoveries at the nanoscale. If one expects science to be critical of received theories and to produce a better understanding of the world, if one expects technology to enhance transparency and control by disenchanting and rationalizing nature, these pragmatic integrations appear regressive rather than revolutionary. If one abandous these expectations and makes the shift from epistemic certainty to systemic robustness, these pragmatic integrations hold the promise of producing socially robust technologies. In the meantime, there is no incentive for researchers and hardly any movement on the side of institutions to consider seriously the question of a disciplinary reorientation and consolidation of the nanosciences and nanotechnologies. A nanotechnological revolution has not happened yet: we may be waiting for it in vain and this is probably a good thing.[39]

38) See footnotes 3 and 29. The term Soziale Naturwissenschaft was coined in the context of the finalization thesis and could be designated more literally as a social natural science – science of a nature that is socially shaped through applied science, technology and human action. It is thus not social science but an integrated approach that acknowledges the social character of the world. Here, this proposal is taken up in two ways. Materials (as opposed to matter) and molecules defined by their history and situation are social entities, as such objects of this socials science of nature. Second, nanotechnoscience is a program for shaping and reshaping, for designing and redesigning, for reforming the world. To the extent that this is also a social reform it is systematically incomplete without societal agenda setting: What are the projects, the problems to be solved, the targets and design norms of nanotechnoscience?

39) Some are waiting, of course, not for a radically new and progressive way of doing science but for the far-off scientific breakthroughs that inspire speculations about human enhancement and mind–machine interfaces. However, in the context not of pure philosophy but of supposed implications of current research, "revolutionary" human enhancement is a non-issue that can only distract from more urgent questions [62].

References

1 Latour, B. (1987) *Science in Action*, Harvard University Press, Cambridge, MA.

2 Haraway, D. (1997) *Modest_Witness@Second_Millenium*, Routledge, New York.

3 Nordmann, A. (2004) Was ist TechnoWissenschaft? – Zum Wandel der Wissenschaftskultur am Beispiel von Nanoforschung und Bionik, in *Bionik: Aktuelle Forschungsergebnisse in Natur-, Ingenieur- und Geisteswissenschaften* (eds T. Rossmann and C. Tropea), Springer, Berlin, pp. 209–218.

4 Hacking, I. (1983) *Representing and Intervening*, Cambridge University Press, New York.

5 Galison, P. (2006) The pyramid and the ring, presented at the conference of the Gesellschaft für analytische Philosophie (GAP), Berlin.

6 Brune, H., Ernst, H., Grunwald, A., Grünwald, W., Hofmann, H., Krug, H., Janich, P., Mayor, M., Rathgeber, W., Schmid, G., Simon, U., Vogel, V. and Wyrwa, D. (2006) *Nanotechnology: Assesssments and Perspectives*, Springer, Berlin.

7 Nanoscience and Nanotechnologies: Opportunities and Uncertainties, Royal Society and Royal Academy of Engineering, London. 2004.

8 Echeverría, J. (2003) *La Revolución Tecnoscientífica*, Fondo de Cultura Económica de España, Madrid.

9 Johnson, A.(3 March 2005) Ethics and the epistemology of engineering: the case of nanotechnology, presented at the conference Nanotechnology: Ethical and Legal Issues, Columbia, SC.

10 Heisenberg, W. (1974) The notion of a "closed theory" in modern science, in *Across The Frontiers* (ed. W. Heisenberg), Harper and Row, New York, pp. 39–46.

11 Bokulich, A. (2006) Heisenberg meets Kuhn: closed theories and paradigms. *Philosophy of Science*, **73**, 90–107.

12 Schäfer, W. (ed.) (1983) *Finalization in Science*, Reidel, Dordrecht.

13 Wilholt, T. (2006) Design rules: industrial research and epistemic merit. *Philosophy of Science*, **73**, 66–89.

14 Roukes, M. (2001) Plenty of room indeed. *Scientific American*, **285** (3), 48–57.

15 Eberhart, M. (2002) Quantum mechanics and molecular design in the twenty first century. *Foundations of Chemistry*, **4**, 201–211.

16 Maynard, A.D. *et al.* (2006) Safe handling of nanotechnology. *Nature*, **444**, 267–269.

17 Hertz, H. (1956) *The Principles of Mechanics*, Dover, New York.

18 Cartwright, N. (1999) *The Dappled World: a Study of the Boundaries of Science*, Cambridge University Press, Cambridge.

19 Morgan, M. and Morrison, M. (eds) (1999) *Models as Mediators*, Cambridge University Press, Cambridge.

20 Nordmann, A. (2004) Molecular disjunctions: staking claims at the nanoscale, in *Discovering the Nanoscale* (eds D., Baird, A. Nordmann and J., Schummer), IOS Press, Amsterdam, pp. 51–62.

21 Chen, J., Reed, M.A., Rawlett, A.M. and Tour, J.M. (1999) Large on–off rations and negative differential resistance in a molecular electronic device. *Science*, **286**, 1550–1552.

22 Winsberg, E. (2006) Handshaking your way to the top: simulation at the nanoscale, in *Simulation: Pragmatic Constructions of Reality* (eds J. Lenhard, G. Küppers and T. Shinn), Springer Dordrecht (*Sociology of the Sciences Yearbook*, Vol. 25), 139–154.

23 Batterman, R.W. (2006) Hydrodynamics versus Molecular Dynamics: Intertheory Relations in Condensed Matter Physics, in *Philosophy of Science*, **73**, 888–904.

23a Nordmann, A. (2006) Unsichtbare Ursprünge: Herbert Gleiter und der Beitrag der Materialwissenschaft, in *Nanotechnologien im Kontext: Philosophische, ethische und gesellschaftliche Perspektiven* (eds A. Nordmann, J.

Schummer and A. Schwarzl), Akademische Verlagsgesellschaft, Berlin, 81–96.

24 Bensaude-Vincent, B. (2001) The Construction of a Discipline: Material Sciences in the United States. *Historical Studies in the Physical and Biological Sciences*, **31**, 223–248.

25 Bensaude-Vincent, B. and Guchet, X. (2007) Nanomachine: one word for three different paradigms. *Techné*, **11** (1), 71–89.

26 Bensaude-Vincent, B., Two cultures of nanotechnology? in *Nanotechnology Challenges: Implications for Philosophy, Ethics and Society* (eds J. Schummer and D. Baird), World Scientific Publishing, Singapore, 7–28.

27 Whitesides, G.M. and Grzybowski, B. (2002) Self-Assembly at all scales. *Science*, **295**, 2418–2421.

28 Scientific Committee on Emerging and Newly Identified Health Risks (SCENIHR) (2005) Opinion on the Appropriateness of Existing Methodologies to Assess the Potential Risks Associated with Engineered and Adventitious Products of Nanotechnologies, European Commission, Health and Consumer Protection Directorate-General, Brussels.

29 Schummer, J. (2001) Ethics of chemical synthesis. *International Journal for the Philosophy of Chemistry*, **7**, 103–124.

30 Kurath, M. and Maasen, S. (2006) Toxicology as a nanoscience? – disciplinary identities reconsidered. *Particle and Fibre Toxicology*, **3**, 6.

31 Nordmann, A. (2006) Collapse of distance: epistemic strategies of science and technoscience. *Danish Yearbook of Philosophy*, **41**, 7–34.

32 Jones, R. (2004) *Soft Machines*, Oxford University Press, Oxford.

33 Mody, C. (2004) How probe microscopists became nanotechnologists, in *Discovering the Nanoscale* (eds D., Baird, A. Nordmann and J. Schummer), IOS Press, Amsterdam, pp. 119–133.

34 Hennig, J. (2006) Lokale Bilder in globalen Kontroversen: Die heterogenen Bildwelten der Rastertunnelmikroskopie, in *The Picture's Image. Wissenschaftliche Visualisierungen als Komposit* (eds I. Hinterwaldner and M., Buschhaus), Fink, Munich, pp. 243–260.

35 Johnson, A. (2008) Modeling molecules: from computational chemistry to computational nanotechnology. Perspectives on Science, submitted.

36 Scerri, E. (2000) Have orbitals really been observed? *Journal of Chemical Education*, **77**, 1492–1494.

37 Vermaas, P.E. (2004) Nanoscale technology: a two-sided challenge for interpretations of quantum mechanics, in *Discovering the Nanoscale* (eds D. Baird, A. Nordmann and J. Schummer), IOS Press, Amsterdam, pp. 77–91.

38 Harré, R. (2003) The materiality of instruments in a metaphysics for experiments, in *The Philosophy of Scientific Experimentation* (ed. H. Radder), University of Pittsburgh Press, Pittsburgh, PA, pp. 19–38.

39 Winsberg, E. (2007) A tale of two methods, unpublished, http://www.cas.usf.edu/~ewinsb/papers.html.

40 Mitchell, W. (2005) *What Do Pictures Want? The Lives and Loves of Images*, University of Chicago Press, Chicago.

41 Belting, H. (2001) *Bildanthropologie. Entwürfe für eine Bildwissenschaft*, Fink, Munich.

42 Nordmann, A. (2006) Noumenal technology: reflections on the incredible tininess of nano, in *Nanotechnology Challenges: Implications for Philosophy, Ethics and Society* (eds J. Schummer and D. Baird), World Scientific, Singapore, pp. 49–72. originally published as Nordmann, A. (2005) *Techné*, **8** (3), 3–23.

43 Nordmann, A. Technology naturalized: a challenge to design for the human scale, in *Philosophy and Design: from Engineering to Architecture* (eds P.E. Vermaas, P. Kroes, A. Light and S.A. Moore), Springer, Dordrecht, 173–184.

44 Nordmann, A. (2007) Entflechtung – Ansätze zum ethisch-gesellschaftlichen

Umgang mit der Nanotechnologie, in *Nano – Chancen und Risiken aktueller Technologien* (eds A. Gazsó, S. Greßler and F., Schiemer), Springer, Berlin, pp. 215–229.

45 ETC Group (2003) The strategy for converging technologies. *ETC Communiqué*, **78**, 1–8.

46 Gaskell, G., Ten Eyck, T., Jackson, J. and Veltri, G. (2004) Public Attitudes to nanotechnology in Europe and the United States. *Nature Materials*, **3**, 496.

47 Heidegger, M. (1977) *The Question Concerning Technology and Other Essays*, Harper and Row, New York.

48 Bensaude-Vincent, B. (1992) The balance: between chemistry and politics. *Eighteenth Century*, **33**, 217–237.

49 Böhme, G. and Schäfer, W. (1983) Towards a social science of nature, in *Finalization in Science* (ed. W. Schäfer), Reidel, Dordrecht, pp. 251–269.

50 Janich, P. (2006) Wissenschaftstheorie der Nanotechnologie, in *Nanotechnologien im Kontext: philosophische, ethische und gesellschaftliche Perspektiven* (eds A. Nordmann, J. Schummer and A. Schwarz), Akademische Verlagsgesellschaft, Berlin, pp. 1–32.

51 Grunwald, A. (2006) Nanotechnologie als Chiffre der Zukunft, in *Nanotechnologien im Kontext: philosophische, ethische und gesellschaftliche Perspektiven* (eds A. Nordmann, J. Schummer and A. Schwarz), Akademische Verlagsgesellschaft, Berlin, pp. 49–80.

52 Baird, D. and Shew, A. (2004) Probing the history of scanning tunneling microscopy, in *Discovering the Nanoscale* (eds D. Baird, A. Nordmann and J. Schummer), IOS Press, Amsterdam, pp. 145–156.

53 Baird, D. (2004) *Thing Knowledge: a Philosophy of Scientific Instruments*, University of California Press, Berkeley, CA.

54 Beck, U. (1986) *Risikogesellschaft. Auf dem Weg in eine andere Moderne*, Suhrkamp, Frankfurt. Beck, U. (1992) *Risk Society: Towards a New Modernity*, Sage, London.

55 Dupuy, A. and Grinbaum, J.-P. (2004) Living with uncertainty: toward the ongoing normative assessment of nanotechnology. *Techné*, **8** (2), 4–25.

56 Nordmann, A. (2007) Design choices in the nanoworld: a space odyssey, in *Nano Researchers Facing Choices, The Dialogue Series* (eds M. Deblonde, L. Goorden, *et al.*), Vol. **10**, Universitair Centrum Sint-Ignatius, Antwerp. pp. 13–30. Nordmann, A. (2007) Gestaltungsspielräume in der Nanowelt: Eine Space-Odyssee, in *Nanotechnologie: Erwartungen, Anwendungen, Auswirkungen* (eds D. Korczak and A. Lerf), Asanger, Kröning, pp. 159–184.

57 Tao, N.J. (2006) Electron transport in molecular junctions. *Nature Nanotechnology*, **1**, 173–181.

58 Clifford, C.A. and Seah, M.P. (2006) Modelling of nanomechanical nanoindentation measurements using an AFM or nanoindenter for compliant layers on stiffer substrates. *Nanotechnology*, **17**, 5283–5292.

59 Galison, P. (1997) *Image and Logic: a Material Culture of Microphysics*, Chicago University Press, Chicago.

60 Staley, T.W., The coding of technical images of nanospace: analogy, disanalogy and the asymmetry of worlds. *Techné*, in press.

61 HLEG (High Level Expert Group "Foresighting the New Technology Wave") (2004) Converging Technologies: Shaping the Future of European Societies, Office for Official Publications of the European Communities, Luxembourg.

62 Nordmann, A. (2007) If and then: a critique of speculative nanoethics. *NanoEthics*, **1** (1), 31–46.

8
Ethics of Nanotechnology. State of the Art and Challenges Ahead

Armin Grunwald

8.1
Introduction and Overview

In view of the revolutionary potentials attributed to nanosciences and nanotechnology with respect to nearly all fields of society and individual life [1, 2], it is not surprising that "nano" has attracted great interest in the media and in the public. Parallel to high expectations, for example in the fields of health, growth and sustainable development, there are concerns about risks and side-effects. Analyzing, deliberating and assessing expectable impacts of nanotechnology on future society are regarded as necessary parts of present and further development. There have already been commissions and expert groups dealing with ethical, legal and social implications of nanotechnology (ELSI) [3, 4]. An ethical reflection on nanotechnology emerged and has already led to new terms such as "nano-ethics" and the recent foundation of a new journal *Nano-Ethics*. The quest for ethics in and for nanotechnology currently belongs to public debate in addition to scientific self-reflection. The ethical aspects of nanotechnology discussed in the (so far few) treatises available show broad evidence of this relatively new field of science and technology ethics.

Nanotechnology has been attracting increasing awareness in practical philosophy and in professional ethics. However, there has been some time delay compared with the development of nanotechnology itself: "While the number of publications on NT [nanotechnology] *per se* has increased dramatically in recent years, there is very little concomitant increase in publications on the ethical and social implications to be found" [5, p. R10]. Certain terms, such as privacy, man–machine interface, the relationship between technology and humankind or equity are often mentioned.

In the last few years, ethical reflection on nanotechnology developed quickly and identified many ethically relevant issues [6, 7]. However, well-justified criteria for determining why certain topics, such as nanoparticles or crossing the border between technology and living systems, should be ethically relevant are mostly not given and there is no consensus yet. In particular, the *novelty* of the ethical questions touched by developments emerging from nanotechnology compared with ethical issues in

Nanotechnology. Volume 1: Principles and Fundamentals. Edited by Günter Schmid

ISBN: 978-3-527-31732-5

well-known fields of technology is often not clear. Furthermore, although ethical aspects of nanotechnology have been identified, their analysis and the elaboration of proposals for how to deal with them in society is still at the beginning. Ethics of nanotechnology is, therefore, still an emerging field, in spite of the thematic broadness and the scientific awareness of the ongoing discussion.

In view of this situation, the purpose of this chapter consists primarily in studying current and foreseeable developments in nanotechnology from the viewpoint of philosophical ethics. Which developments are ethically relevant? Are there ethical questions which have already been tackled by current or recent discussions in the ethics of technology or in bioethics? Could the analysis of ethical aspects of nanotechnology benefit from other ethical discussions? Are there developments which pose completely new ethical questions? To this end, it is necessary to clarify the understanding of the notion of "ethics" to be used in order to look for criteria for deciding when something is ethically relevant in a transparent way (Section 8.2). These criteria are then applied to the field of nanotechnology and the ethical challenges are "mapped out" and described briefly, in order to give a broad overview (Section 8.3). This overview is complemented by two in-depth case studies of ethical aspects in nanotechnology: the challenge of dealing with possible risks of nanoparticles and the role of the precautionary principle (Section 8.4) and the human enhancement case, which is directly related to nanotechnology via the debate on "converging technologies" (Section 8.5). Dealing constructively and in a rational way with these ethical challenges requires specific conceptual and methodical developments. In particular, some effort has to be invested into handling the dimension of the future in normative as well as in epistemological regard in a non-partisan way (Section 8.6).

8.2
The Understanding of Ethics[1)]

In modern discussion, the distinction between factual morals on the one hand and ethics as the reflective discipline in cases of moral conflicts or ambiguities on the other has widely been accepted [10]. This distinction takes into account the plurality of morals in modern society. As long as established traditional moral convictions (e.g. religious ones) are uncontroversial and valid among all relevant actors and as long as they are sufficient to deal with the respective situation and do not leave open relevant questions, ethical reflection is not in place. Morals are, in fact, the action-guiding maxims and rules of an individual, of a group or of society as a whole. Ethical analysis, on the other hand, takes these morals as its subjects to reflect on. Ethics is concerned with the justification of moral rules of action, which can lay claim to validity above and beyond the respective, merely particular morals [8]. In

1) This chapter summarises general work of the author in the field of ethics of technology [8, 9] in order to introduce the basic notions to be used in the following in a transparent way. See also [1, Section 6.2].

particular, ethics serves the resolution of conflict situations which result out of the actions or plans of actors based on divergent moral conceptions by argumentative deliberation only, which is grounded in philosophical ideas such as the Categorical Imperative by Immanuel Kant, the Golden Rule or the Pursuit of Happiness (Utilitarianism).

Normative aspects of science and technology lead, in a morally pluralistic society, unavoidably to societal debates at the least and often also to conflicts over technology. We can witness recent examples in the fields of nuclear power and radioactive waste disposal, stem cell research, genetically modified organisms and reproductive cloning. As a rule, what is held to be desirable, tolerable or acceptable is controversial in society. Open questions and conflicts of this type are the point of departure for the *ethics of technology* [10, 11]. Technology conflicts are, as a rule, not only conflicts over technological means (e.g. in questions of efficiency), but also include diverging ideas over visions of the future, of concepts of humanity and on views of society. Technology conflicts are often controversies about *futures*: present images of the future – which are considerably influenced by our illustrations of the scientific and technological advance – are highly contested [12]. The role of the ethics of technology consists of the analysis of the normative structure of technology conflicts and of the search for rational, argumentative and discursive methods of resolving them. In this "continental" understanding, ethics is part of the philosophical profession. In ethical reflection in the various areas of application, however, there are close interfaces to and inevitable necessities for interdisciplinary cooperation with the natural and engineering sciences involved as well as with the humanities. Even transdisciplinary work might be included in cases of requests for broad participation, for example in the framework of participatory technology assessment [13].

Technology is not nature and does not originate of itself, but is consciously produced to certain ends and purposes – namely, to bring something about which would not happen of itself. Technology is therefore always embedded in societal goals, problem diagnoses and action strategies. In this sense, there is no "pure" technology, that is a technology completely independent of this societal dimension. Therefore, research on and development of new technologies always refer to normative criteria of decision-making including expectations, goals to be reached and values involved [14].

But even if technology is basically beset with values, this does not imply that every decision in research and technology development must be scrutinized in ethical regard. Most of the technically relevant decisions can, instead, be classified as a "standard case" in moral respect in the following sense [6, 9]: they do not subject the normative aspects of the basis for the decision (criteria, rules or regulations, goals) to specific reflection, but assume them to be given for the respective situation and accept the frame of reference they create. In such cases, no explicit ethical reflection is, as a rule, necessary, even if normative elements self-evidently play a vital role in these decisions – the normative decision criteria are clear, acknowledged and unequivocal. It is then out of the question that this could be a case of conflict with moral convictions or a situation of normative ambiguity – the information on the normative framework can be integrated into the decision by those affected and by those deciding on the basis of axiological information, without analyzing it or deliberating on it. The (national and international) legal regulations, the rules of the relevant institutions (e.g. corporate guidelines), where applicable, the code of ethics of the professional group concerned, as well as general societal usage, are elements of this normative framework [14]. The requirements to be fulfilled in the affected normative framework in order that the respective moral situation can be assumed to be a "standard case" can be operationalized according to the following criteria [6, 11]:

- *Pragmatic completeness:* the normative framework has to treat the decision to be made fully with regard to normative aspects.
- *Local consistency:* there must be a "sufficient" measure of consistency between the normative framework's elements.
- *Unambiguity:* among the relevant actors, there must be a sufficiently consensual interpretation of the normative framework.
- *Acceptance:* the normative framework must be accepted by those affected as the basis for the decision.
- *Compliance:* the normative framework also has to be complied with in the field concerned.

Standard situations in moral respect in this sense are governing decision-making in many fields (e.g. in many cases of laboratory work, in public administration or in private businesses). Technical innovations and scientific progress, however, can challenge such situations by presenting new questions or by shaking views previously held to be valid. This is then the entry point for ethical reflection in questions of science and engineering, for the explicit confirmation, modification or augmentation of the normative framework [15] but also for influencing the direction of further development. This conceptual framework provides a point of departure to identify ethically relevant aspects in new technological developments. Whether there are new challenges for ethics in nanotechnology and which ethical questions will be addressed will be investigated against this background.

8.3
Ethical Aspects of Nanotechnology – an Overview

Ethical challenges of nanotechnology are, following the preceding section, challenges of the existing normative frameworks including the social values represented by them by emerging nanotechnological innovations. The task of identifying ethical challenges emerging from nanotechnology, therefore, can be transformed into the search for affected normative frameworks and social values. Because of the novelty of nanotechnology, which still consists mostly of nanoscience [1, 2], corresponding technology development will frequently take place in form of "radical design processes" rather than as "normal design processes" which may be characterized by more incremental approaches (following [14]). Therefore, we can expect that ethical challenges will be relevant in many cases [15]. Establishing a "map" of the "landscape" of ethical questions caused by nanotech innovation means answering the following questions:

- Which are the ethical aspects of nanotechnology and related innovations in the sense defined above?
- Which of the identified ethical aspects of nanotechnology are *specific* for nanotechnology and *novel* to ethics?
- Where are relations to recent or ongoing ethical debates in other technology fields, if any?

The resulting "map" of ethical aspects described in the following is organized with reference to existing ethical debates (e.g., debates on privacy, equity or human nature). This classification has the advantage that it automatically allows one to refer to normative frameworks and therefore enables us to investigate whether (a) existing frameworks are sufficient to deal with the value problems involved and (b) if not, whether there are new challenges to the frameworks which are specifically caused by nanotechnology. In this way, it is possible to arrive at a structured and well-founded picture of ethical aspects in nanotechnology.[2)]

8.3.1
Equity: Just Distribution of Opportunities and Risks

A first type of ethical aspects of nanotechnology might result from considerations of equity and distributional justice. Ethical questions concern the *distribution* of the benefits of nanotechnology among different groups of the population or among different regions of the world, as well as the spatial and temporal distribution of the risks of nanotechnology [16]: "Nanotech offers potential benefits in areas such as biomedicine, clean energy production, safer and cleaner transport and environmental remediation: all areas where it would be of help in developing countries. But it is at present mostly a very high-tech and cost-intensive science and a lot of the current research is focused on areas of information technology where one can imagine the result being a widening of the gulf between the haves and the have-nots" [5].

Problems of distributive justice are inherent to many fields of technical innovation. Because scientific and technical progress requires considerable investment, it usually takes place where the greatest economic and human resources are already available. Technical progress does, by its nature, tend to increase existing inequalities of distribution. For example, nanotechnology-based medicine will, in all probability, be rather expensive. Questions of equity and of access to (possible) medical treatments could become urgent in at least two respects: *within* industrialized societies, existing inequalities in access to medical care could be exacerbated by a highly technicized medicine making use of nanotechnology and – with regard to *less developed societies* – because likewise, already existing and particularly dramatic inequalities between industrialized and developing nations could be further increased. Apprehensions

2) This analysis builds on earlier work in the field [1, 6] and goes beyond the state reached. In particular, the categorical classification in fields of ethical interest has been improved.

with regard to both of these types of a potential "nano-divide" (after the well-known "digital divide") are based on the assumption that nanotechnology can lead not only to new and greater options for individual self-determination (e.g. in the field of medicine), but also to considerable improvement of the competitiveness of national economies. Current discussions on distributive justice on both national and international levels (in the context of sustainability as well) are therefore likely to gain increased relevance with regard to nanotechnology.

A specific future field of debate with respect to equity will be the "human enhancement" issue ([17]; see Section 8.5 of this chapter). If technologies of improving human performance were to be available then the question arises of who will have access to those technologies, especially who will be able to pay for them and what will happen to persons and groups excluded from the benefits. There could develop a separation of the population into "enhanced" and "normal" people where a situation is imaginable with "normal" to be used as a pejorative attribute [18] and a coercion towards enhancement might occur: "Merely competing against enhanced co-workers exerts an incentive to use neuro-cognitive enhancement and it is harder to identify any existing legal framework for protecting people against such incentives to compete" [19, p. 423]. Special problems can be expected for disabled persons [20].

Equity aspects, however, are not really new ethical aspects caused by nanotechnology, but are rather intensifications of problems of distribution already existing and highly relevant. Problems of equity belong indispensably to modern technology in general and are leading to ongoing and persistent debates in many fields. The digital divide [21] is, perhaps, the best-known example. But also in military respects or with regard to access to medical high-tech solutions these inequalities exist and are debated in many areas. There is no new ethical question behind them but there might be new and dramatic cases emerging driven by nanotechnological R&D. This point has already arrived at the international level of the nanotech debate [22].

8.3.2 Environmental Issues[3)]

Technology is of major importance for the sustainability of humankind's development. On the one hand, technology determines to a large extent the demand for raw materials and energy, needs for transport and infrastructure, mass flows of materials, emissions and amount and composition of waste. Technology is, on the other hand, also a key factor of the innovation system and influences prosperity, consumption patterns, lifestyles, social relations and cultural developments. Therefore, the development, production, use and disposal of technical products and systems have impacts on the ecological, economic and social dimensions of sustainable development. In most cases, these impacts are ambivalent with regard to sustainable development [23]: there are both positive contributions and negative consequences. The overall concept of

3) A detailed analysis of nanotech potential with regard to sustainable development can be found in [26] and the respective Special Issue of the *Journal of Cleaner Production*.

sustainable development requires production and consumption patterns to be shaped in a manner such that the needs of the current generations can be satisfied in a way that they do not limit or threaten opportunities of future generations for satisfying their needs [24, 25]. Of special relevance in this respect are (following [26]):

- The *limited availability of many natural resources* such as clean water, fossil fuels and specific minerals highlights the importance of the efficiency of their use, of recycling and of substituting non-renewable resources by renewable ones.
- The *limited carrying capacity of the environment* (atmosphere, groundwater and surface water, oceans and pedosphere, ecosystems) emphasizes the necessity for limiting or reducing emissions and for regenerating damaged environments.
- The postulate of *intergenerational equity* as a core part of the idea of sustainable development requires consideration of the distribution of risks and benefits of new technologies among the population and in the global dimension ([22]; see also the preceding section of this paper).
- The sustainability issue of *participation* leads to consequences for the processes of opinion-forming and decision-making in shaping technology and its interfaces with the public (e.g., by means of participatory technology assessment [13]).

Many scientists and engineers claim that nanotechnology promises less material and energy consumption and less waste and pollution from production. Nanotechnology is also expected to enable new technological approaches that reduce the environmental footprints of existing technologies in industrialized countries or to allow developing countries to harness nanotechnology to address some of their most pressing needs [22]. Nanoscience and nanotechnology may be a critical enabling component of sustainable development when they are used wisely and when the social context of their application is considered [26]. There are a lot of high expectations concerning positive contributions of nanotechnology to sustainable development.

However, all the potential positive contributions to sustainable development may come at a price. The ambivalence of technology with respect to sustainable development also applies to nanotechnology [27]. The production, use and disposal of products containing nanomaterials may lead to their appearance in air, water, soil or even organisms [1, Chapter 5, 28]. Nanoparticles could eventually be transported as aerosols over great distances and be distributed diffusely. Despite many research initiatives throughout the world, only little is known about the potential environmental and health impacts of nanomaterials. This situation applies also and above all for substances which do not occur in the natural environment, such as fullerenes or nanotubes. The challenge of acting under circumstances with high uncertainties but with the nanoproducts already at the marketplace is the heart of the ethical challenges by nanoparticles (because of the high relevance and because nanoparticles and their possible risks are under intensive public observation today [29, 30], this topic will be dealt with in-depth (see Section 8.4).

Questions of eco- or human toxicity of nanoparticles, on nanomaterial flow, on the behavior of nanoparticles in spreading throughout various environmental media, on their rate of degradation or agglomeration and their consequences for the various conceivable targets are, however, not ethical questions (see Section 8.2). In these cases, empirical–scientific disciplines, such as human toxicology, eco-toxicology or environmental chemistry, are competent. They are to provide the knowledge basis for practical consequences for working with nanoparticles and for disseminating products based on them. However, as the debate on environmental standards of chemicals or radiation has shown [31, 32], the results of empirical research do not determine how society should react. Safety and environmental standards – in our case for dealing with nanoparticles – are to be based on sound knowledge but cannot logically be derived from that knowledge. In addition, normative standards, for example concerning the intended level of protection, the level of public risk acceptance and other societal and value-laden issues enter the field. Because of this situation, it is not surprising that frequently conflicts about the *acceptability* of risks occur [33, 34] – and this is obviously a non-standard situation in moral respect (see Section 8.2). Therefore, the field of determining the acceptability and the tolerability of risks of nanoparticles is an ethically relevant issue.

In particular, there are a lot of sub-questions where ethical investigation and debate are asked for in the field of nanoparticles. Such questions are [1, Section 6.2]:

- What follows from our present lack of knowledge about the possible side-effects of nanoparticles? This is a challenge to acting rationally under the condition of high uncertainty – a common problem in practical ethics.
- Is the precautionary principle [35] relevant in view of a lack of knowledge and what would follow from applying this principle [28, 31, 32] (see Section 8.4)?
- Which role do the – doubtlessly considerable – *opportunities* of nanoparticle-based products play in considerations of this sort? According to which criteria may benefits and hazards be weighed against each other, especially in cases when the benefits are (relatively) concrete, but the hazards are hypothetical?

A further question would be by which means such balancing could be performed in an inter-subjectively valid and commonly acceptable way. The quantification of risks and benefits by using utility values might be very difficult or even impossible in cases of high uncertainty about the probability, the kind and the extent of a possible damage as well as in cases of ethical problematic procedures of quantification (such as expressing the value of human life in monetarian units).

Are comparisons of the possible risks of nanoparticles with other types of risk possible, in order to learn from them? Can criteria for assessing the nanoparticle risks be gained from experience in developing new chemicals or medicines? Are we allowed to look at risks of our daily life in order to determine what nanoparticle risk should be acceptable in general [31]? Which normative premises enter into such comparisons and by which argumentative means could they be morally justified?

The questions of the acceptability and comparability of risks, the advisability of weighing up risks against opportunities and the rationality of action under uncertainty are, without doubt, of great importance in nanotechnology. A new field of application is developing here for the ethics of science and technology. The type of questions posed, however, is well known from established discussions on risks (e.g. risks by exposure to radiation or by new chemicals). Really novel ethical questions are not to be expected in spite of the high practical relevance of the field.

From an ethical point of view this situation is well known: there are, on the one hand, positive expectations with regard to sustainable development which legitimate a moral postulate to explore further and to exhaust those potentials. On the other hand, there are risks and uncertainties. This situation is the basic motivation of technology assessment (TA) as an operationalization of ethical reflections on technology [36]. The basic challenge with strong ethical support is *shaping* the further development of nanotechnology in the direction of sustainable development [25, 37].

8.3.3 Privacy and Control

Another field regularly mentioned among the ethical aspects of nanotechnology is the threat to privacy through new monitoring and control technologies. Nanotechnology offers a range of possibilities for gathering, storing and distributing personal data to an increasing extent. In the course of miniaturization, a development of sensor and memory technology is conceivable which, unnoticed by its "victim", drastically increases the possibilities for acquiring data. Furthermore, miniaturization and networking of observation systems (e.g., in the framework of "pervasive" or "ubiquitous" computing) could considerably impede present control methods and data protection regulations or even render them obsolete [38]. Passive observation of people could, in the distant future, be complemented by actively manipulating them – for instance, if it would be possible to gain direct technical access to their nervous system or brain [19, 39].

Within the private sphere, health is a particularly sensitive area. The development of small analyzers – the "lab on a chip" – can make it possible to compile comprehensive personal diagnoses and prognoses on the basis of personal health data. This technology can facilitate not only medical diagnoses, but can also make fast and economical comprehensive screening possible. Everyone could let him- or herself be tested, for example, for genetic dispositions for certain disorders – or could be urged by his/her employer or insurance company to do so. In this manner, individual persons could find themselves put under social pressure. Without sufficient protection of their private sphere, people are rendered manipulable, their autonomy and freedom of action are called into question. Stringent standards for data protection and for the protection of privacy therefore have to be set.

Questions of privacy, of monitoring and controlling people are doubtlessly ethically relevant. But all of these questions of monitoring and of data and privacy protection are not posed exclusively by nanotechnology. Even without nanotechnology, observation technologies have reached a remarkable stage of development which poses questions on the preservation of the private sphere. Even today, so-called smart tags, based on RFID technology (radiofrequency identification), are being employed for access control, such as ticketing, for example, in public transportation and in logistics. These objects have at present a size of several tenths of a millimeter in each dimension, so that they are practically unnoticeable to the naked eye. Further miniaturization will permit further reductions in size and the addition of more functions – without nanotechnology being needed – but nanotechnology will promote and accelerate these developments.

The ethically relevant questions on a right to know or not to know, on a personal right to certain data, on a right to privacy, as well as the discussions on data protection and on possible undesirable inherent social dynamisms and, in consequence, of a drastic proliferation of genetic and other tests, have been a central point in bio- and medical-ethical discussions for some time. Nanotechnological innovations can accelerate or facilitate the realization of certain technical possibilities and therefore increase the urgency of the problematics of the consequences; in this area, however, they do not give rise to qualitatively new ethical questions.

8.3.4 Military Use of Nanotechnology

Nanotechnology can improve not only multiple peaceful uses but also military and future arms systems. The foundation of the so-called Institute for Soldier Nanotechnologies (http://web.mit.edu/isn/) to enhance soldier survivability makes it clear that nanotechnology has applications in the military field. It is predicted that nanotechnology will bring revolutionary changes in this areas as well [40–42]. Nevertheless, progress in a military technology will not only improve survival and healing, it always implies its use to enhance the efficacy of weapons, surveillance systems and other military equipment. As nanotechnology will provide materials and products that are stronger, lighter, smaller and more sensitive, there may be projectiles with greater velocity and smaller precision-guidance systems. Moreover, nanotechnology will influence the processing in energy generation and storage, displays and sensors, logistics and information systems, all being important elements of warfare. A particular point of interest is the use of BCI (brain–computer interaction) for navigation support for jet pilots, which is being researched by many projects funded by the US defense agency DARPA [43]. In several countries the Departments or Ministries of Defense have arranged nanotechnological programs. No one wants to be left behind [42].

The ethical concerns related to these developments mentioned in the available literature may be classified into the following points [41, 42]:

- An arms race similar to that of nuclear weapons cannot be excluded.
- Present asymmetric power relation could be increased or intensified, for instance to the disadvantage of developing countries [7].
- Some nanobased weapons might be much smaller and, perhaps, cheaper that traditional ones. This could increase the risks of proliferation and of terrorist usage of those weapons.

As long as the military in general is regarded as ethically allowed – and this is the case in most concepts of ethics as far as the military is used for legitimate purposes such as self-defense of democratic states or legitimate interventions into totalitarian states or for humanitarian reasons – there is little reason to investigate ethical aspects of the use of new technologies in the military areas as a specific topic. At any time, technology has been an important factor in military affairs and had a decisive influence on power relations. Nations with highly developed industrial capacities will be able to exploit the military possibilities of scientific and technological advances to a greater extent than less developed nations.

However, the developments must be observed carefully from the standpoint of peace-keeping and arms control. Possibly areas of international agreements must be reconsidered [41], above all the arms control agreements (e.g. Biological Weapons Convention; limits on conventional forces by new weapons types outside of treaty definitions) or the international laws of warfare (e.g. through the introduction of autonomous fighting systems not reliably discriminating between combatants and non-combatants). There is a certain risk that military "facts" could be created before an open debate about such developments has been launched. In this way, awareness in ethical regard is required in this field but novel ethical questions are not in view at present.

8.3.5 Health

Miniaturization is an essential means of progress in many medical areas. Smaller samples for *in vitro* analysis allow less invasive and less traumatic methods of extraction. Better interfaces and biocompatible materials provide new occasions for implants and restoring damaged organic facilities. The field of medical application seems to be the largest area of future nanotech applications [2]. Nanomedicine is defined as "(1) the comprehensive monitoring, control, construction, repair, defense and improvement of all human biological systems; working from the molecular level, using engineered nanodevices and nanostructures; (2) the science and technology of diagnosing, treating and preventing disease and traumatic injury, of relieving pain and of preserving and improving human health, using molecular tools and molecular knowledge of the human body; (3) the employment of molecular machine systems to address medical problems, using molecular knowledge to maintain and improve human health at the molecular scale" [44, p. 418].

With the help of nanotechnology-based diagnostic instruments, diseases or predispositions for diseases could possibly be discovered earlier than at present [45]. Through the development of "lab-on-a-chip"-technology, the emerging tendency towards personalized medicine would be further promoted. In therapy, there is the prospect, with the help of nanotechnology, of developing targeted treatments free of side-effects. The broad use of nanoparticle dosage systems could lead to progress in medicinal treatment. Through nanotechnological methods, the biocompatibility of artificial implants can be improved. Drug delivery systems could considerably enhance the efficiency of medication and minimize side-effects: "Although many of the ideas developed in nanomedicine might seem to be in the realm of science fiction, only a few more steps are needed to make them come true, so the 'time-to-market' of these technologies will not be as long as it seems today. Nanotechnology will soon allow many diseases to be monitored, diagnosed and treated in a minimally invasive way and it thus holds great promise of improving health and prolonging life. Whereas molecular or personalized medicine will bring better diagnosis and prevention of disease, nanomedicine might very well be the next breakthrough in the treatment of disease" [46, p. 1012).

Addressing symptoms more efficiently or detecting early onsets of diseases is without doubt recommended by ethics. These potentials are so remarkable that ethical reflection almost seems to be superfluous – if one looks solely at the potentials. A comprehensive analysis, however, has to include – as noted above – also possible side-effects, especially risks [48–50]. New types of responsibility and new tasks for weighing up pros and cons might occur. For example, new forms of drug delivery based on nanotechnology (using fullerenes, nanostructured membranes, gold nanoshells, dendrimers [1, Section 3.3]) could also have consequences which are not expected and might not be welcome. Careful observations of the advance of knowledge and early investigations of possible side-effects have to be conducted. HTA (Health Technology Assessment) offers several established approaches for early warning. The ethically relevant issues are [7]:

- the gulf between diagnostic and therapeutic possibilities and the problem of undesirable information;
- data protection and the protection of privacy (see Section 8.3.3), especially the danger of genetic discrimination;
- preventive medicine and screening programs;
- increases in costs through nanomedicine and problems of access and equity (see Section 8.3.1);
- deferment of illness during the lifetime of humans;
- changes in the understanding of illness and health [52].

However, there is probably no field of science in which dealing with risks is so well established as in medicine and pharmaceutics. Advances in medicine (diagnosis and therapy) are evidently related to risks and there are a lot of established mechanisms such as approval procedures for dealing with them. There is nothing new about this situation. Therefore, using nanotechnology for medical purposes is a standard situation in moral respects (following the notion introduced in Section 8.2). Against this background, it seems to be improbable that direct applications of nanotechnology for medical purposes might lead to completely new ethical questions [47, 51]. The ethical topics to be aware of are not specific for the use of nanotechnology but are also valid for a lot of other advances in medical science and practice.

The boundaries of such a standard situation in moral respects would, however, be transgressed in some more visionary scenarios as in the vision of longevity or the abolition of aging. Nanotechnology could, in connection with biotechnology and, perhaps, neurophysiology, build the technological basis for realizing such visions. In this respect, the idea has been proposed that nanomachines in the human body could permanently monitor all biological functions and could, in case of dysfunction, damage or violation, intervene and re-establish the "correct" status. In this way, an optimal health status could be sustained permanently [53] which could considerably enlarge the human lifespan. Such methods, however, would require dramatic technological progress [2, Section 7.2.3]. According to the state of present knowledge, neither the prediction of the time needed for such developments nor an assessment of their feasibility at all can seriously be given.

A new area that is both practically and ethically interesting and much closer to realization consists of creating direct connections between technical systems and the human nervous system [39, 54, 55]. There is intensive current work on connecting the world of molecular biology with that of technology. An interesting field of development is nanoelectronic neuro-implants (neurobionics), which compensate for damage to sensory organs or to the nervous system [43]. Micro-implants could restore the functions of hearing and eyesight. Even today, simple cochlear or retina implants, for example, can be realized. With progress in nano-informatics, these implants could approach the smallness and capabilities of natural systems. Because of the undoubtedly positive goals of healing and restoring damaged capabilities, ethical reflection could, in this case, concentrate above all on the definition and prevention of misuse. Technical access to the nervous system, because of the possibilities for manipulation and control which it opens up, is a particularly sensitive issue. A more complex ethical issue would be neuro-cognitive *enhancement* of functions of the brain [19] (see Section 8.5).

Summing up, in the field of medical applications of nanotechnology there are, considered from the standpoint of today's knowledge, no ethical concerns which are specifically related to the use of nanotechnology. There are a lot of positive potentials which probably also will bear risks – these risks, however, might be dealt with by "standard" measures of risk analysis and management established in medical practice [48]. However, things might change in the more distant future (e.g., if there were to be a shift from the classical medical viewpoint to the perspective of "enhancing" human performance; see Section 8.5).

8.3.6
Artificial Life

Basic life processes take place on a nanoscale, because life's essential building blocks (such as proteins) have precisely this size. By means of nanobiotechnology, biological processes will, due to frequently expressed expectations, be made technologically controllable. Molecular "factories" (mitochondria) and "transport systems", which play an essential role in cellular metabolism, can be models for controllable bio-nanomachines. Nanotechnology on this level could permit the "engineering" of cells and allow a "synthetic biology" constructing living systems "from the bottom" or modifying existing living systems (such as viruses) by technical means. An intermeshing of natural biological processes with technical processes seems to be conceivable. The classical barrier between technology and life is increasingly being breached and crossed.

The technical design of life processes on the cellular level, direct links and new interfaces between organisms and technical systems portend a new and highly dynamic scientific and technological field. Diverse opportunities, above all, in the field of medicine, but also in biotechnology, stimulate research and research funding [1, Section 3.3]. New ethical aspects are certainly to be expected in this field. They will possibly address questions of artificial life and of rearranging existing forms of living systems by technical means, for example the reprogramming of viruses. Their concrete specification, however, will only be possible when research and development can give more precise information on fields of application and products.

Without knowing much about products and systems emerging from the mentioned developments, there is one thing which seems clear already today. We can surely expect discussions about risks because technically modifying or even creating life is morally and with respect a very sensitive field [56]. The corresponding discussions of risks will have structural similarities to the discussion on genetically modified organisms (GMOs) because in both cases the "source code of life" is attached by technical means. It could come to discussions about safety standards for the research concerned, about containment strategies, about "field trials" and release problems. The danger of misuse will be made a topic of debate, such as technically modifying viruses in order to produce new biological weapons that could possibly be used by terrorists. In this area of nanotechnology, opposition, rejection and resistance in society could be feared, comparable to the GMO case. There will be a demand for early dealing with possible ethical and risk problems and for public dialogue and involvement.

In spite of the partly still speculative nature of the subject, ethical reflection of the scientific advance on crossing the border between technology and life does not seem to be premature [49, 50]. There are clear indications that scientific and technical progress will intensify the – at present non-existent – urgency of these questions in the coming years. However, against the questions to be answered in this section, it has to be stated that these ethical questions are not really specific to nanotechnology. The slogan "shaping the world atom by atom" [57] does not make a difference between technology and life and is, therefore, background to crossing the border between technology and life. But the ethical debates to be expected can rely on preceding investigations. Since the 1980s, these subjects have been repeatedly discussed in the debates on GMO, on artificial intelligence and on artificial life. There is a long tradition of ethical thought which has to be taken into account when facing the new challenges in the field.

8.3.7 Human Enhancement[4)]

Within the tradition of technical progress, that has, at all times, transformed conditions and developments – which, until then, had been taken as given, as unalterable fate – into influenceable, manipulable and formable conditions and developments, the human body and its psyche are rapidly moving into the dimension of the formable. The vision of "enhancing human performance" has been conjured

4) Because of the high relevance in current debates and the challenge to traditional thinking involved, this topic has been selected for an in-depth investigation (see Section 8.5).

up, above all, in the field of “converging technologies” [17]. Nanotechnology, in combination with biotechnology and medicine, opens up perspectives for fundamentally altering and rebuilding the human body. At present, research is being done on tissue and organ substitution, which could be realized with the help of nano- and stem cell technologies. Nano-implants would be able to restore human sensory functions or to complement them, but they would also be able to influence the central nervous system (for an overview on types of enhancement, see [20]).

While the examples of medical applications of nanotechnology cited above [45] remain within a certain traditional framework – because the purpose consists of “healing” and “repairing” deviations from an ideal condition of health, which is a classical medical goal –chances (or risks) of a remodeling and “improvement” of the human body are opened up. This could mean extending human physical capabilities, for example to new sensory functions (e.g., broadening the electromagnetic spectrum that the eye is able to perceive). It could, however, also – by means of the direct connection of mechanical systems with the human brain – give rise to completely new interfaces between humans and machines. Even completely technical organs and parts of the body (or even entire bodies) are being discussed, which, in comparison with biological organisms, are supposed to have advantages such as – perhaps – increased stability against external influences [17].

There are initial anthropological questions of our concept of humanity and of the relationship between humanity and technology. With them, however, and at the same time, the question poses itself of how far human beings *can*, *should* or *want to* go in remodeling the human body and to what end(s) this should or could be done. The practical relevance of such ethical questions in view of a possible technical improvement of human beings (with the substantial participation of nanotechnology) may, at first sight, seem limited. Three considerations, however, contest this estimation. First, the vision of the technical enhancement of human beings is actually being seriously advocated. Research projects are being planned in this direction and milestones for reaching this goal are being set up, whereby nanotechnology takes on the role of an “enabling technology”. Second, the visions of human enhancement show consequences by merely communicating them – they currently modify the *conditio humana* [58]. Third, technical enhancements are by no means completely new, but are – in part – actually established, as the example of plastic surgery as a technical correction of physical characteristics felt to be imperfections shows, and as is the case in the practice of administering psycho-active substances. It is not difficult to predict that the possibilities and the realization of technical improvements of human beings will increase; demand is conceivable. In view of the moral questions connected with this development and of their conflict potential, ethical reflection is needed in this field already today although the feasibility of the enhancement technologies at all and the time scale of their availability cannot be assessed with any certainty. Because nanotechnology is seen as the “enabling technology” of these developments, the emerging ethical questions can be related to nanotech developments. However, we have to keep in mind that there is a tradition of ethical thought in anthropology, bioethics, medicine ethics and ethics of technology which provides a lot of argumentations and analyses for the field of “human enhancement”.

8.4 Nanoparticles and the Precautionary Principle[5)]

Nanoparticles and nanomaterials are among the first outcomes of nanotechnology to enter the marketplace and can already be found in everyday products [59]. From the "imperative of responsibility" [60] or the quest for "projected futures" [61, 62], philosophical conclusions have been drawn for slowing the process of bringing more and more nanoparticles into the environment and the human body without knowing much about possible side-effects. Ethical reflection is asked for concerning the question of how to deal responsibly with this situation involving high uncertainties of knowledge.

8.4.1 The Risk Debate on Nanoparticles

Currently, special attention in public risk debate is being paid to synthetic nanoparticles. A vast potential market for nano-based products is seen in this field. By means of admixtures or specific applications of nanoparticles, for example, new properties of materials can be brought about, for instance, in surface treatment, in cosmetics and in sunscreens. Consumers and citizens could easily come in contact with nanoparticles already today and the probability of directly having contact with synthetic nanoparticles will increase considerably in the next few years because of the expanding market for the respective products [59]. In spite of this situation, there is still little knowledge about possible impacts of nanoparticles on human health and the environment. In order to allow rational risk management strategies, knowledge about ways of spreading, behavior in the atmosphere or in fluids, lifetime of nanoparticles *as* nanoparticles until agglomeration to form other (larger) particles, their behavior in the human body and in the natural environment, and so on, would be needed. Such knowledge, however, is currently not available to an extent which would allow for classical risk management strategies [1, Section 5.2, [27, 28]].

The growing awareness of this risk issue in combination with the fact of having nearly no knowledge available about side-effects of nanotechnology led to severe irritations and to a kind of helplessness in the early stages of that debate. Some statements from that time will illustrate that situation. A re-assurance company stated: "The new element with this kind of loss scenario is that, up to now, losses involving dangerous products were on a relatively manageable scale whereas, taken to extremes, nanotechnology products can even cause ecological damage which is permanent and difficult to contain. What is therefore required for the transportation

5) Parts of this work have been performed within the expert group "Nanotechnology. Assessment and Perspectives" of the European Academy Bad Neuenahr-Ahrweiler [1]. I would like to thank the group members for many fruitful discussions. More specifically, I am deeply indebted to Harald Krug, who introduced the knowledge available in human and eco-toxicology about impacts of nanoparticles into those discussions. An extended version of this chapter is currently being published [66].

of nanotechnology products and processes is an organizational and technical loss prevention programme on a scale appropriate to the hazardous nature of the products" [64, p. 13]. The ETC group postulated a ban on the commercial use of nanoparticles until more knowledge was available: "At this stage, we know practically nothing about the possible cumulative impact of human-made nanoscale particles on human health and the environment. Given the concerns raised over nanoparticle contamination in living organisms, the, ETC group proposes that governments declare an immediate moratorium on commercial production of new nanomaterials and launch a transparent global process for evaluating the socio-economic, health and environmental implications of the technology" [29, p. 72]. A completely different but also far-reaching recommendation aims at "containing" nanotech research: "CRN has identified several sources of risk from MNT (molecular nanotechnology), including arms races, gray goo, societal upheaval, independent development and programmers of nanotech prohibition that would require violation of human rights. It appears that the safest option is the creation of one – and only one – molecular nanotechnology programme and the widespread but restricted use of the resulting manufacturing capability" [65, p. 4]. This containment strategy would imply a secret and strictly controlled nanotech development, which seems to be unrealistic and unsafe as well as undemocratic. These different proposals have enriched (and heated) public and scientific debate on possible nanoparticle risks. Nanotechnology, itself still in an embryo state, experienced itself, at that time, more or less suddenly, as a subject of public risk debate. The actors in the field seemed not to be prepared for this case. Against this background, it is understandable that the first years of the nanotech risk debate may be characterized mainly by mere suspicions, speculations and uncertainties rather than by knowledge-based and rational deliberation.

In analyzing this situation with respect to conclusions for risk management, careful normative reflection is required [35]. More precisely, thinking about the precautionary principle implies the absence of a standard situation in moral, in epistemic and in risk respect ([6]; see also Section 8.2 of this chapter). Ethical reflection is needed to shed light on the normative premises of the options at hand as well as on the criteria of decision-making. Such an ethical "enlightenment" is a necessary precondition of deliberative procedures within which society could identify adequate levels of protection, threshold values or action strategies. Questions of the acceptability and comparability of risks, the advisability of weighing up risks against opportunities and the rationality of action under uncertainty are, without doubt, of great importance in the field of nanoparticles (for the general challenge related to rationality and risk, see [33]). A new field of application is developing here for the ethics of technology, where close cooperation with toxicology, social sciences and jurisprudence is necessary [63].

8.4.2 The Precautionary Principle

Risk management strategies accompanying the implementation of new technologies and the introduction of new materials are standing in a long tradition. In earlier times, often a "wait-and-see" approach had been taken. New substances have been introduced assuming that either probably no negative developments and impacts at all would occur or that, in case of adverse effects, *ex post* repair and compensation strategies would be appropriate. The asbestos case is one of the best-known experiences where this approach failed and where the failure had dramatic consequences [67].

Such experiences with hazards caused by new materials, by radiation or by new technologies (see [68] for impressive case studies) led to risk regulations in different fields, in order to prevent further negative impacts on health and the environment. Important areas are [1, Section 5.1]:

- regulations for working places with specific risk exposures (nuclear power plants, chemical industry, aircraft, etc.) to protect staff and personnel;
- procedural and substantial regulations for nutrition and food (concerning conservation procedures, maximum allowed concentrations of undesirable chemicals such as hormones, etc.) to protect consumers;
- environmental standards in many areas to sustain environmental quality (concerning ground water quality, maximum allowed rate of specific emissions from fabrication plants, power plants, heating in households, etc.);
- safety standards and liability issues to protect users and consumers (in the field of automobile transport, for power plants, engines, technical products used in households, etc.).

There are established mechanisms of risk analysis, risk assessment and risk management in many areas of science, medicine and technology, for example in dealing with new chemicals or pharmaceuticals. Laws such as the Toxic Substance Control Act in the USA [69] constitute an essential part of the normative framework governing such situations. This "classical" risk regulation is adequate if the level of protection is defined and the risk can be quantified as the product of the probability of the occurrence of the adverse effects multiplied by the assumed extent of the possible damage. In situations of this type, thresholds can be set by law, by self-commitments or following participatory procedures, risks can be either minimized or kept below a certain level and also precautionary measures can be taken to keep particular effects well below particular thresholds by employing the ALARA (as low as reasonably achievable) principle [63]. Insofar as such mechanisms are able to cover challenges at hand to a sufficient extent, there is a standard situation in moral respect (see Section 8.2) regarding the risk issue. As the ongoing debate shows (which will be explained in more detail below), the field of risks of nanoparticles will not be a standard situation in this sense.

As soon as the conditions for the classical risk management approach are no longer fulfilled, uncertainties and ambivalent situations are the consequence. This is the case if, on the one hand, scientific knowledge concerning possible adverse effects is not available at all or is controversial or highly hypothetical or if empirical evidence is still missing. On the other, classical risk management might not be applicable if adverse affects could have catastrophic dimensions with respect to the extent of possible damage, also in case of (nearly) arbitrary small probabilities of their occurrence. In the field of nuclear power plants, scenarios of this type have been used as counter-arguments against that technology. The catastrophic dimension of possible, at least thinkable, accidents should, in the eyes of opponents, be a decisive argument even in case of an extreme low probability of such events. This type of situation motivated, for example, Hans Jonas [60] to postulate a "heuristics of fear" and the obligation to use the worst scenario as orientation for action.

In the philosophical debate, however, it became clear that Jonas's approach – besides inherent philosophical problems of the naturalistic and teleological approach – might be very appropriate to raise awareness with regard to precautionary situations but completely inadequate to be operationalized by regulation. Jonas's approach missed completely a legitimate procedure for deciding about the applicability and adequacy of precautionary strategies. What can still be learned from Jonas's work is the high relevance of normative reflection in cases where classical risk management would no longer be adequate [35]. Such situations often are welcomed entry points for ideology and interest-driven statements in arbitrary directions. In fact, it is very difficult to identify what a "rational" approach to dealing with non-classical situations could be and in which way it could be proven to be rational [33].

The observation that in many cases severe adverse effects in the course of the introduction of new materials had not been detected in an early stage but rather led to immense damage to human health, the environment and also the economy [68] motivated debates about precautionary regulation measures which could be applied *in advance* of having certain and complete knowledge – because it might then be too late to prevent damage. Wide international agreement on the precautionary principle was reached during the Earth Summit [United Nations Conference on Environment and Development (UNCED)] in Rio de Janeiro in 1992 and became part of Agenda 21: "In order to protect the environment, the precautionary approach should be widely applied by States according to their capabilities. Where there are threats of serious or irreversible damage, lack of full scientific certainty shall not be used as a reason for postponing cost-effective measures to prevent environmental degradation" [70]. The precautionary principle was incorporated in 1992 in the Treaty on the European Union: "Community policy on the environment shall aim at a high level of protection taking into account the diversity of situations in the various regions of the Community. It shall be based on the precautionary principle ..." (Article 174).

The precautionary principle substantially lowers the (threshold) level for action of governments (see [35] for the following). It considerably alters the situation in comparison with the previous context in which politicians could use (or abuse) a persistent dissent among scientists as a reason (or excuse), simply *not to take action at all*. In this way, political action simply could come much too late. It is, however, a difficult task to establish legitimate decisions about precautionary measures without either to run into the possible high risks of a "wait-and-see" strategy or to overstress precautionary argumentation with the consequence of no longer to be able to act any more or to cause other types of problems (e.g., for the economy) without need. The following characterization of the precautionary principle shows – in spite of the fact that it still does not cover all relevant aspects – the complex inherent structure of the precautionary principle:

> *"Where, following an assessment of available scientific information, there is reasonable concern for the possibility of adverse effects but scientific uncertainty persists, measures based on the precautionary principle may be adopted, pending further scientific information for a more comprehensive risk assessment, without having to wait until the reality and seriousness of those adverse effects become fully apparent" (modified from [35]).*

Thinking about applying the precautionary principle generally starts with a *scientific examination.*

An assessment of the state of the knowledge available in science and of the types and extents of uncertainties involved is needed. In assessing the uncertainties involved, *normative* qualifiers come into play [35]. It has to be clarified whether there is *reasonable concern* in this situation of uncertainty. The qualifier "reasonable concern" as employed by the EC guidelines makes no prejudice about the degree of likelihood, but this qualifier relates to a judgment on the quality of the available information [35]. Therefore, the assessment of the knowledge available including its uncertainties is crucial to precautionary reflections.

8.4.3
The Precautionary Principle Applied to Nanoparticles

Following the preceding analysis, the central questions in the current situation concerning the use of nanoparticles and the knowledge about possible impacts are as follows [66]:

1. Is there a precautionary situation at all, characterized by *epistemic uncertainty* or *unquantifiable risk?*
2. Is there "*reasonable concern* for the *possibility* of adverse effects" in the field of nanoparticles to legitimate the application of the precautionary principle?
3. If yes, what would follow from this judgment with respect to adequate precautionary measures?

The first question has to be answered *positive* (e.g. [1, Section 5.2, 27]). There are unknown or uncertain cause-effect relations and still unknown probabilities of risks resulting from the production, use and proliferation of nanoparticles. The scope of possible effects, their degrees and the nature of their "seriousness" (in relation to the chosen level of protection) can currently, even in the best cases, only be estimated in qualitative terms. Therefore, we are witnessing a typical situation of uncertainty where established risk management strategies are not to be applied [3, p. 4f, 27].

The answer to the second question, whether there is "*reasonable concern* for the *possibility* of adverse effects" in the field of synthetic nanoparticles is also to be answered positive. There are first toxicological results from the exposure of rats to high concentrations of specific nanoparticles which showed severe up to lethal consequences. Because the exposure concentrations were extremely high and the transfer of knowledge achieved by the exposure of rats to the situation of humans is difficult, these results do not allow for the conclusion of evidence of harm to human health – but they support assuming "reasonable concern for the *possibility* of adverse effects" caused by synthetic nanoparticles. There is a strict difference from classical risk management where *evidence of adverse effects* is required, not evidence of their *possibility* (only).

The third question is the most difficult one. In the precautionary situation given by positive answers to the first two questions, the challenge is to identify a "rational" course of action. The suspicion of adverse effects of nanoparticles could serve as a legitimating reason for very strict measures such as a moratorium on nanoparticle use [29, 30]. However, a mere suspicion – this would be sufficient in the above-mentioned Jonas scenario – does not legitimate strict precautionary measures. Instead, this would depend on a scientific assessment of the state of the art and that the quality of the information available [35]. This scientific assessment of the knowledge available for the nanoparticle case has recently been performed [1, Section 5.2] in a comprehensive manner. It resulted that there are indications for nanoparticle risks for health and the environment in some cases but that there is, based on the state of the art, *no reason for serious concern* about adverse effects but well-grounded serious concern about the *possibilities* of such effects. In the same direction: "Taking into account our present-day knowledge, there is, with regard to nano-specific effects (excluding self-organization effects and cumulative effects of mass production), no reason for particularly great concern about global and irreversible effects of the specific technology 'per se', with it being on a par with the justifiable apprehension concerning nuclear technology and genetic engineering" [27, p. 16].

The mere *possibility* of serious harm implied by a wider use of nanoparticles, however, *does not legitimize* using the precautionary principle as an argument for a moratorium or other prohibitive measures. However, because the state of knowledge permanently changes, continuous monitoring and assessment of the knowledge production concerning impacts of nanoparticles on human health and the environment are urgently required. One of the most dramatic lessons which can be learned from the asbestos story is exactly the lesson about the crucial necessity for such a systematic assessment [67].

The next question is what other (and weaker) types of measures are required in the present (precautionary) situation. These could, for example, be self-organization measures in science, such as the application of established codes of conduct or the adaptation of existing regulation schemes to special features of nanoparticles. It seems helpful to reconsider, as an example, the implementation of the precautionary principle in the genetically modified organisms (GMOs) case: "The European Directive 2001/18 (superseding Directive 90/220), concerning the deliberate release of GMOs into the environment, is the first piece of international legislation in which the precautionary principle is translated into a substantial precautionary framework. ... In the framework of this Directive the precautionary principle is translated into a regulatory framework which is based on a so-called *case-by-case* and a *step-by-step* procedure. The case-by-case procedures facilitates a mandatory scientific evaluation of every single release of a GMO. The step by step procedure facilitates a progressive line of development of GMOs by evaluating the environmental impacts of releases in decreasing steps of physical/biological containment (from greenhouse experiments, to small-scale and large-scale fields tests up to market approval). This procedural implementation of the precautionary principle implies an ongoing scientific evaluation and identification of possible risks" [35]. The application of the precautionary principle would, in this argument, imply "that there is a need for a cautious step-by-step diffusion of risk-related activities or technologies until more knowledge and experience is accumulated" [27, p. 6].

Against this background there is a wide consensus that no really new regulatory measures are needed: ". . . the existing regulative framework is adequate to deal with the introduction of new substances in the nanotechnology sector" [27, p. 27]. There are particular points of criticism concerning existing regulatory systems [69]; these points may be taken into account by modifications and supplements. Treating nanoparticles as new chemicals seems to be the adequate risk management approach [1].

In the framework of the precautionary principle, there are urgent tasks to be done by science [66], especially in order to close the still large knowledge gaps [27]. A second line of activities required consists of systematically collecting all relevant knowledge and of regularly conducting comprehensive evaluations of the respective state of knowledge. A third field of activities should be directed to contributing to public debate and to improving communication at the interface between science and society. The following steps for further action, thereby meeting several of recommendations by other groups [3, 27, 71], have been proposed [1]:

- develop a nomenclature for nanoparticles and assign a new Chemical Abstracts Service (CAS) registry number to engineered nanoparticles;
- group and classify nanomaterials with respect to categories of risk, toxicity and proliferation;
- treat nanoparticles as new substances and develop and approve tools for screening and testing;
- improve the knowledge base in toxicology;
- develop guidelines and standards for good practices in cautiously dealing with nanoparticles;
- avoid or minimize production and unintentional release of waste nanoparticles;
- establish comprehensive evaluation of the state of knowledge and its evaluation with respect to implications for risk management as well as for identifying knowledge gaps which should urgently be closed;
- create institutions to monitor nanotechnologies and the knowledge about possible risks;
- establish a permanent and open dialogue with the public and with industry.

This step-by-step approach conforms to present knowledge and seems to be an adequate way of applying the precautionary principle in a pragmatic way. However, there is no guarantee of preventing all kinds of possible risks by applying these steps. To repeat an important lesson from the asbestos story, the present situation which may be characterized by "no evidence of harm" must not be reinterpreted as "evidence of no harm" [67].

With respect to the "reasonable concern for the possibility of hazards", specific caution in dealing responsibly with synthetic nanoparticles is required. Such particles should be handled analogously to *new* chemicals even in the case that the chemical composition is well known beyond the nano character. Dealing with new nanoparticles is still based on a case-by-case approach because established nomenclature and classification schemes are not well prepared to be applied to nanoparticles. This situation should be changed as soon as possible [1, 63].

Beyond risk management and regulation, public risk communication has to be observed carefully because irritations in this communication could have a dramatic impact on public acceptance and political judgment. Denying the existence of risks of new technologies often causes mistrust and suspicion instead of creating optimism. To speak frankly about possible risks increases the chance of a trusty relationship between science and society: Building trust in public debate needs an open debate about chances and risks and often requires schemes for comparing different types and amounts of risk to be available [56, 63]. Transparency about the premises of different risk assessment exercises is urgently required. Knowledge about risks includes knowledge about the validity and the limits of that knowledge. Communicative and participative instruments of technology assessment [13, 72] could help improve mutual understanding and public risk assessment.

8.5 Human Enhancement by Converging Technologies

The report "Converging Technologies for Improving Human Performance" [17] caused a worldwide wave of intensive debate about the future of human nature and on the current agenda of science [71]. It was based on earlier debates in the context of biotechnical and genetic improvement strategies [73, 74]. The idea of NBIC (Nano–Bio–Info–Cogno) convergence [1], however, led to a new degree of intensity. Nanotechnology has been seen as the "enabling technology" of such new strategies of enhancement which allows the "converging technologies" to be given a place in a book on nanotechnology. Without any doubt the question for human enhancement will not be covered by a standard situation in moral respect.

In this section, we will use a topic-oriented approach, starting by discussing the "human enhancement" topic first in general and then mentioning the contributions to this old topic of nanotechnologies and the converging technologies (Section 8.5.1). The implications of the human enhancement idea by NBIC technologies for the need for orientations will be discussed (Section 8.5.2), followed by addressing specific ethical questions (Section 8.5.3). As a result, a rather probable scenario of the further development will be given and discussed in an ethical respect (Section 8.5.4).

8.5.1 Human Enhancement: Visions and Expectations

Human enhancement is a very old theme. Mankind's dissatisfaction with itself has been known from ancient times – discontent with mankind's physical endowments, its physical and intellectual capabilities, with its vulnerability to exogenic eventualities such as disease, with the inevitability of aging and, finally, of death, dissatisfaction with its moral capacities or – and this will probably be particularly frequent – with one's physical appearance. Various methods have been developed and established in order to give certain wishes for improvement a helping hand. Today's esthetic surgery, as a kind of business with considerable and further growing returns, is at present the probably most widespread method of human enhancement. Pharmaceuticals enhancing the performance of the mind or preventing tiredness are increasingly used [19]. Extending the physical limits of human capabilities through intensive training in competitive sports can also be understood as enhancement. Making use of *technical means* for improving performance in sport (doping), however, is being practiced as we often can read about in newspapers but is still held to be unsportsmanlike and unethical.

If the types of enhancement just listed apply to individuals (high athletic performance, individual beauty), *collective* human enhancement is, in its turn, also no new topic. Humankind's often deplored defects in terms of morals or civilization led, for example, in the – with regard to morality as well – progress-optimistic European Enlightenment to approaches towards trying to improve humankind *as a whole* through education. Beginning with the individual, above all, in school education, far-reaching processes towards the advancement of human civilization were to be stimulated and supported.

In the twentieth century, the totalitarian regimes in the Soviet Union and in the German Nazi Reich also applied strategies for improvement – due to the respective ideologies, either related to ideas of socio-Darwinist racism and anti-Semitism or to the orthodox communist and anti-bourgeois ideology. These historical developments show that strategies of improvement must be scrutinized very careful with respect to possibly underlying totalitarian ideologies.

In the current discussion of human enhancement, it is neither a question of an improvement through education and culture nor by indoctrination or power, but of *technical* improvement. Initiated by new scientific visions and utopias which are under discussion, completely new possibilities of human development have been proposed. The title of the report of an American research group to the National Science Foundation conveys its program: "Converging Technologies for Improving Human Performance" [17]. Nanotechnology and the Converging (NBIC) Technologies offer, according to this report, far-reaching perspectives for perceiving even the human body and mind as formable, to improve them through precisely targeted technical measures, and, in this manner, also to increase their societal performance. Strategies of enhancement start at the individual level but are aiming, in the last consequence, at the societal level: "Rapid advances in convergent technologies have the potential to enhance both human performance and the nation's productivity. Examples of payoff will include improving work efficiency and learning, enhancing individual sensory and cognitive capacities, revolutionary changes in healthcare, improving both individual and group efficiency, highly effective communication techniques including brain to brain interaction, perfecting human–machine interfaces including neuromorphic engineering for industrial and personal use, enhancing human capabilities for defense purposes, reaching sustainable development using NBIC tools and ameliorating the physical and cognitive decline that is common to the aging mind" [17, p. 1]. Among these proposed strategies of enhancement are the following:

- *The extension of human sensory faculties:* the capabilities of the human eye can be augmented, for example, with respect to visual acuity ("Eagle Eye") or with regard to a night vision capability by broadening the electromagnetic spectrum visible in the infrared direction; other sensory organs, such as the ear, could likewise be improved, or completely new sensory capabilities, such as the radar sense of bats, could be made accessible to human beings.
- *Expanding memory through technical aids:* it would be conceivable, by means of a chip which could be directly connected to the optic nerve [43], to record all of the visual impressions perceived in real time and to store them externally. In this manner, all of the visual impressions which accumulate in the course of a lifetime could be recalled at any time. In view of our forgetfulness, this could be an attractive idea for many people. Also the human memory could be enhanced by technical means (neuro-cognitive enhancement [19]).
- *Retardation of aging:* according to our present knowledge, aging can, roughly speaking, be interpreted as a form of degradation on the cellular level. If one could succeed in discovering and repairing immediately all forms of such degradation, aging could be greatly delayed or even abolished. For example, the famous nanorobots acting as "submarines" in our bodies could detect and eliminate pathogens: "In the hunt for pathogens, doctors send tiny machines into the furthest recesses of the body. These 'mini-submarines' are so small that they are not visible to the naked eye even as a speck of dust. Rotating cutting devices from the same dwarf world burrow their way through blocked blood vessels to eliminate the causes of heart attacks and strokes" [64, p. 3]. Overcoming aging with the help of nanotechnology would, then, in the sense of medical ethics, be nothing other than fighting epidemics or other diseases.

This thinking aims at broadening human capabilities in comparison with those we traditionally ascribe to a healthy human being. It is obvious that an entire series of ethical or anthropological questions are associated with these visionary expectations (or even just possibilities), which increase the contingency of the *conditio humana* [58, 76]. These questions pertain to the moral *permissibility* or *forbiddenness* of enhancement, to a possible *duty* to enhancement (if such a duty were possible), to the consequences of enhancement with regard to distributive justice (who can afford to have him- or herself enhanced [18]), to the *consequences* for our *concept of humankind* and for the *society of the future,* to the questions of the possible limits of technical enhancement and to legitimizing criteria for drawing such a boundary line. Without any doubt it seems clear that these challenges demarcate a non-standard situation in moral respect (see Section 8.2) and that, therefore, this is a field where ethical reflection is required in spite of the partially speculative status (see Section 8.3.7).

8.5.2 Occasions of Choice and Need for Orientation

Scientific and technical progress leads to an increase in the options for human action and has therefore at first sight an *emancipatory function*: an augmentation of the possibilities for acting and deciding and a diminution of the conditions which have to be endured as unalterable takes place. Whatever had been inaccessible to human intervention, whatever had to be accepted as non-influenceable nature or as fate becomes an object of technical manipulation or shaping. This is an increase of contingency in the *conditio humana*, a broadening of the spectrum of choices possible among various options [58].

Influencing the faculties of the "healthy" human body in the form of an improvement can be shown to be the *logically consistent continuation of* scientific and technical progress [76]. Up to now the physical capabilities of healthy humans have to be taken as given, as a heritage of the evolution of life. The sensoric capabilities of the eye or the ear, for example, cannot be extended (except by technical means outside the human body, such as microscopes). It would be an act of emancipation from Nature to be able to influence these capabilities intentionally by technical means. New occasions of choice would appear: in which directions would an enhancement be sensible, which additional functions of the human body should be realized, and so on? Also, more individuality could be the result if decisions on specific enhancements could be made at the individual level. In this way, human enhancement seems to contribute further to realizing grand normative ideas of the Era of Enlightenment.

However, there is also another side of the coin. The first price to be paid is the dissolution of established self-evidence about human beings and their "natural" capabilities. The increase of contingency is also an increased need for orientation and decision-making [58]. Second, the outcomes of those decisions about technical improvements can then be attributed to human action and decision – increased accountability and responsibility are the consequence [77]. Third, not only are new options opened but existing ones might be closed. For example, it is rather probable that disabled persons in a society using enhancement technologies to a large extent would have greater problems of conducting their life [20]. Furthermore, problems of access and equity would arise in increasing amount [18]. This ambivalent situation is characteristic of many modern technologies and does not per se legitimize arguments against enhancement but calls for attention and awareness and also for rational debates about dealing with the "dark side" of enhancement technologies.

8.5.3 Human Enhancement – No Simple Answers from Ethics

Increasing humankind's occasions of choice and its emancipation from the givenness of its biological and intellectual constitution also bring about uncertainties, a loss of security which have been unquestioned up to now and the need for new orientation to answering the above-mentioned general questions, as well as other, more specific questions, for example concerning neuro-implants: "However, ethical dilemmas regarding the enhancement of the human brain and mind are potentially more complex than, for example, genetically enhancing one's growth rate, muscle mass or even appearance because we primarily define and distinguish ourselves as individuals through our behavior and personality" [55]. This diagnosis poses the question of how far humans *may, should, want* to go in the (re-)construction of the human body and mind with the aim of enhancement [18, 74]. In advance of identifying the ethical questions involved, it seems appropriate to highlight an anthropological aspect affected: the difference between healing and enhancement.

Legitimate interventions into the human body and mind are at present carried out with the aims of healing or preventing disease or deficiencies. *Improving* human beings is, as yet, not a legitimate aim of medicine. Although the borderline between healing and enhancing interventions can hardly be drawn unambiguously [74] and although the terms "health" and "illness" are not well clarified [20, 52], there is obviously a categorical difference between the intentional enhancement of human beings and healing disorders [20]. Healing orients itself conceptually on a condition of health held to be ideal. This can either be explicitly defined or merely implicitly understood – in both cases, healing means closing the gap between the actual condition and the assumed ideal condition. What is to be understood under the ideal condition has certainly been defined culturally in different manners in the course of history. In each individual case, however, this is, at least in context, obvious enough. The ophthalmologist who subjects his patient to an eye-test has a conception of what the human eye should be able to perceive. He or she will propose technical improvements of the current state (e.g. a pair of spectacles) only for deviations from this conception and only from a certain degree of deviation on. The purpose of such measures is restoring the normal state, which succeed may more or less well. Traditional medical practice is probably unimaginable without the manner of thinking that a normal or ideal state serves in the background as a normative criterion for defining deviation. Medical treatment does not extend beyond this normal or ideal state. Just this – for medical practice, essential – way of thinking would, in view of the possible technical improvement of human beings, probably become meaningless. A situation would develop where normative frameworks for technical interventions into the human body and mind would be needed but will not be available because this is a completely new situation in moral respects. This poses the question of how far humans *may, should, want* to go in the (re-)construction of the human body and mind with the aim of improving them.

Ethical issues raised by this situation have been characterized with regard to neuro-cognitive enhancement [19]. This classification can be extended to other forms of enhancement also:

- *Safety:* risks of enhancement might consist of unintended side-effects for the brain of the person to be enhanced (e.g., by unsuccessful technical interventions or in the long term).

- *Coercion:* the use of enhancement technologies by parts of the population might exert force on others also to enhance themselves. Otherwise disadvantages in professional or private life might be the consequence.
- *Distributive justice:* access to enhancement technologies might be distributed over the population or between different regions of the world in a very inequitable way, for example because of high costs of enhancement ([18]; for the topic of equity in general, see Section 8.3.2).
- *Personhood and intangible values:* the way in which we see ourselves as humans and our imagination of a "good life" could be changed.

First we have to draw our attention to the fact that the spontaneous rejection with which the concept of human enhancement is often confronted in the population is, in itself, no ethical argument *per se*. The fact that we are not accustomed to dealing with the enhancement issue and the cultural alien-ness of the idea of technically enhanced human beings are social facts and are quite understandable – but they have only limited argumentative force as such. Spontaneous rejection might occur only because of feeling uncomfortable and unfamiliar in thinking about technical enhancement and could be changed by getting more familiar with it. Therefore, feeling and intuition are factual but still have to be scrutinized for whether there are ethical arguments hidden behind them at all and how strong these arguments would be.

The often-mentioned assertion that a human being's "naturalness" would be endangered or even be eliminated by technical improvement is also no strong argument *per se*. Humankind's naturalness or culturality are competing and partially linked patterns of interpretation of the human condition. Using humankind's naturalness as an argument in the sense that we should not technically improve the evolutionarily acquired faculties of sight, hearing, thinking, and so on, just because they are naturally developed and evolutionarily adapted, would be a naive naturalistic fallacy: out of the fact that we find ourselves to be human beings, for instance, with eyes which function only within a certain segment of the electromagnetic spectrum, nothing follows – normatively –directly at all. Limiting human capabilities to the naturally given properties would reduce humans to museum pieces and would blind out the cultural aspects of being humans, to which also belongs transcending the status quo, that is thinking beyond what is given, as has been the thesis of many writers on philosophical anthropology such as Arnold Gehlen.

From these considerations it cannot be concluded that human enhancement is permitted or even imperative. It merely follows that one should not make it too easy on oneself with an ethical repudiation. Strong imperative arguments are, in fact, not in sight [18]. However, argumentatively, the repudiation front also is not very strong. It points to a great extent to the *consequences* of enhancement – consequences which, like the fears of an increasing separation of society [18], are, to a great extent, hypothetical and which can therefore provide only very general and provisional orientation. In the final analysis, the ethical debate seems to narrow itself down to single-case argumentation: which concrete improvement is meant, which goals and purposes are connected with it, which side-effects and risks are to be apprehended and the question of weighing up these aspects against the background of ethical theories, such as Kantianism or utilitarianism. *Universally* applicable verdicts such as a strong imperative duty or a clear rejection of any improvement whatsoever seem at present to be scarcely justifiable. What follows from this situation for the future is the responsibility to reflect on the *criteria* for the desirability or acceptability of concrete possibilities for enhancement. A lot of work is still in front of ethics [75, 78].

Recently, it has been proposed to structure the field of possible ethical standpoints and perspectives with regard to enhancement technologies ([79, p. 9], applied to cognitive enhancement, but the scheme seems to be more generally usable) in the following, instructive way:

- *Laissez-faire* – emphasizes freedom of individuals to seek and employ enhancement technologies based on their own judgment.
- *Managed technological optimism* – believes that although these technologies promise great benefits, such benefits cannot emerge without an active government role.
- *Managed technological skepticism* – views that the quality of life arises more out of society's institutions than its technologies.
- *Human essentialism* – starts with the notion of human essence (whether God-given or evolutionary in origin) that should not be modified.

The decision to take one of these perspectives and to work with it in conceptualizing future developments in this field depend on the assessment of the strength of ethical arguments pro and con and also on images of "human nature".

8.5.4
Enhancement Technologies – A Marketplace Scenario Ahead?

What would follow from the description of the current debate given above? In order to give an answer we should take a brief look at the field of human cloning, which is not a technical enhancement of humans but would be a deep technical intervention into human life. Human cloning is regarded against the background of many ethical positions, for example Kantian ethics [74] as a form of instrumentalization of humans. The genetic disposition of an individual would be intentionally fixed by cloning. The persons affected would not be able to give their agreement to be cloned in advance. Cloning means an intentional and external determination of the genome of a later individual. An "informed consent", the information of the affected person about the cloning process and its impacts as well as the agreement of that person, would not be possible because the cloning has to be done at the early embryo phase. Therefore, human (reproductive) cloning and research to this end were banned in many countries soon after the clone sheep "Dolly" had been presented.

The ethical debate about human enhancement is, so far, completely different from the debate about human cloning. A restrictive regulation for human enhancement technologies and the research which could enable it, similar to the ban on human cloning technologies, has not been postulated yet. What makes the difference in debating about regulation while the intuitive and spontaneous public reactions are similar in both fields? The thesis is that the situation in ethical respect itself is completely different.

In the debate on human cloning, there is a strong ethical argument which motivated severe concern. The postulate of human dignity implies, in many interpretations, the idea that humans must not be instrumentalized without their consent. But cloning means determining *intentionally* the genome of an individual without any chance of consent in advance. In contradiction to the usual fertilization, which includes a high degree of statistical influence, cloning would double or multiply a well-known genome. This would imply a determination of the developing persons without any chance of reversibility [74, 80, 81]. This point seems to be the ethical reason behind the strong regulatory measures which have been taken extremely quickly.

In the field of human enhancement, however, things are completely different in this respect. Human enhancement in the fields mostly mentioned [1] would be applied to adults. Additional or improved sensory capabilities, for example, would be implemented in the same way as other medical procedures today: the candidates would be informed about the operational approach, about costs, about the process of adapting the new features, as well as about possible risks and side-effects. After having received such information, the enhancement candidates would either leave the "enhancement station" (in order not to use the term "hospital") or would sign a letter of agreement and, thereby, would give their "informed consent".

Ethical argumentation, therefore, will be performed by using different types of arguments. The arguments given so far are mostly concerned with possible *impacts* of technical enhancement. Expectable problems of distributive justice and equity are often mentioned [18, 50, 75]. Such argumentations, however, work with highly uncertain knowledge about future impacts of enhancement technologies. As can be learned from the history of technology assessment, it is very difficult to assess the consequences of completely new technologies where no or only little experience would be available as a basis for projections. In such cases, normative biases, pure imagination and ideology are difficult to separate from what could be derived from scientific knowledge. Often, an argument can be countered by a contradicting argument – for example, in the case of coercion mentioned above: "The straightforward legislative approach of outlawing or restricting the use of neurocognitive enhancement in the workplace or school is itself also coercive. It denies people the freedom to practice a safe means of self-improvement, just to eliminate any negative consequences of the (freely taken) choice not to enhance" [19, p. 423]. Another example would be the problem of equity: "Unequal access is generally not grounds for prohibiting neurocognitive enhancement, any more than it is grounds for prohibiting other types of enhancement, such as private tutoring or cosmetic surgery .... In comparison with other forms of enhancement that contribute to gaps in socioeconomic achievement, from good nutrition to high-quality schools, neuro-cognitive enhancement could prove easier to distribute equitably" [19, p. 423]. Or take arguments which make use of a possible change of the personhood of people by enhancement: "And if we are not the same person on Ritalin as off, neither we are the same person after a glass of wine as before or on vacation as before an exam"

19, p. 424]. Therefore, ethical analysis building on such highly uncertain assumptions about possible impacts could only constitute weak ethical arguments, pro enhancement as well as contra.

In this situation, it is rather probable that the introduction of enhancement technologies could happen according to the *economic market model*. Enhancement technologies would be researched, developed and offered by science and transferred to the marketplace. History shows (see Section 8.5.1) that a demand for such enhancement technologies is imaginable, as is currently the case in the field of esthetic surgery. Public interest would be restricted to possible situations of market failure. Such situations would be of the types discussed above: problems of equity, prevention of risks, questions of liability or avoidance of misuse. This scenario could be perceived as defensive with respect to current ethical standards, as cynical or as poor in ethical respects. Many people might feel uneasy about it. However, at the moment, such a scenario would fit the state of ethical analysis of technical enhancement of the human body and mind.

Such a scenario, however, has not to be equivalent to a *laissez-faire* model of the use of enhancement technologies [79; see the quotation at the end of Section 5.3]. Also, a marketplace scenario would have to be embedded in a societal environment, consisting of normative frameworks [15], ethical standards and regulation. To clarify the ethical issues involved in enhancement technologies and their relations to social values and to regulation is, to a large extent, still a task ahead.

8.6 Conceptual and Methodical Challenges

It is a characteristic trait of modern societies that they draw the orientation needed for opinion formation and decision-making increasingly from debates about future developments and less and less from existing traditions and values [77, 82]. Modern secular and scienticized society generally orients itself, instead of on the past, more on wishes and hopes, but also on fears with regard to the future. The notion of a "risk society" [77] and the global movement towards sustainable development [25] are examples of this approach. Ethical reflection, therefore, has to be related to communication about those future prospects. The necessity of providing orientation in questions such as human enhancement leads to the methodical challenge of applying the "moral point of view" on non-existing but only *projected* ideas with their own uncertainties and ambiguities.

8.6.1 Ethical Assessments of Uncertain Futures

Ethical analysis on nanotech issues is to a large extent confronted with the necessity to deal with, in part far-reaching, future projections [58, 76]. Ethical inquiry in fields such as artificial life or human enhancement takes elements of future communication like visions as its subject. Providing orientation then means to analyze, assess and judge those far-reaching future prospects on the basis of today's knowledge and seen from the today's moral point of view.

Frequently, the "futures" used in the debate on nanotech and society differ to a large extent and, sometimes, waver between expectations of paradise and fears of apocalyptic catastrophes. Expectations which regard nanotech as the anticipated solution of all of humanity's problems standing in Drexler's [53] technology-optimistic tradition contrast radically with fears expressed in the technology-skeptical tradition of Joy's [83] line of argumentation where self-reproducing nanorobots are no longer simply a vision which is supposed to contribute to the solution of humanity's gravest problems [53], but are communicated in public partially as a nightmare. The visionary pathos in many technical utopias is extremely vulnerable to the simple question of whether everything could not just be completely different – and it is as good as certain that this question will also be asked in an open society. But as soon as it is posed, the hoped effect of futuristic visions evaporates and can even turn into its opposite [1, Section 5.3].

The general problem is that the spectrum of proposed future projections often seems to be rather arbitrary, for example between expectations of paradise and apocalypse, warning against catastrophes in completely diverging directions. One example will be given: the uncertainty of our knowledge about future developments of nanotechnology and its consequences in connection with the immense imagined potential for damage, of possibly catastrophic effects, are taken as an occasion for categorizing even the precautionary principle (see Section 8.4) as insufficient for handling these far-reaching future questions [61, 62]. Instead, the author's view of society's future with nanotechnology leaves open solely the *existential renunciation* of nanotechnology as the only solution, going beyond even Hans Jonas' *Imperative of Responsibility* [60], inasmuch as they formulate a "duty to expect the catastrophe" in order to prevent the catastrophe (an analysis of the inherent contradictory structure of this argument is included in [76]). It seems interesting that the argumentative opponents, the protagonists of human enhancement by converging technologies, also warn against catastrophes, but only in the opposite sense: "If we fail to chart the direction of change boldly, we may become the victims of unpredictable catastrophe" [17, p. 3]. If, however, the ultimate catastrophe is cited in both directions as a threat, this leads to an arbitrariness of the conclusions. Ethics, therefore, is confronted not only with problems of judging states or developments by applying well-justified moral criteria but also with the necessity to assess the "futures" used in these debate with respect to their "rationality".

This situation leads to new methodical challenges for ethical inquiry. Especially the uncertainty of the knowledge available standing behind the future prospects makes it difficult to assess whether a specific development at the human–machine interface or concerning the technical improvement of the human body should be regarded as science fiction (SF), as technical potential in a far future, as a somewhat realistic scenario or as probably becoming part of reality in the near future. Ethical analysis has to take into account this uncertainty and has to be combined with an "epistemology" of future projections [76] and with new types of a future knowledge assessment [84]. In the following, we will propose an "Ethical

Vision Assessment" and an "Ethical Foresight Approach" which would be parts of a concomitant ethical reflection parallel to the ongoing scientific advance, in close relation to that advance but also in a distance which allows independent assessments and judgments.

8.6.2 Ethical Vision Assessment

Far-reaching visions have been put forward in the nanotech debate since its very beginning. A new paradise has been announced since the days of Eric Drexler's book *Engines of Creation* [53]. This line of story-telling about nanotech has been continued in the debate on human enhancement by converging technologies: "People will possess entirely new capabilities for relations with each other, with machines and with the institutions of civilization. . . . Perhaps wholly new ethical principles will govern in areas of radical technological advance, such as the routine acceptance of brain implants, political rights for robots and the ambiguity of death in an era when people upload aspects of their personalities to the Solar System Wide Web" [17, p. 19]. It might seem that such speculations should not or could not be subject to ethical inquiry at all.

Because of the "power of visions" in the societal debates – for examples for research funding, motivating young scientists and for public acceptance – it is of great importance to scrutinize such visions carefully instead of denoting them as obscure and fantastic. In spite of the "futuristic" and speculative nature of those visions (see [85] for the notion of "futuristic visions" and [76] and [1] for first steps toward an epistemological analysis of future knowledge), an early ethical dealing with nanotech visions would be an important and highly relevant task in order to allow more transparent debate, especially about science's agenda [1, 71]. There is a need to make those visions more specific, to extract their epistemic and normative key assumptions and ideas and to relate them to other key issues in public debate. Visions in nanotechnology are very different in nature and often refer to more general ideas, such as [49]

- the notion of free and curiosity-driven research as justification and paradigm in itself;
- the Enlightenment ideology of human-centered and emancipatory progress;
- a transhumanist dogma driving forward human evolution by physically changing humans and transforming them into technology;
- the aims and emphases of different religious belief systems including different ideas about the "quality of life";
- the promotion of global development, health and sustainable development, partially following goals of environmentalism, movements for social justice or particular campaign groups in the framework of the civil society;
- a neo-liberal winner-takes-all capitalist system;
- compassion, motivated by the desire to alleviate human suffering;
- a "medical success" culture driven by the felt obligation or goal to leave no disease or condition without a technical solution and to correct every physical disadvantage.

Scrutiny of nanotech visions could be done in the framework of an "ethical vision assessment" [1, Section 8.6]. Vision assessment can be analytically divided in several steps which are not sharply separated and not linearly ordered but which serve different sub-objectives and involve different methods [85]:

1. With respect to analysis (*Vision Analysis*), it would be a question of disclosing the *cognitive and normative* contents of the visions and of judging epistemologically the extent of their reality and practicability, self-evidently on the basis of current knowledge. Then, an important aspect, the prerequisites for the visions' realizability and the time spans involved would have to be investigated. In both analytical steps, observing the language used on the one hand and the question of the antecedents of the predictive statements play a special role: " ... the nanoethics researcher must be attentive to the twists and turns of language which can be symptoms bringing light to the most hidden layers of the scientific or technological imagination" [61]. Further, the visions' *normative* contents have to be reconstructed analytically: the visions of a future society or of the development of human beings, as well as possible diagnoses of current problems, to the solution of which the visionary innovations are supposed to contribute. For a "rational" discussion, the transparent disclosure of the stocks of knowledge, uncertainties and values involved is necessary, above all, with regard to the relationship of fact to fiction [86]. The contribution of such reflective analyses could consist in this respect in the "clarification" of the pertinent communication: the partners in communication should know explicitly what they are talking about as a prerequisite for more rational communication. It is a matter of society's "self-enlightenment" and of supporting the appropriate learning processes.
2. Vision Assessment would, further, include evaluative elements (*Vision Evaluation*). These are questions of how the cognitive aspects are to be categorized, how they can be judged according to the degree of realization or realizability, according to plausibility and evidence and which status the normative aspects have, for example relative to established systems of values or to ethical standards. The purpose is the transparent disclosure of the relationship between knowledge and values, between knowledge and the lack of it and the evaluation of these relationships and their implications. On the one hand, one can draw upon the established evaluation methods of technology assessment, which often include a participative component [13, 72]. On the other, in the field of human enhancement there are some far-reaching questions in a normative respect which stand to discussion and which require ethical and philosophical reflection (see Section 8.5 and [18, 73, 74]).
3. Finally, it is a matter of deciding and acting (*Vision Management*). The question is how the public, the media, politics and science can be advised with regard to a "rational" use of visions. The question of alternatives, either already existing or to be developed, to the visions already in circulation, stands here in the center of interest, in accordance with the basic position of technology assessment, of always thinking in terms of alternatives and options. In this manner, visions

based on technology can be compared with one another or with non-technological visions.

In particular, it would be the assignment of Vision Assessment in an ethical respect to confront the various and, in part, completely divergent normative aspects of the visions of the future directly with one another. This can, on the one hand, be done by ethical analysis and desk research; on the other, however, the representatives of the various positions should discuss their differing judgments in workshops directly with and against one another, in order to lay open their respective premises and assumptions (e.g., using technology assessment procedures [72]).

8.6.3 Ethical Reflection in Technology Foresight

In the last consequence, it would be necessary to transform the future prospects of nanotech visionarists [17] into more negotiable and communicable, transparent scenarios of the future. This could be realized by using the toolbox of (technology) foresight [87]. According to the above-mentioned ideas and challenges, ethical reflection of far-ranging future visions should be combined with research on and procedures for making the various future prospects more transparent. A close relation between an epistemology of future knowledge [84] ethical deliberation emerges. This situation motivates having a brief look into the field of "technology foresight" which has been developed in the last 10 years, especially at the European level [87]. It seems to be possible to benefit from experiences which have been made there in dealing with different kinds of future knowledge and assessments concerning the agenda-setting processes. In particular, it adds to the previously mentioned dimensions of epistemology and ethics possibilities for public debate and involving actors other than researchers and philosophers.

Foresight is the process of looking systematically into the longer-term future of science, technology, the economy and society with the aim of identifying the areas of strategic research (agenda setting) and the emerging generic technologies likely to yield the greatest economic and social benefits at low risk [87]. As the term implies, these approaches involve thinking about emerging opportunities and challenges, trends and breaks from trends, future risks and ways of dealing with them, and so on. Foresight involves bringing together key actors of change and sources of knowledge, in order to develop *strategic visions* and to establish networks of knowledgeable actors who can respond better to policy challenges in awareness of each others' knowledge resources, values, interests and strategic orientations. The contexts in which foresight can be employed are equally wide-ranging: much work to date has focused on national competitiveness and especially the prioritization and development of strategic goals for areas of research in science and technology. Foresight thus occupies the space in which planning, future studies, technology assessment, strategic deliberation and policy development overlap. It is not a matter of academic- or consultancy-based forecasts of the future (though it has to take these into account as necessary knowledge grounds).

As can easily be seen, there are some similarities to the situation described above in the field of nanotech: far-reaching visions and expectations, uncertainty of knowledge, diverse positions of different actors and the aim at influencing agenda setting. Obviously, there are also differences: usually, in foresight processes ethical reflection is not part of the game whereas regional cooperation, creation of new networks, mobilization and contributing to economic welfare by exploiting chances of new technologies and of regional resources are major issues. The reason for bringing the more philosophical issue of assessing and reflecting nanotech visions with respect to epistemological and ethical questions with established foresight methodologies lies in the common challenge of being confronted with the necessity to deal "rationally" with highly uncertain future knowledge and the inseparably interwoven normative elements of expectations, desires and fears often involved. Giving advice to the scientific agenda, for example via contributing to the processes of defining issues and priorities of the public funding of science and technology, following an open and democratic debate, is, therefore, difficult to achieve.

In various foresight exercises, it has been a common experience that in order to arrive at a workable view on the future in the respective field it is crucial to focus on a concrete and tangible topic. Compared with this experience, it seems impossible to apply a foresight exercise to such a broad and grand topic like nanotechnology. On the contrary, more specific subtopics should be addressed such as the use of nanoparticles in cosmetics, developments towards artificial life or opportunities or threats concerning the equity of access to nanotech benefits. Then it is imaginable to set up a foresight process which would

- include research and reflection parts as the vision assessment activities described above;
- involve other societal groups (stakeholders, customers, policymakers, regulators, business people, etc.);
- provide a balanced and ethically as well as epistemologically reflected view on the respective part of the nanotech field which then could be used as a valuable input for debates about science's agenda in this field.

The task of such a foresight exercise would consist, first, of a "rationalization" of diffuse future prospects in a double manner: rationalization concerning the epistemic contents as well as concerning the normative aspects involved. Second, the resulting images of the future should be communicated to and debated with a broader audience in society. In this way, there could be a success in transforming the rather diffuse visionary prospects of nanotechnology into more specific scenarios in specific areas, for instance of the human–machine interface.

8.6.4
Concomitant Ethical Reflection on Nanotechnology

Since the very beginning of ethical reflection in science and technology, there has been an ongoing discussion about an adequate relation in time between scientific–technological advances and ethics. Ethics often seems to pant helplessly behind technical progress and to fall short of the occasionally great expectations [88]. The rapid pace of innovation in technicization has the effect that ethical deliberations often come too late: when all of the relevant decisions have already been made, when it is long since too late for shaping technology. Technological and scientific progress sets facts which, normatively, can no longer be revised [74]: "It is a familiar cliché that ethics does not keep pace with technology" [38]. This "ethics last" model means that first there have to be concrete technological developments, products and systems which then could be reflected by ethics. Ethics in this perspective, could, at best, act as a repair service for problems which are already on the table.

In contrast, the "ethics first" model postulates comprehensive ethical reflection on possible impacts already *in advance* of technological developments. Ethics actually can provide orientation in the early phases of innovation, for example because future projections and visions emerging on the grounds of scientific and technical advances may be subject to ethical inquiry. Because there are early ideas available about the scientific and technical knowledge and capabilities as well as about their societal impacts – risks as well as chances – long before market entry, it is possible to reflect and discuss their normative implications. For example, Jonas worked on ethical aspects of human cloning long before cloning technology was available even in the field of animals. Obviously, ethical reflection in this model has to deal with the situation that the knowledge about technology and its consequences is uncertain and preliminary.

This does not necessarily mean that ethical deliberations have to be made for absolutely every scientific or technical idea. The problems of a timely occupation with new technologies appear most vividly in the diverse questions raised by the visions of salvation and horror as regards nanotechnology and human enhancement. What sense is there in concerning oneself hypothetically with the ethical aspects of an extreme lengthening of the human life span or with self-replicating nanorobots [38]? The "ethics first" perspective is exaggerated in these cases to such an extent that any relevance will be lost. Most scientists are of the opinion that these are speculations which stem much rather from the realm of science fiction than from problem analysis which is to be taken seriously: "If discussions about the ethics and dangers of nanotechnology become fixated on a worry that exists only in science fiction, we have a serious problem" [89]. We should not forget that ethical reflection binds resources and there should therefore be certain evidence for the "validity" of these visions, if resources are to be invested in them which could then be lacking elsewhere. Therefore, methods of assessing visions of human enhancement are required which allow for an epistemological investigation of the visions under consideration (see the sections above).

Ethical judgment in very early stages of development could provide orientation for shaping the *process* of scientific advance and technological development (e.g., with regard to questions of equity or of risks of misuse). In the course of the continuing concretization of the possibilities for application of nanotechnologies, it is then possible continuously to concretize the – at first abstract – estimations and orientations on the basis of newly acquired knowledge and finally to carry out an ethically reflected technology assessment. In this way, ethical analysis is an ongoing process, accompanying the scientific and technological advance.

Due to nanotechnology's and the converging technologies' early stage of development, we have here a rare case of an advantageous opportunity: there is the chance and also the time for concomitant reflection, as well as the opportunity to integrate the results of reflection into scientific agenda and technology design and thereby to contribute to the further development of science and technology [38]. In view of the visionary nature of many the prospects in nanotechnology and of long and longer spans of time within which the realization of certain milestones can be expected, there is, in all probability, enough time to analyze the questions posed. In general, it applies in this case that this reflective discussion should take place already in the early phases of development, because then the greatest possibilities for influencing the process of scientific development are given. The chances are good that, in the field of nanotechnology, ethical reflection and the societal discussion do not come too late, but can accompany scientific–technical progress critically and can, in particular, help to influence science's agenda by ethically reflected advice.

References

1 Schmid, G., Brune, H., Ernst, H., Grünwald, W., Grunwald, A., Hofmann, H., Janich, P., Krug, H., Mayor, M., Rathgeber, W., Simon, B., Vogel, V. and Wyrwa, D. (2006) *Nanotechnology – Assessment and Perspectives*, Springer, Berlin.

2 Paschen, H., Coenen, C., Fleischer, T., Grünwald, R., Oertel, D. and Revermann, C. (2004) *Nanotechnologie. Forschung und Anwendungen*, Springer, Berlin.

3 Royal Society (2004) *Nanoscience and Nanotechnologies: Opportunities and Uncertainties*, Royal Accademy, London.

4 Nanoforum (2004) *Nanotechnology. Benefits, Risks, Ethical, Legal and Social Aspects of Nanotechnology*, http://www.nanoforum.org, accessed 2 October 2006.

5 Mnyusiwalla, A., Daar, A.S. and Singer, P.A. (2003) Mind the gap. Science and ethics in nanotechnology. *Nanotechnology*, **14**, R9–R13.

6 Grunwald, A. (2005) Nanotechnology – a new field of ethical inquiry? *Science and Engineering Ethics*, **11**, 187–201.

7 Ach, J.S. and Jömann, N. (2006) Size matters. Ethical and social challenges of nanobiotechnology – an overview. LIT, Münster.

8 Grunwald, A. (1999) Ethische Grenzen der Technik? Reflexionen zum Verhältnis von Ethik und Praxis, in *Ethik in der Technikgestaltung. Praktische Relevanz und Legitimation* (eds A. Grunwald and S. Saupe), Springer, Berlin, pp. 221–252.

9 Grunwald, A. (2003) Methodical reconstruction of ethical advises, in *Expertise and Its Interfaces* (eds G. Bechmann and I. Hronszky), Edition Sigma, Berlin, pp. 103–124.

10 Gethmann, C.F. and Sander, T. (1999) Rechtfertigungsdiskurse, in *Ethik in der Technikgestaltung. Praktische Relevanz und Legitimation* (eds A. Grunwald and S. Saupe), Springer, Berlin, pp. 117–151.

11 Grunwald, A. (2000) Against over-estimating the role of ethics in technology. *Science and Engineering Ethics*, **6**, 181–196.

12 Brown, N., Rappert, B. and Webster, A. (eds) (2000) *Contested Futures. A Sociology of Prospective Techno-science*, Ashgate, Burlington.

13 Joss, S. and Belucci, S. (eds), (2002) *Participatory Technology Assessment – European Perspectives*, Centre of the Study of Democracy, London.

14 van Gorp, A. (2005) Ethical Issues in Engineering Design: Safety and Sustainability, *Simon Stevin Series in the Philosophy of Technology*, Delft.

15 van Gorp, A. and Grunwald, A. (2008) Ethical responsibilities of engineers in design processes, risks, regulative frameworks and societal division of labour, in preparation.

16 Gethmann, C.F. (1994) Die Ethik technischen Handelns im Rahmen der Technikfolgenbeurteilung, in *Technikbeurteilung in der Raumfahrt – Anforderungen, Methoden, Wirkungen* (eds A. Grunwald and H. Sax), Edition Sigma, Berlin, pp. 146–159.

17 Roco, M.C. and Bainbridge, W.S. (eds) (2002) *Converging Technologies for Improving Human Performance*, National Science Foundation, Arlington, VA.

18 Siep, L. (2005) *Die biotechnische Neuerfindung des Menschen, presented at the XX. Deutscher Kongress für Philosophie*, Berlin.

19 Farah, M.J., Illes, J., Cook-Deegan, R., Gardner, H., Kandel, E., King, P., Parens, E., Sahakian, B. and Wople, P.R. (2004) Neurocognitive enhancement: what can we do and what should we do? *Nature Reviews. Neuroscience*, **5**, 421–425.

20 Wolbring, G. (2005) The Triangle of Enhancement Medicine, Disabled People and the Concept of Health: A New Challenge for HTA, Health Research and Health Policy. Research Paper, Alberta; http://www.cspo.org/ourlibrary/documents/HTA.pdf, accessed 27 July 2007.

21 Krings, B.-J. and Riehm, U. (2006) Die Nutzung und Nichtnutzung des Internets. Eine kritische Reflexion der Diskussion zum "Digital Divide", in *Netzbasierte Kommunikation, Identität und Gemeinschaft. Net-based Communication, Identity and Community* (eds U. Nicanor and A. Metzner-Szigeth), Berlin, pp. 233–251.

22 Salamanca-Buentello, F., Persad, D.L., Court, E.B., Martin, D.K., Daar, A.S. and Singer, P. (2005) Nanotechnology the Developing World. *PLoS Medicine*, 2, e97.

23 Fleischer, T. and Grunwald, A. (2002) Technikgestaltung für mehr Nachhaltigkeit – Anforderungen an die Technikfolgenabschätzung, in *Technikgestaltung für eine nachhaltige Entwicklung* (ed. A. Grunwald), Edition Sigma, Berlin, pp. 95–146.

24 World Commission on Environment and Development (1987) Our Common Future, World Commission on Environment and Development, Oxford.

25 Grunwald, A. and Kopfmüller, J. (2006) *Nachhaltigkeit*, Campus, Frankfurt, New York.

26 Fleischer, T. and Grunwald, A. (2008) Making nanotechnology developments sustainable. A role for technology assessment? *Journal of Cleaner Production* (forthcoming).

27 Haum, R., Petschow, U., Steinfeldt, M. and von Gleich, A. (2004) *Nanotechnology and Regulation within the Framework of the Precautionary Principle*, Institut für ökologische Wirtschaftsforschung, Berlin.

28 Colvin, V. (2003) *Responsible Nanotechnology: Looking Beyond the Good News*, Center for Biological and Environmental Nanotechnology at Rice University, http://www.eurekalert.org, accessed 2 October 2006.

29 ETC Group, *The Big Down. Atomtech: Technologies Converging at the Nanoscale*, (2003) http://www.etcgroup.org, accessed 2 October 2006.

30 Friends of the Earth, *Nanomaterials, Sunscreens and Cosmetics. Small Ingredients*, (2006) http://www.foe.org/camps/comm/nanotech/nanocosmetics.pdf, accessed 19 November 2006.

31 Gethmann, C.F. and Mittelstrass, J. (1992) Umweltstandards. *GAIA*, **1**, 16–25.

32 Gethmann, C.F., Pinkau, K., Renn, O., Decker, K., Levi, H.W., Mittelstrass, J., Peyerimhoff, S., Putlitz, G. zu, Randelzhofer, A., Streffer, C. and Weinert, F.E. (1998) *Environmental Standards.*

Scientific Foundations and Rational Procedures of Regulation with Emphasis on Radiological Risk Management, Boston.
33 Shrader-Frechette, K.S. (1991) *Risk and Rationality – Philosophical Foundations for Populist Reforms*, University of California Press, Berkeley, CA.
34 Grunwald, A. (2005) Zur Rolle von Akzeptanz und Akzeptabilität von Technik bei der Bewältigung von Technikkonflikten. *Technikfolgenabschätzung Theorie Praxis*, **14**, 54–60.
35 von Schomberg, R. (2005), The precautionary principle and its normative challenges, in *The Precautionary Principle and Public Policy Decision Making* (eds E. Fisher, J. Jones and R. von Schomberg), Edward Elgar, Cheltenham, UK, Northampton, pp. 141–165.
36 Grunwald, A. (1999) Technology assessment or ethics of technology? Reflections on technology development between social sciences and philosophy. *Ethical Perspectives*, **6**, 170–182.
37 Fleischer, T. (2003) Technikgestaltung für mehr Nachhaltigkeit: Nanotechnologie, in *Nachhaltigkeitsprobleme in Deutschland. Analyse und Lösungsstrategien* (eds R. Coenen and A. Grunwald), Edition Sigma, Berlin, pp. 356–373.
38 Moor, J. and Weckert, J. (2003) *Nanoethics: Assessing the Nanoscale from an Ethical Point of View*, Technical University of Darmstadt, Darmstadt.
39 Abbott, A. (2006) In search of the sixth sense. *Nature*, **442**, 125–127.
40 Altmann, J. and Gubrud, A.A. (2002) Risks from military uses of nanotechnology – the need for technology assessment and preventive control, in *Nanotechnology – Revolutionary Opportunities and Societal Implications* (eds M. Roco and R. Tomellini), European Commission, Luxembourg.
41 Altmann, J. (2004) *Military Nanotechnology. Potential Applications and Preventive Arms Control*, Routledge, New York.
42 Haper, T. (2002) *Nanotechnology Arms Race: Why Nobody Wants To Be Left Behind* http://www.nanotechweb.org/articles/column/1/1/11/1, accessed 2 December 2006.
43 Stieglitz, T. (2006) Neuro-technical interfaces to the central nervous system. *Poiesis Praxis*, **4**, 95–110.
44 Freitas, J.A. (1999) *Nanomedicine. Volume I: Basic Capabilities*, Landes Biosciences, Georgetown, TX.
45 European Technology Platform (2006) Nanomedicine – Nanotechnology for Health, European Commission, Luxembourg.
46 Kralj, M. and Pavelic, K. (2003) Medicine on a small scale. How molecular medicine can benefit from self-assembled and nanostructured materials. *EMBO Reports*, **4**, 1008–1012.
47 Farkas, R. and Monfeld, C. (2004) Ergebnisse der Technologievorschau Nanotechnologie pro Gesundheit 2003. *Technikfolgenabschätzung Theorie Praxis*, **13**, 42–51.
48 Baumgartner, C. (2004) Ethische Aspekte nanotechnologischer Forschung und Entwicklung in der Medizin. *Parlament*, **B23–24**, 39–46.
49 Bruce, D. (2006) Ethical and social issues in nanobiotechnologies. *EMBO Reports*, **7**, 754–758.
50 Ach, J. and Siep, L. (eds) (2006) *Nano-bio-ethics. Ethical Dimensions of Nanobiotechnology*, Berlin.
51 MacDonald, C. (2004) Nanotech is novel; the ethical issues are not. *Scientist*, **18**, 3.
52 Gethmann, C.F. (2004) Zur Amphibolie des Krankheitsbegriffs, in *Wissen und Verantwortung. Band 2: Studien zur medizinischen Ethik* (eds A. Gethmann-Siefert and K. Gahl), Alber, Freiburg.
53 Drexler, K.E. (1986) *Engines of Creation – The Coming Era of Nanotechnology*, Anchor Books, Oxford.
54 Scott, S.H. (2006) Converting thoughts into action. *Nature*, **442**, 141–142.
55 Turner, D.C. and Sahakian, B.J. (2006) Ethical questions in functional neuroimaging and cognitive enhancement. *Poiesis Praxis*, **4**, 81–94.

56 Grunwald, A. (2004) The case of nanobiotechnology. Towards a prospective risk assessment. *EMBO Reports*, **5**, 32–36.

57 National Nanotechnology Initiative (1999) Shaping the World Atom by Atom, Washington, DC.

58 Grunwald, A. (2007) Converging technologies: visions, increased contingencies of the conditio humana and search for orientation, *Futures*, **39**, pp 380–392.

59 Luther, W. (ed.) (2004) *Industrial Application of Nanomaterials – Chances and Risks*, http://www.techportal.de, accessed 2 October 2006.

60 Jonas, H. (1979) Das Prinzip Verantwortung, Suhrkamp Frankfurt/Main, English edn.: *The Imperative of Responsibility*, London 1984.

61 Dupuy, J.-P. (2005) The Philosophical Foundations of Nanoethics. Arguments for a Method, Lecture at the University of South Carolina, 3rd March.

62 Dupuy, J.-P. and Grinbaum, A. (2004) Living with uncertainty: toward the ongoing normative assessment of nanotechnology. *Techné*, **8**, 4–25.

63 Renn, O. and Roco, M. (2006) Nanotechnology and the need for risk governance. *Journal of Nanoparticle Research*, **8** (2), pp. 153–191.

64 Münchener Rückversicherung, (2002) Nanotechnology – What is in Store for Us? Münchener Rückversicherungsgesellschaft, http://www.munichre.com/publications/302-03534_en.pdf, accessed 12 November 2006.

65 Phoenix, C. and Treder, M. (2003) *Applying the Precautionary Principle to Nanotechnology*, http://www.crnano.org/Precautionary.pdf, accessed 2 October 2006.

66 Grunwald, A. (2008) Nanotechnology and the precautionary principle, in Nanotechnology and Nanoethics: Framing the Field (ed. F. Jotterand), Berlin, in press.

67 Gee, D. and Greenberg, M. (2002) Asbestos: from 'magic' to malevolent mineral, in *The Precautionary Principle in the 20th Century. Late Lessons from Early Warnings* (eds P. Harremoes, D. Gee, M. MacGarvin, A. Stirling, J. Keys, B. Wynne and S. Guedes Vaz), Sage, London, pp. 49–63.

68 Harremoes, P., Gee, D., MacGarvin, M., Stirling, A., Keys, J., Wynne, B. and Guedes Vaz, S. (eds) (2002) *The Precautionary Principle in the 20th Century. Late Lessons from Early Warnings*, Sage, London.

69 Wardak, A. (2003) Nanotechnology & Regulation, a Case Study Using the Toxic Substance Control Act (TSCA), Woodrow Wilson International Center, Foresight and Governance Project, Paper 2003–2006.

70 United Nations (2002) Report of the United Nations Conference on Environment and Development, A/CONf.151/26 (Vols. I–III), United Nations, New York.

71 Nordmann, A. (2004) *Converging Technologies – Shaping the Future of European Societies*, European Commission, Brussels.

72 Decker, M. and Ladikas, M. (eds) (2004) *Bridges Between Science, Society and Policy, Technology Assessment – Methods and Impacts*, Springer, Berlin.

73 Fukuyama, F. (2002) Our Posthuman Future: Consequences of the Biotechnology Revolution, Farrar, Strauss and Giraux.

74 Habermas, J. (2001) *Die Zukunft der menschlichen Natur*, Suhrkamp, Frankfurt.

75 Khushf, G. (2004) Systems theory and the ethics of human enhancement: a framework for NBIC convergence. *Annals of the New York Academy of Sciences*, **1013**, 124–149.

76 Grunwald, A. (2006) Nanotechnologie als Chiffre der Zukunft, in *Nanotechnologien im Kontext* (eds A. Nordmann, J. Schummer and A. Schwarz), Akademie Verlag, Berlin, pp. 49–80.

77 Beck, U. (1992) *Risk Society, Towards a New Modernity*, Sage, London.

78 Khushf, G. (2004) The ethics of nanotechnology – visions and values for a new generation of science and engineering, in *Emerging Technologies and Ethical Issues in Engineering* (ed National Academy of Engineering), Washington, DC, pp. 29–55.

79 Sarewitz, D. and Karas, T.H. (2006) Policy Implications of Technologies for Cognitive Enhancement, Arizona State University, Consortium for Science, Policy and Outcomes, 3–5 May.

80 Hermerén, G.(1 March 2004) Nano ethics primer, presented at the conference "Mapping Out Nano Risks", Brussels.

81 Habermas, J. (2002) Replik auf Einwände, *Deutsche Zeitschrift für Philosophie*. **50**, pp. 214–226.

82 Stehr, N. (2004) *The Governance of Knowledge*, Sage, London.

83 Joy, B. (2000) Why the future does not need us. *Wired Magazine*, 2nd April, pp. 238–263.

84 Pereira, A.G.,von Schomberg, R. andFuntowicz, S. (2007) Foresight knowledge assessment. *International Journal of Foresight and Innovation Policy*, **4**, pp. 44–59.

85 Grunwald, A. (2004) Vision assessment as a new element of the technology futures analysis toolbox. Proceedings of the EU–US Scientific Seminar: New Technology Foresight, Forecasting and Assessment Methods, Seville. http://www.jrc.es/projects/fta/index.htm, accessed 2 December 2006.

86 Schmidt, J. (2003) Zwischen Fakten und Fiktionen: NanoTechnoScience als Anfrage an prospektive Wissenschaftsbewertung und Technikfolgenabschätzung, in *Zukunftsorientierte Wissenschaft* (eds W. Bender and J. Schmidt), LIT Münster, pp. 207–220.

87 FOREN (2001) A Practical Guide to Regional Foresight, http://foren.jrc.es, accessed 2 December 2006.

88 Ropohl, G. (1995) Die Dynamik der Technik und die Trägheit der Vernunft, in *Neue Realitäten – Herausforderung der Philosophie* (eds H. Lenk and H. Poser), Akademie Verlag, Berlin, pp. 221–237.

89 Ball, P. (2003) *Nanoethics and the Purpose of New Technologies*, Royal Society for Arts, London, http://www.whitebottom.com/philipball/docs/Nanoethics.doc, accessed 2 December 2006.

9
Outlook and Consequences

Günter Schmid

"Nanotechnology could become the most influential force to take hold of the technology industry since the rise of the Internet. Nanotechnology could increase the speed of memory chips, remove pollution particles in water and air and find cancer cells quicker. Nanotechnology could prove beyond our control and spell the end of our very existence as human beings. Nanotechnology could alleviate world hunger, clean the environment, cure cancer, guarantee biblical life spans or concoct super weapons of untold horror . . . . Nanotechnology could spur economic development through spin-offs of the research. Nanotechnology could harm the opportunities of the poor in developing countries . . . . Nanotechnology could change the world from the bottom up. Nanotechnology could become an instrument of terrorism. Nanotechnology could lead to the next industrial revolution . . . . Nanotechnology could change everything."

This only incompletely cited collection of meaningful as well as senseless predictions is listed in a brochure on "The Ethics and Politics of Nanotechnology" by UNESCO published in 2006 [1]. It impressively demonstrates how the understanding of nanotechnology depends from the observer's individual opinion, the spreading medium or from the different political trends. Depending on the respective author's personal attitude, hopes are raised or catastrophes are predicted.

In spite of the variations of the definition of nanotechnology (see Chapters 1 and 2), from a scientific point of view, strictly observed in this and the other books in this series, we must refrain from unserious promises, but also from scenarios drawing the decline of mankind. Both belong to the field of science fiction and are not based on scientific findings. Indeed, there are enough scientifically substantiated facts making nanotechnology one of the most influential technologies we have ever had, in agreement with the very first of the UNESCO headlines cited above. This can be followed from the few examples given in Sections 2.2.1–2.2.7 in Chapter 2, but even more from the following volumes that deal with information technology, medicine, energy and instrumentation on the nanoscale, making observations in the nanoworld possible. Without entering science fiction fields, our present knowledge in nanoscience allows the prognosis that nano-based data storage systems will offer

Nanotechnology. Volume 1: Principles and Fundamentals. Edited by Günter Schmid

ISBN: 978-3-527-31732-5

capacities to store the whole of the world's literature on a single chip and to construct notebooks with the capacity of a computer center of today. Low-priced solar cells, moldable accumulators of high capacities, highly efficient fuel cells and hydrogen storage systems will become available. Last but not least, progress to be expected in medical diagnosis and therapy will revolutionize health care, beginning with enduringly working drug delivery systems, diagnostic potentials, improved by orders of magnitude, up to individual cancer therapies, based on the personal genome of a patient.

These few examples indeed demonstrate the innovative power of nanoscience and -technology. On the other hand, we should not close our eyes to possible dangers linked with the expansion of nanotechnology. From experiences in the past, for instance in the cases of nuclear energy and genetically modified organisms, we should avoid similar mistakes in the case of nanotechnology. Especially scientists are asked to contribute to an objective discussion in public, free of ideologies and pre-opinions.

An indispensable condition to discuss chances and risks of nanotechnology, free from any prejudice, is a minimum knowledge about nanotechnology by the public. This is not easy to achieve, due to the very complex nature of nanotechnology extending from physics to the life sciences. Therefore, education processes have to start as early as possible in primary and secondary schools, in colleges and universities. This is especially necessary in order to protect people from wrong prophets, in a positive as well as in a negative sense. There is no doubt that nanotechnology will change our lives, but it depends on us what this change will look like.

References

1 The Ethics and Politics of Nanotechnology United Nations Educational, Scientific and Cultural Organization (2006), Paris.

Index

Nanotechnology. Volume 1: Principles and Fundamentals. Edited by Günter Schmid

ISBN: 978-3-527-31732-5